Lecture Notes in Mathematics

Edited by A. Dold and B. Eckmann

Subseries: Department of Mathematics, University of Maryland
Adviser: M. Zedek

1222

Anatole Katok
Jean-Marie Strelcyn

with the collaboration of
F. Ledrappier and F. Przytycki

Invariant Manifolds, Entropy and Billiards; Smooth Maps with Singularities

Springer-Verlag
Berlin Heidelberg New York London Paris Tokyo

Authors

Anatole Katok
Mathematics 253-37, California Institute of Technology
Pasadena, CA 91125, USA

Jean-Marie Strelcyn
Université Paris-Nord, Centre Scientifique et Polytechnique
Département de Mathématiques
Avenue J.-B. Clément, 93430 Villetaneuse, France

François Ledrappier
Laboratoire de Probabilités, Université Paris VI
4 Place Jussieu, 75230 Paris, France

Feliks Przytycki
Mathematical Institute of the Polish Academy of Sciences
ul. Sniadeckich 8, 00-950 Warsaw, Poland

Mathematics Subject Classification (1980): Primary: 28D20, 34F05, 58F11, 58F15
Secondary: 34C35, 58F08, 58F18, 58F20, 58F22, 58F25

ISBN 3-540-17190-8 Springer-Verlag Berlin Heidelberg New York
ISBN 0-387-17190-8 Springer-Verlag New York Berlin Heidelberg

Printing and binding: Druckhaus Beltz, Hemsbach/Bergstr.
2146/3140-543210

TABLE OF CONTENTS

1. INTRODUCTION

During the past twenty-five years the hyperbolic properties of smooth dynamical systems (i.e. of diffeomorphisms and flows) were studied in the ergodic theory of such systems in a more and more general framework (see $[\text{Ano}]_{1,2}$, [Sma], [Nit], [Bri], $[\text{Kat}]_1$, $[\text{Pes}]_{1,3}$, $[\text{Rue}]_{2,3}$). The detailed historical survey of the hyperbolicity and its role in the ergodic theory up to 1967 is given in $[\text{Ano}]_2$, Chapter 1.

One of the most important features of smooth dynamical systems showing behavior of hyperbolic type is the existence of invariant families of stable and unstable manifolds and their so called "absolute continuity". The most general theorem concerning the existence and the absolute continuity of such families has been proved by Ya. B. Pesin ($[\text{Pes}]_{1,2}$).

The final results of this theory give a partial description of the ergodic properties of a smooth dynamical system with respect to an absolutely continuous invariant measure in terms of the Lyapunov characteristic exponents. One of the most striking of the many important consequences of these results described in $[\text{Pes}]_{2,3}$ is the so called Pesin entropy formula which expresses the entropy of a smooth dynamical system through its Lyapunov characteristic exponents.

Our first main purpose is to generalize Pesin's results to a broad class of dynamical systems with singularities and at the same time to fill gaps and correct errors in Pesin's proof of absolute continuity of families of invariant manifolds ($[\text{Pes}]_1$, Sec. 3). We followed Pesin's scheme very closely and this may at least partly explain the length of our presentation and heaviness of details, especially in Part II. Parts I and II contain the theory of stable (and unstable) invariant manifolds in our more general situation and correspond to the context of $[\text{Pes}]_1$. At the end of Part II we also prove an infinite dimensional counterpart of Pesin's results from $[\text{Pes}]_1$.

The motivation for our generalization lies in the fact that some important dynamical systems occurring in classical mechanics (for example, the motion of the system of rigid balls with elastic collisions) do have singularities. Some of these systems (including the example mentioned) can be reduced to so-called billiard systems. Briefly speaking, a billiard system describes the motion of a point mass within a Riemannian manifold with boundary with reflection from the boundary. Our general conditions on the singularities formulated in Sec. 1 of Part I grew out of an attempt to understand the nature of singularities in the billiard problem.

Since a Poincaré map (first-return map on a section) for a smooth flow usually has singularities, considering transformations with singularities may also provide a unified treatment of discrete time and continuous time dynamical systems.

In Part III we prove the below estimate for the Pesin entropy formula. This part reproduces with minor changes the paper $[\mathrm{Led}]_1$, whose idea goes back to $[\mathrm{Sin}]_1$ and $[\mathrm{Pes}]_{2,3}$. This proof uses in an essential way the results of Parts I and II. Recently R. Mañe ($[\mathrm{Mañ}]_1$) gave in smooth case an alternative very ingenious proof of the estimation from below. His proof avoids completely the use of invariant manifolds and it is substantially simpler than the proof along the Sinai-Pesin line. It seems that Mañe's method can be applied to our case.

The above entropy estimate proved in Part IV is largely independent of the rest of the book.

In $[\mathrm{Pes}]_3$ Pesin derives from his results on invariant stable and unstable manifolds the description of ergodic properties of a smooth dynamical system on the invariant set with non-zero Lyapunov exponents. All his arguments with the sole exception of his proof of Bernoulli property literally apply to our case. It seems that the proof of Bernoulli property requires a somewhat stronger estimate of the Jacobian of the Poincaré map than the one obtained in Part II.

Results from $[\mathrm{Kat}]_2$ concerning the connection between entropy and the growth of periodic points also hold in our situation assuming that the measure satisfies the conditions from Sec. 1 of Part I.

In Part V we study in great detail the singularities of the Poincaré map for plane billiards and show that the conditions from Sec. 1 of Part I are satisfied with respect to the natural absolutely continuous invariant measure for a broad class of such systems. This class includes all compact regions bounded by a finite number of convex and concave arcs of class C^3 and straight line intervals, with the extra assumption: every convex arc has the tangency of finite order with all its tangents. By the results of Parts III and IV the Pesin entropy formula is satisfied for such billiards. We do not know whether the above entropy estimate through the Lyapunov exponents holds for an arbitrary invariant measure for such a billiard. Let us notice that recently M. Wojtkowski ($[\mathrm{Woj}]_{1,2}$) found an easy proof that for so-called Sinai-Bunimovich billiards the Lyapunov exponents are non-zero.

Resuming, one can say that in the present book we completed the lower right corner of the following diagram,

The theory of Anosov systems and of the related systems as Axiom A systems, etc.	⟶	The theory of billiards of Sinai and Bunimovich
↓		↓
Pesin Theory of diffeomorphisms of compact manifolds	⟶	Pesin Theory of mappings with singularities

A concise résumé of the main results of the present book can be found in [Str].

Other presentations of Pesin's theorem concerning the existence of invariant manifolds were given later by D. Ruelle ($[\text{Rue}]_1$) and A. Fathi, M. Herman and J.-C. Yoccoz ([Fat]). D. Ruelle has developed several generalizations of that theorem (non-invertible smooth maps, a class of infinite-dimensional maps ($[\text{Rue}]_{2,3}$)). R. Mañe has found another infinite-dimensional version of Pesin's theorem ($[\text{Mañ}]_2$).

The authors would like to point out their unequal participation in the preparation of this book. Almost all the text was actually written by the second author. The first author suggested the general plan of the work and worked out the arguments which allow us to overcome the presence of singularities in the construction of invariant manifolds and in the above entropy estimate. Naturally, we discussed together numerous questions concerning practically all subjects treated in the text.

The first draft of the theory described in the present book was presented by the second author in December 1978 at the Seminar of Mathematical Physics at IHES (Bures-sur Yvette, France). The material of this book represents a part of the "Thèse d' Etat" of the second author, defended 30 April 1982 at University Paris VI (France).

Our notations are very similar to those used by Pesin, but they are not the same.

Concerning the enumeration of formulas, theorems, etc, the first number indicates the section in which the given formula, theorem, etc., is contained. The lower Roman numeral indicates the part of the book. In the interior of the same parts, the Roman numerals are not marked.

Despite all our efforts, some mistakes can remain. We will be grateful to the readers kind enough to point them out.

Acknowledgments. This book owes very much to Dr. F. Ledrappier (CNRS, University Paris VI, France) and to Dr. F. Przytycki (Mathematical Institute of Polish Academy of Sciences, Warsaw).

Besides being a co-author of Part III, F. Ledrappier made numerous useful remarks concerning other topics treated in the book. In particular he played a very important role in the elaboration of the infinite dimensional case.

The role of F. Przytycki can hardly be overestimated. We owe him the final formulation of conditions characterizing our class of maps with singularities. In the previous versions conditions on the growth of the first derivative as well as of the growth of the two first derivatives of the inverse mapping near the singularities were assumed. Using ideas of F. Przytycki we were able to dispose of these conditions in Parts I-III and consequently to extend the class of mappings under consideration. We thank sincerely both of them.

We also thank Dr. G. Benettin (University of Padova, Italy), Dr. M. Brin (University of Maryland, USA), Dr. P. Collet (Ecole Polytechnique, Palaiseau, France), Dr. M. Misiurewicz (University of Warsaw, Poland), Dr. Ya. B. Pesin (Moscow, URSS), Dr. R. Roussarie (University of Dijon, France), Dr. J.-P. Thouvenot (CNRS, University Paris VI, France) and Dr. L.-S. Young (Michigan State University, U.S.A) for very useful discussions. In particular the first author discussed the early version of the theory described in this book with Ya. B. Pesin who made several useful remarks. G. Benettin communicated to us important formula $(4.10)_V$. M. Misiurewicz found the counterexample described in Sec. 7.8_V.

We thank also Dr. R. Douady (Ecole Polytechnique, Palaiseau, France), Dr. M. Levi (Boston University, U.S.A.) and Dr. T. Nadzieja (University of Wroclaw, Poland) for their help in the final editing of the text.

We would especially like to acknowledge the advice and gentle criticism of Dr. R. Douady, whose careful reading of the manuscript enabled us to make many corrections and improvements.

It is our pleasant obligation to express our gratitude to the Department of Mathematics of the University of Maryland and the National Science Foundation for the support given to the second author for his trips to the U.S.A. (NSF Grant MCS79-030116). The second author also thanks the Department of Mathematics of the Centre Scientifique et Polytechnique of University Paris XIII, for excellent working conditions.

PART I

EXISTENCE OF INVARIANT MANIFOLDS FOR SMOOTH MAPS WITH SINGULARITIES

A. Katok and J.-M. Strelcyn

1. CLASS OF TRANSFORMATIONS WITH SINGULARITIES

1.1 In Section 1 we describe the general setting for the problem and formulate basic assumptions (1.1)-(1.3) on maps under consideration. These conditions insure that the set of singularities of the map under consideration is "thin" and that the second derivatives of the map grow moderately near this set.

Let M be a compact metric space with metric ρ satisfying the following conditions (A) and (B).

(A) M contains as an open dense subset an open smooth (at least of class C^4) Riemannian manifold V of finite dimension m.

(B) There exist positive constant $C < 1$, $R < 1$, $g \geq 1$ and q such that for every point $x \in V$

a) $R(x,V) \geq C(\min(R,[\rho(x,M\setminus V)]^g)) \overset{\text{def}}{=} R_V(x)$ where by $\rho(x,X)$ we denote the distance from the point x to the set $X \subset M$, and by $R(x,V)$ the radius of injectivity of the exponential map $\exp_x\colon T_xV \to V$.

b) for every $y \in V$ such that $\rho(x,y) < R_V(x)$, one has $\|d\exp_x(w)\| \leq q$, where $w = \exp_x^{-1}y$ and $\|d(\exp_x^{-1})(y)\| \leq q$. Here $\|\cdot\|$ denotes the operator norm of the linear maps $d\exp_x(w)\colon(T_xV,\|\cdot\|_x) \to (T_yV,\|\cdot\|_y)$ and $d(\exp_x^{-1})(y)\colon(T_yV,\|\cdot\|_y) \to (T_xV,\|\cdot\|_x)$,

respectively, where $\|\cdot\|_x$ (or shortly $\|\cdot\|$) denotes the norm in T_xV induced by the Riemannian metric.

c) For every $\varepsilon > 0$ there exists $r_\varepsilon > 0$ such that for every $x, y \in V$ satisfying $\rho(x,y) < \min(r_\varepsilon, R_V(x)) \overset{\text{def}}{=} R_\varepsilon(x)$ one has $\|d\exp_x(w)\| \leq 1+\varepsilon$ where $w = \exp_x^{-1}(y)$ and $\|d(\exp_x^{-1}(y)\| \leq 1+\varepsilon$.

Obviously condition (B) is always satisfied when M is a smooth compact Riemannian manifold and V is a smooth open submanifold of M.

We will say that the metric space Y with metric d has a capacity equal to m if

$$\limsup_{\varepsilon \to 0} \frac{\log s(\varepsilon)}{\log \frac{1}{\varepsilon}} = m$$

where $s(\varepsilon)$ denotes the minimal cardinal of a covering of Y by open balls of radius ε.

As follows from Sec. 3 of [Ko 1], every compact manifold (with border) of dimension m has a capacity equal to m.

C) The space M with metric ρ is of finite capacity.

Let us note that condition (Bc) will only play a role in Parts II and III and the condition (C) only in Part IV.

Let N be an open subset of V and let $\Phi : N \to V$ be a mapping which is a $C^p (p \geq 2)$ diffeomorphism between N and its image, i.e., a diffeomorphic embedding of N into V.

Let μ be a Borel probability measure on M invariant with respect to Φ, i.e., $\mu(\Phi^{-1}(B)) = \Phi(B)$ for any Borel set $B \subset M$. We will always assume the following conditions (1.1)-(1.3) concerning M, Φ and μ.

Let $A = M \setminus N$. Let $U_\varepsilon(L)$ be the open ε-neighborhood of the subset L of M with respect to the metric ρ.

(1.1) <u>There exist positive constants</u> c_1 <u>and</u> a <u>such that for every positive</u> ε,

$$\mu(U_\varepsilon(A)) \leq c_1 \varepsilon^a .$$

Obviously, condition (1.1) implies that $\mu(A) = 0$ or equivalently that $\mu(N) = 1$. From the Φ-invariance of the measure μ one obtains that $\mu(\tilde{N}) = 1$, where $\tilde{N} = \bigcap_{r=-\infty}^{\infty} \Phi^r(N)$ is the set where all iterates of Φ are defined. Thus $\tilde{N}$ is always non empty.

We will denote by $d^k \Phi^\ell_x$ or $d^k \Phi^\ell(x)$ the k-th derivative of Φ^ℓ at the point x. We will suppose that

$$\left.\begin{aligned} &\int_M \log^+ \|d\Phi_x\| d\mu < +\infty \\ &\int_M \log^+ \|d\Phi_x^{-1}\| d\mu < +\infty \end{aligned}\right\} \qquad (1.2)$$

where, by definition, $\log^+ x = \max(\log x, 0)$ and the norm $\|\cdot\|$ is the operator norm induced by the Riemannian metric ρ.

These conditions quarantee the applicability of the Oseledec Multiplicative Ergodic Theorem (see Appendix 2, cf. also [Ose], $[\text{Pes}]_2$, [Rag], $[\text{Led}]_1$ and $[\text{Rue}]_{2,3}$).

For $x \in N$, like in condition (Ba) let us denote by $R(x,N)$ the radius of the injectivity of the exponential map $\exp_x : T_x N \to N$. It is clear that $R(x,N) = \min(R(x,V), \rho(x,A))$. Thus in particular

$R(x,N) \geq C \min(R, [\rho(x,A)]^g) \overset{\mathrm{def}}{=} R_N(x)$.

Let us denote for $x \in N$, $\Phi_{0x} = \exp^{-1}_{\Phi(x)} \circ \Phi \circ \exp_x$. The mapping Φ_{0x} is well defined in a neighborhood U of $0 \in T_xN$ and $\Phi_{0x}(U)$ is a neighborhood of $0 \in T_{\Phi(x)}N$. Nevertheless, in general, $\Phi_{0x}(h)$ can not be defined for all $h \in T_xN$, $\|h\| < R_N(x)$.

We will now formulate the main condition concerning the growth of Φ.

(1.3) <u>There exist positive constants</u> $c_2 \geq 1$ <u>and</u> b <u>such that if</u> $x \in N$, $h \in T_xN$, $\|h\| < R_N(x)$ and $\Phi_{0x}(h)$ is defined then one has

$$\|d^2\Phi_{0x}(h)\| < c_2\rho(\exp_x(h),A)^{-b}.$$

Without loss of generality one can always assume that $b > 2g$ where the constant g comes from the condition (Ba).

The conditions (A), (B) and (1.1)-(1.3) are sufficient for all purpose of Parts I - III, in particular for the construction of stable manifolds of Φ as well as unstable ones (i.e., stable for Φ^{-1}).

It is remarkable that the counterparts of conditions (1.1) and (1.3) with respect to Φ^{-1} are not necessary.

To prove the estimation of the entropy from above we will need the following extra assumption.

(1.4) <u>There exist positive constants</u> c_3 <u>and</u> $d \geq 1$ <u>such that for every</u> $x \in N$

$$\|d\Phi_x\| < c_3\rho(x,A)^{-d}.$$

Compared to condition (1.4), condition (1.3) looks somewaht artificial because we want to work with the "quadratic part of the second derivative. This is only possible for the maps between linear spaces. To achieve that, we consider $d^2\Phi_{0x}$ instead of $d^2\Phi_x$.

Let us note also that when M is a compact smooth manifold, the validity of conditions (1.1)-(1.4) is independent of the choice of the Riemannian metric ρ on M. In this case, the choice of metric influences only the constants c_1 and c_3.

It was supposed above that M is a compact metric space. In fact, for the purpose of Parts I - III it is sufficient to suppose M to be a complete metric space. In this more general setting the condition (1.3) is somewhat restrictive. It can be replaced by the following more general condition.

(1.3') There exists positive constants C^1, c_2 and b such that for every $x \in N$ and $h \in T_xN$ such that $\|h\| < R_N(x)$ and such that $\Phi_{0x}(h)$ make a sense one has

$$\|d^2\Phi_{0x}(h)\| \le \max(C^1, c_2\rho(x,A)^{-b}).$$

Clearly, in case of compact M, the conditions (1.3) and (1.3') are equivalent.

Nevertheless, to avoid additional purely formal complications in an already sufficiently complicated text, we will restrict ourselves to the case of compact M.

It is natural to consider A as a "singular set" of Φ. However, in some examples (see Part V), the set A has to be chosen substantially larger than the singular set in the usual sense, in order to quarantee conditions (1.3) and (1.4).

Now we describe the most important example of a situation satisfying condition (1.1). Namely, M is the union of a finite number of smooth compact manifolds $M_1,\ldots,M_r$ (perhaps with boundary and with angles) all of dimension $m \ge 2$ glued together along a finite number of C^1 submanifolds of positive codimensions; A is the union of a finite number of C^1 submanifolds of positive codimensions of $M_1,\ldots,M_r$; μ is a measure such that its restriction to $N = M\backslash A$ is an absolutely continuous measure with a bounded density, i.e., a measure of the form $h\nu$, where ν is the Riemannian volume induced by ρ and h is a bounded function. In this case we can set $a = 1$ in (1.1). Even if the density h has "moderate growth" near A it is still possible for (1.1) to be true for some $a < 1$.

Finally, let us note that as it will be proved in Sec. 12_{II}, for the purposes of Parts I-III the second condition of (1.2) can be dropped. Moreover, in what concern Part I, the invertibility of ϕ is also inessential.

1.2 The class of transformations with singularities described above is derived from the very detailed study of singularities of plane billiards discussed in Part V. In Part V it is shown that all above conditions are satisfied for an appropriate Poincaré map of the billard system in a region whose boundary is not too degenerate, e.g. a compact region in the Euclidean plane bounded by a finite number of real-analytic curves (i.e. of real-analytic images of closed intervals).

For the class of transformations introduced in Sec. 1.1 we will prove the counterpart of results of Pesin from $[\mathrm{Pes}]_{1-3}$. Nevertheless there is no doubt that for many other classes of transformations with singularities which may arise from others examples, the Pesin's results are still true since the proofs from $[\mathrm{Pes}]_{1-3}$ are based on very few general ideas which can be adapted to various settings.

Let us note that for any map which is the composition of a finite number of maps satisfying the assumptions (1.1)-(1.3), all results of Parts I-III remain true (see Sec. 4.2). This class of maps is actually larger than the previous one because, for example, one can construct a map Φ satisfying the assumptions (1.1)-(1.3) such that for $\Phi \circ \Phi$, condition (1.3) fails.

2. PRELIMINARIES

2.1 Recall that if V is a smooth manifold, $x \in V$, then T_xV denotes the tangent space to V at the point x. If $X \subset V$, we denote by T_XV the set of all tangent vectors to V with support in X. TV denotes the tangent bundle of V.

Let E,F be two linear manifolds in a Euclidean space H. We denote by $\gamma(E,F)$ and $\Gamma(E,F)$ the minimal angle and the aperture between E and F respectively, i.e.

$$\gamma(E,F) = \inf_{\substack{0\neq e\in E' \\ 0\neq f\in F'}} \gamma(e,f) \quad \text{and} \quad \Gamma(E,F) = \sup_{\substack{e\in E' \\ \|e\|=1}} (\inf_{v\in F'} \|e-v\|),$$

where $\gamma(e,f)$ denote the angle between vectors e and f and where E', F' are linear subspaces of H parallel to E and F respectively. Let us note that in general $\Gamma(E,F) \neq \Gamma(F,E)$. It is clear that $2\Gamma(E,F) \geq \gamma(E,F)$.

2.2 The set $\tilde{N}$ defined in Section 1.1 is Φ-invariant, and we can consider the restriction of the derivative $d\Phi$ to the set $T_{\tilde{N}}N$ as a linear extension of Φ. Condition (1.2) shows that the assumptions of the Oseledec Multiplicative Ergodic Theorem (see Appendix 2) are satisfied for this extension. To see this, one can construct a measurable family of linear isometries

$$\{\sigma_x : (T_xN, \|\cdot\|_x) \to (\mathbb{R}^m, \|\cdot\|)\}_{x\in\tilde{N}},$$

where $\|\cdot\|$ is the standard Euclidean norm in $\mathbb{R}^m$, and transfer the restriction of $d\Phi$ to $T_{\tilde{N}}N$ to the direct product $N \times \mathbb{R}^m$. Then condition (1.2) becomes the condition of integrability for the matrix cocycle in the Oseledec theorem.

To obtain the family $\{\sigma\}$ one can proceed as follows. First choose one point x_i in every connected component V_i of V, and for every point $x \in V_i$ choose a smooth path γ_x connecting x with the point x_i such that V_x depends on x measurably. Then identify $T_x V$ with $T_{x_i} V$ by the parallel translation along γ_x (which is an isometric map). Finally choose isometries

$$\sigma_i : (T_{x_i} V, \|\cdot\|_{x_i}) \to (\mathbb{R}^m, \|\cdot\|) \quad i = 1,\ldots,k,\ldots .$$

By the Oseledec Theorem there exists a Φ-invariant measurable subset $\Lambda \subset \tilde{N}$ such that

(i) $\mu(\Lambda) = 1$,

(ii) for every point $x \in \Lambda$ there exists a $d\Phi$-invariant μ-measurable decomposition

$$T_x M = \bigoplus_{i=1}^{s(x)} H_i(x),$$

such that uniformly for all vectors $v \in H_i(x), \|v\| = 1$, one has the finite limit

$$\lim_{n\to\pm\infty} \frac{\log\|d\Phi_x^n(v)\|}{n} = \chi_i(x),\ i = 1,\ldots,s(x),$$

$\chi_1(x) < \chi_2(x) <\ldots< \chi_{s(x)}(x)$,

The numbers $\chi_i(x)$ are called the Lyapunov characteristic exponents of Φ at the point x and $d_i(x) \stackrel{\text{def}}{=} \dim H_i(x)$ is called the multiplicity of the exponent $\chi_i(x)$.

(iii) for every point $x \in \Lambda$, let us denote by $\lambda_1(x) \le \lambda_2(x) \le\ldots \le \lambda_m(x)$ all Lyapunov characteristic exponents at x, where any exponent $\chi_i(x)$ is repeated $k_i(x)$ times; i.e. the first $k_1(x)$ of them coincides with $\chi_1(x)$, the following $k_2(x)$ with $\chi_2(x)$, etc. Now, for $x \in \Lambda$, let $f_1(x),\ldots,f_m(x)$ be an arbitrary family of bases of $\mathbb{R}^m$ such that for every $i, 1 \leqq i \leqq m$,

$$\lim_{n\to\infty} \frac{1}{n}\|\log d\phi_x^n(f_i(x))\| = \lambda_i(x).$$

Then, if $m \ge 2$, for every two non-empty disjoint subsets $P,Q \subset \{1,2,\ldots,m\}$ we have:

$$\lim_{n\to\infty} \frac{1}{n} \log \gamma(d\phi_x^n(E_P), d\phi_x^n(E_Q)) = 0,$$

where E_P and E_Q denote the linear subspaces of $\mathbb{R}^m$ spanned by the vectors $\{f_i(x)\}_{i\in P}$ and $\{f_i(x)\}_{i\in Q}$ respectively.

(ii) and (iii) imply easily

(iv) for every point $x \in \Lambda$ and every $k, 1 \leq k < s(x)$, let us note $E_1 = \bigoplus_{i=1}^{k} H_i(x)$. Let $E_2 \subset T_xN$ be an arbitrary linear subspace such that $E_1 \oplus E_2 = T_xN$. Then

$$\lim_{n\to\pm\infty} \frac{\log \gamma(d\Phi_x^n(E_1), d\Phi_x^n(E_2))}{n} = 0.$$

The points x such that (ii) and (iii) hold are called regular. If (ii) and (iii) hold but only for $n \to \infty$, the point is called regular in the future.

2.3 Let us fix $T \le 0$ (usually, in the applications, one takes $T = 0$). Let $\ell(x)$ be the largest positive integer such that

$$\chi_{\ell(x)}(x) < T \le \chi_{\ell(x)+1}(x).$$

Let us define

$$\lambda(x) = \exp \chi_{\ell(x)}(x) \tag{2.1}$$

$$\mu(x) = \exp \chi_{\ell(x)+1}(x). \tag{2.2}$$

Clearly, $\lambda(x) < 1$. Let furthermore

$$\left.\begin{aligned} E_{1x} &= \bigoplus_{i=1}^{\ell(x)} H_i(x) \\ E_{2x} &= \bigoplus_{i=\ell(x)+1}^{s(x)} H_i(x). \end{aligned}\right\} \tag{2.3}$$

For the positive numbers r and s, $r < 1$, $r < s$, let us define $\Lambda_{r,s} \subset \Lambda$ by

$$\Lambda_{r,s} = \{x \in \Lambda: \lambda(x) \le r < s \le \mu(x)\}. \tag{2.4}$$

Obviously, every set $\Lambda_{r,s}$ is measurable and ϕ-invariant.

For $x \in \Lambda_{r,s}$, let $\varepsilon(x) = \varepsilon_{r,s}$ be a so small positive number depending only on r and s that

$$r \exp(100\varepsilon_{r,s}) < 1 \quad \text{and} \quad r \exp(100\varepsilon_{r,s}) < s. \tag{2.5}$$

Furthermore, let

$$\lambda'(x) = \lambda(x) \exp(3\varepsilon(x)) \tag{2.6}$$

$$\mu'(x) = \mu(x) \exp(-3\varepsilon(x)) \tag{2.7}$$

$$\nu(x) = \exp(-\varepsilon(x)). \tag{2.8}$$

From (2.5) it follows that

$$0 < \lambda(x) < \lambda'(x) < \mu'(x) < \mu(x), \ \lambda'(x) < 1. \tag{2.9}$$

The following theorem is proved in Sec. 1.3 of $[\mathrm{Pes}]_1$ for diffeomorphisms of compact manifolds. The proof remains valid also in our case.

THEOREM 2.1. *There exists a measurable positive function* L *defined on* $\Lambda_{r,s}$ *such that if* $v \in d\Phi^m E_{1x}$, *then for every* $n \in \mathbb{Z}$

$$(\|d\Phi^n_{\Phi^m x} v\|)^{\operatorname{sign} n} \le L(x) r^n \exp(\varepsilon_{r,s}(n + |m|))\|v\|; \tag{2.1.1}$$

if $v \in d\Phi^m E_{2x}$, *then for every* $n \in \mathbb{Z}$

$$(\|d\Phi^n_{\Phi^m x} v\|)^{\operatorname{sign} n} \ge (L(x))^{-1} s^n \exp(-\varepsilon_{r,s}(n + |m|))\|v\|; \tag{2.1.2}$$

and the angle $\gamma(\Phi^m(x))$ *between the subspaces* $d\Phi^m(E_{1x})$ *and* $d\Phi^m(E_{2x})$ (*i.e., the smallest angle between the vectors* $0 \ne u \in d\Phi^m(E_{1x})$ *and* $0 \ne v \in d\Phi^m(E_{2y})$) *satisfies for every* $m \in \mathbb{Z}$

$$(L(x))^{-1} \exp(-\varepsilon_{r,s}|m|) \le \gamma(\Phi^m(x)). \tag{2.1.3}$$

For $\ell > 0$, let us define now the set

$$\Lambda^{\ell}_{r,s} = \{x \in \Lambda_{r,s};\ L(x) \le \ell\}.$$

obviously, if $\ell_1 > \ell_2$ then $\Lambda^{\ell_1}_{r,s} \supset \Lambda^{\ell_2}_{r,s}$ and $\bigcup_{\ell>0} \Lambda^{\ell}_{r,s} = \Lambda_{r,s}$.

Let us denote for $0 \le k \le m$

$$\left.\begin{aligned} &\dim E_{1x} = k(x),\ \Lambda_k = \{x \in \Lambda; k(x) = k\},\ \Lambda_{k,r,s} = \Lambda_k \cap \Lambda_{r,s}, \\ &\text{and } \Lambda^{\ell}_{k,r,s} = \Lambda_k \cap \Lambda^{\ell}_{r,s} \end{aligned}\right\} \tag{2.10}$$

The sets Λ_k and $\Lambda_{k,r,s}$ are clearly Φ-invariant.

LEMMA 2.1. *For any integer* q *and* $\ell \ge 1$ *there exists a positive integer* $\alpha = \alpha(q,\ell,s)$ *such that* $f^q(\Lambda^{\ell}_{r,s}) \subset \Lambda^{\alpha}_{r,s}$.

Theorem 2.2 formulated below coincides almost verbatium with Theorem 1.5.1 formulated in $[\text{Pes}]_1$ for diffeomorphisms and the proof uses only the definition of the sets $\Lambda^{\ell}_{r,s}$ and works in our case without any change.

THEOREM 2.2. *There exists a measurable Riemannian metric on the set* Λ (*i.e., a family of scalar products* $\langle \cdot,\cdot \rangle'_x$ *on the spaces* $T_x N$, $x \in \Lambda$, *which depends on* x *in a measurable way*) *and a measurable function* $A(x)$ *such that for every* $x \in \Lambda$ *and any integer* n *the following is true:*

$$A(\Phi^n x) \le A(x)\exp(\varepsilon(x)|n|), \tag{2.2.1}$$

$$\sup_{x \in \Lambda^{\ell}_{r,s}} A(x) = A^{\ell}_{r,s} < +\infty, \tag{2.2.2}$$

$$\left.\begin{aligned} &\|d\Phi_x|E_{1x}\|' < \lambda'(x) \\ &\left(\|d\Phi_x^{-1}|E_{2x}\|'\right)^{-1} \ge \mu'(x), \end{aligned}\right\} \tag{2.2.3}$$

where $\|A\|'$ *denotes the usual operator norm of an operator* A *acting between two tangent spaces (or their subspaces) provided with norms* $\|\cdot\|'_y$ *generated by the scalar products* $\langle \cdot,\cdot \rangle'_y$.

(2.2.4) *The subspaces* E_{1x} *and* E_{2x} *are orthogonal with respect to the scalar product* $\langle \cdot,\cdot \rangle'_x$.

(2.2.5) *The norm* $\|\cdot\|'_x$ *is related to the norm* $\|\cdot\|_x$ *generated by the Riemannian metric* ρ *on* N *by the following inequalities:*

$$\frac{1}{\sqrt{2}}\|\cdot\|_x \le \|\cdot\|'_x \le A(x)\|\cdot\|_x.$$

The norm $\|\cdot\|'_x$ is sometimes called Lyapunov norm.

Recall that $\Lambda_k \subset \Lambda$ is the set of all points $x \in \Lambda$ such that $\dim E_{1x} = k$.

For points $x \in \Lambda_k$, we can define a measurable family of invertible linear transformations

$$\tau'_x\colon T_xM \to \mathbb{R}^m,\ \tau'_x = \tau'_{1x} \times \tau'_{2x},\ \tau'_{1x}\colon E_{1x} \to \mathbb{R}^k,\ \tau'_{2x}\colon E_{2x} \to \mathbb{R}^{m-k}$$

such that τ'_x transforms the scalar product $\langle \cdot,\cdot \rangle'_x$ into the standard scalar product in $\mathbb{R}^m$.

3. OVERCOMING INFLUENCE OF SINGULARITIES

3.1 Let us fix numbers α and γ such that $0 < \alpha < 1$, $\gamma > 0$.

The following sets $\Omega_{\alpha,\gamma}$ will play an important role in subsequent considerations.

(3.1) $\Omega_{\alpha,\gamma} = \{x \in \tilde{N}; \rho(\Phi^n(x), A) \ge \gamma\alpha^{|n|}$ for all integer $n\}$.

LEMMA 3.1. $\lim_{\gamma\to 0} \mu(\Omega_{\alpha,\gamma}) = 1.$

Proof. Since

$$\tilde{N}\backslash\Omega_{\alpha,\gamma} = \{x \in \tilde{N}; \rho(\Phi^n(x),A) < \gamma\alpha^{|n|} \text{ for some integer } n\}$$

we have

$$\mu(\tilde{N}\backslash\Omega_{\alpha,\gamma}) \leq \sum_{n=-\infty}^{\infty} \mu(\{x \in \tilde{N}; \rho(\Phi^n(x),A) < \gamma\alpha^{|n|}\}).$$

Evidently $\{x \in \tilde{N}; \rho(\Phi^n(x),A) < \gamma\alpha^{|n|}\} = \Phi^{-n}(U_{\gamma\alpha^{|n|}}(A)) \cap \tilde{N}$.

Since Φ preserves μ, we have by (1.1),

$$\mu(\Phi^{-n}(U_{\gamma\alpha^{|n|}}(A))\cap\tilde{N}) = \mu(\Phi^{-n}(U_{\gamma\alpha^{|n|}}(A)) = \mu(U_{\gamma\alpha^{|n|}}(A)) < c_1\gamma^a\alpha^{a|n|}$$

and consequently,

$$\mu(M\backslash\Omega_{\alpha,\gamma}) = \mu(\tilde{N}\backslash\Omega_{\alpha,\gamma}) \leq \frac{2c_1\gamma^a}{1-\alpha^a}. \qquad \square$$

In the future we will consider exclusively $\gamma \leq R$, where the constant R come from condition (B a) from Sec. 1.1.

PROPOSITION 3.1.

(3.2.1) $\Phi(\Omega_{\alpha,\gamma}) \subset \Omega_{\alpha,\alpha\gamma}$,

(3.2.2) *for every integer* n *the restriction*

$$d\Phi^n\big|_{T_{\Omega_{\alpha,\gamma}}N}: T_{\Omega_{\alpha,\gamma}}N \to TN$$

is a continuous map.

Proof.

(3.2.1) If $y = \Phi(x)$, $x \in \Omega_{\alpha,\gamma}$, then by (3.1) $\rho(\Phi^n y,A) = \rho(\Phi^{n+1}x,A) \geq \gamma\alpha^{n+1} \geq \gamma\alpha\cdot\alpha^n$.

(3.2.2) Let $x \in \Omega_{\alpha,\gamma}$. The derivative $d\Phi$ is defined and continuous in some neighborhood of every point $\Phi^i x$, $i = 0,\dots,n-1$. Therefore $d\Phi^n$ is defined and continuous in a neighborhood of x.

It follows immediately that $d\Phi^n$ is continuous on the set $\Phi^m\Omega_{\alpha,\gamma}$ for any integers m,n. □

3.2 We proceed to the proof of some basic estimates.

For $x \in N$ and $x \in \mathbb{Z}$ let us denote

$$\left.\begin{aligned} \Phi_{nx} &= \exp^{-1}_{\Phi^{n+1}x} \circ \Phi \circ \exp_{\Phi^n x}, \\ \Phi^{-1}_{nx} &= \exp^{-1}_{\Phi^n x} \circ \Phi^{-1} \circ \exp_{\Phi^{n+1}x}. \end{aligned}\right\} \tag{3.2}$$

The mapping Φ_{nx} is well defined in some small neighborhood of the origin in $T_{\Phi^n x}N, \Phi_{nx}(0) = 0 \in T_{\Phi^{n+1}x}$. Our first aim is to define for a fixed α the small neighborhoods $\Lambda_n(x)$ of $0 \in T_{\Phi^n x}N$ such that the good estimations of $\sup_{h\in\Lambda_n(x)} \|d\Phi_{nx}(h)\|$, $\sup_{h\in\Lambda_{n+1}(x)} \|d\Phi^{-1}_{nx}(h)\|$, $\sup_{h\in\Lambda_n(x)} \|d^2\Phi_{nx}(h)\|$, $\sup_{h\in\Lambda_{n+1}(x)} \|d^2\Phi^{-1}_{nx}(h)\|$ can be obtained. Unfortunately, the kind of estimation that we need can not be obtained for all $x \in N$ or even for all $x \in \Lambda^{\ell}_{k,r,s} \cap \Omega_{\alpha,\gamma}$. Thus now we will give the careful description of the set $\Lambda^{\ell}_{k,s,r,\alpha,\gamma} \subset \Lambda^{\ell}_{k,r,s} \cap \Omega_{\alpha,\gamma}$ on which the needed estimations can be obtained.

First we will prove the following easy lemma.

LEMMA 3.2. *There exists a Φ invariant measurable subset $\Lambda_0 \subset \tilde{N}$, $\mu(\Lambda_0) = 1$ such that for every δ, $0 < \delta < 1$, there exists a positive measurable function $C(x,\delta)$ defined on Λ_0 such that for every $x \in \Lambda_0$, $n \in \mathbb{Z}$, $v \in T_{\Phi^n x}N$ one has*

$$\frac{1}{C(x,\delta)} e^{-\delta|n|} \|v\| \le \|d\Phi_{nx}(0)(v)\| \le C(x,\delta)e^{\delta|n|} \|v\|.$$

Proof. The condition (1.2) implies that $\left|\int_M \log\|d\Phi_x\|d\mu\right| < +\infty$. Thus by Birkhoff Ergodic Theorem applied to $\log\|d\Phi_x\| - \log\|d\Phi_{\Phi(x)}\|$ one deduces that there exists a measurable Φ invariant subset $\Lambda_0 \subset \tilde{N}$, $\mu(\Lambda_0) = 1$, such that for every $x \in \Lambda_0$

$$\lim_{n\to\pm\infty} \frac{\log\|d\Phi_{\Phi^n x}\|}{n} = 0. \tag{3.3}$$

Let $x \in N$. Let us denote $\|d\Phi_x(v)\|$, where $v \in T_x N$ and $\|v\| = 1$ by $\phi(x,v)$. Since

$$\frac{1}{\|d\Phi_x^{-1}\|} \le \phi(x,v) \le \|d\Phi_x\|$$

then from (3.3) one obtains immediately that for $x \in \Lambda_0$

$$\lim_{n\to\pm\infty} \frac{1}{n}\Big(\sup_{\substack{v\in T_{\Phi^n}N \\ \|v\|=1}} \phi(\Phi^n(x),v)\Big) = 0$$

Thus for every $\delta > 0$ there exists $N(x)$ such that for $|n| \ge N(x)$ and all $v \in T_{\Phi^n x}N$, $\|v\| = 1$, one has

$$e^{-n\delta} \le \phi(x,v) \le e^{n\delta}. \tag{3.4}$$

Now, for (3.4) one deduces that for every $\delta > 0$ and every $x \in \Lambda_0$ there exists a measurable positive function $C(x,\delta)$ such that for every $n \in \mathbb{Z}$, $v \in T_{\Phi^n x}N$ one has

$$\frac{1}{C(x,\delta)} e^{-\delta|n|} \|v\| \le \|d\Phi_{\Phi^n x}(v)\| \le C(x,\delta)e^{\delta n}\|v\|. \tag{3.5}$$

From the definition (3.2) of Φ_{nx}, as $d\exp_x(0)$ is the identity mapping, one sees that $d\Phi_{\Phi^n x}(v) = d\Phi_{nx}(0)(v)$ and thus Lemma 3.2 follows from (3.5) □

For every $t > 2$ let us define the subset $\Lambda_0^{t,\delta}$ of Λ_0 by

$$\Lambda_0^{t,\delta} = \{x \in \Lambda_0;\ \frac{1}{t} < C(x,\delta) < t\}.$$

Clearly, if $t_1 < t_2$ then $\Lambda_0^{t_1,\delta} \subset \Lambda_0^{t_2,\delta}$ and $\lim_{t\to\infty} \mu(\Lambda_0^{t,\delta}) = 1$.

For a given α, $0 < \alpha < 1$, let us define $\delta_\alpha > 0$ by

$$\delta_\alpha = \log \frac{1}{\alpha^b} \quad \text{(cf. condition (1.3))} \tag{3.6}$$

As follows from Lemma 3.3, for every $x \in \Lambda_0^{t,\delta_\alpha}$, $n \in \mathbb{Z}$, $v \in T_{\Phi^n x}N$ one has

$$\frac{\alpha^{b|n|}}{t}\|v\| \le \|d\Phi_{nx}(0)(v)\| \le \frac{t}{\alpha^{b|n|}}\|v\|. \tag{3.7}$$

For $x \in \Omega^t_{\alpha,\gamma}$, where

$$\Omega^t_{\alpha,\gamma} = \Omega_{\alpha,\gamma} \cap \Lambda_0^{t,\delta_\alpha}, \tag{3.8}$$

and integer n let us introduce the following neighborhoods $V_n(x)$ and $W_n(x)$, $V_n(x) \subset W_n(x)$, of the point $\Phi^n(x)$:

$$V_n(x) = \{y \in M;\ \rho(y,\Phi^n(x)) < (Q(x)\alpha^{2b|n|})^2\}, \tag{3.9}$$

$$W_n(x) = \{y \in M;\ \rho(y,\Phi^n(x)) < Q(x)\alpha^{2b|n|}\}, \tag{3.10}$$

where $Q(x) = Q(t,\alpha,\gamma)$ is defined by

$$Q(x) = \min\left(\frac{\gamma}{4}, \frac{CR\alpha^g}{6tq^2}, \frac{\gamma^b}{2tc_2(\frac{4}{3})^b}\right). \tag{3.11}$$

Here the constants $0 < R, C < 1$, $q \ge 1$, b and $c_2 \ge 1$ arise from condition (B) and from condition (1.3) respectively.

Let us recall that for $x \in N$, we denote by $R(x,N)$ the radius of injectivity of the exponential map $\exp_x: T_xN \to N$. Since N is an open subset of V

$$R(x,N) = \min(\rho(x,A), R(x,V)). \tag{3.12}$$

From (3.11), from the inequality $b > 2g$ (see Sec. 1.1), and the condition (Ba), as $x \in \Omega_{\alpha,\gamma}$ and $\gamma \le R$, for every $n \in \mathbb{Z}$ one has

$$Q(x)\alpha^{2b|n|} \le \min(CR, C(\gamma\alpha^{|n|})^g) \le C\min(R, [\rho(\Phi^n x, A)]^g) \le$$

$$\le C\min(R, [\rho(\Phi^n x, M\setminus V)]^g) \le R(\Phi^n x, V)$$

We used the fact that $Q(x) \le CR < C$(cf. (3.11)).

On the other hand, since $x \in \Omega_{\alpha,\gamma}$ so that by (3.11) $Q(x) \le \frac{\gamma}{4}$ one has

$$Q(x)\alpha^{2b|n|} \le \frac{1}{4}\gamma\alpha^{|n|} \le \rho(\Phi^n x, A).$$

Consequently $Q(x)\alpha^{2b|n|} \le R(\Phi^n x, N)$ and finally

$$V_n(x) \subset W_n(x) \subset \{y \in M;\ \rho(y,\Phi^n x) \le R(\Phi^n x, N)\}$$

Thus (see Condition (Ba) the mapping $\exp^{-1}_{\Phi^n(x)}$ restricted to $W_n(x)$, maps this neighborhood diffeomorphically onto the ball $M_n(x)$ around the origin in $T_{\Phi^n(x)}N$ of radius $Q(x)\alpha^{2b|n|}$ (cf. e.g., Section 2 of [Che] or the lemma from Section 5.3 in [Gro]). Similarly $\exp^{-1}_{\Phi^n(x)}$ maps the neighborhood $V_n(x)$ diffeomorphically onto the ball $\Lambda_n(x)$ around the origin in $T_{\Phi^n(x)}N$ of radius $(Q(x)\alpha^{2b|n|})^2$.

We will now formulate a lemma saying that for $x \in \Omega^t_{\alpha,\gamma}$ the mappings Φ_{nx} and $\Phi^{-1}_{(n-1)x}$ (cf. (3.2)) are well defined on $\Lambda_n(x)$. This lemma provides also good estimations of $\sup_{h\in\Lambda_n(x)} \|d\Phi_{nx}(h)\|$,

$$\sup_{h\in\Lambda_{n+1}(x)} \|d\Phi^{-1}_{nx}(h)\|, \quad \sup_{h\in\Lambda_n(x)} \|d^2\Phi_{nx}(h)\|, \quad \sup_{h\in\Lambda_{n+1}(x)} \|d^2\Phi^{-1}_{nx}(h)\|.$$

The proof of this lemma is given in the next section.

LEMMA 3.3. *There exist constants* $c_i > 1$, $5 \le i \le 8$, *such that if* $x \in \Omega^t_{\alpha,\gamma}$ *for some* $t > 2$, *then for every* $n \in \mathbb{Z}$ *one has*

$$\Phi(W_n(x)) \subset \{y \in M;\ \rho(y,\Phi^{n+1}(x)) < R(\Phi^{n+1}(x),N)\} \tag{3.3.1}$$

and thus the mapping Φ_{nx} *is well defined on* $M_n(x)$.

$$\sup_{h\in M_n(x)} \|d\Phi_{nx}(h)\| \le \frac{c_5 t}{\alpha^{b|n|}} \tag{3.3.2}$$

$$\sup_{h\in M_n(x)} \|d^2\Phi_{nx}(h)\| \le \frac{c_6}{\gamma^b\alpha^{b|n|}} \tag{3.3.3}$$

$$V_{n+1}(x) \subset \Phi(W_n(x)) \tag{3.3.4}$$

and thus the mapping Φ^{-1}_{nx} *is well defined on* $\Lambda_{n+1}(x)$.

$$\sup_{h\in\Lambda_{n+1}(x)} \|d\Phi^{-1}_{nx}(h)\| \le \frac{c_7 t}{\alpha^{b|n|}} \tag{3.3.5}$$

$$\sup_{h\in\Lambda_{n+1}(x)} \|d^2\Phi^{-1}_{nx}(h)\| \le \frac{c_8 t^3}{\gamma^b\alpha^{4b|n|}} \tag{3.3.6}$$

Remark. Although (3.3.2) is true, one can construct the examples of mappings satisfying all our conditions for which no global estimation of the form $\|d\Phi_x\| \le \frac{C}{[\rho(x,A)]^d}$ for some constants $C > 0$ and $d > 0$ holds.

3.3. Let us define $B_n(x) = \tau'_{\Phi^n x} \exp^{-1}_{\Phi^n x}(V_n(x)) = \tau'_{\Phi^n x}(\Lambda_n(x))$. $B_n(x)$ is a neighborhood of the origin in $\mathbb{R}^m$.

Suppose that $x \in \Lambda_k \cap \Omega_{\alpha,\gamma}$. We can define a map

$$\Phi'_{nx}: B_n(x) \to \mathbb{R}^m$$

in the following way

$$\Phi'_{nx} = \tau'_{\Phi^{n+1}x} \circ \exp^{-1}_{\Phi^{n+1}x} \circ \Phi \circ \exp_{\Phi^n x} \circ (\tau'_{\Phi^n x})^{-1}. \tag{3.13}$$

It follows from (3.3.1) that Φ'_{nx} is a diffeomorphic embedding of $B_n(x)$ into $\mathbb{R}^m$. Moreover, the linear part of Φ'_{nx} at the origin preserves the decomposition

$$\mathbb{R}^m = \mathbb{R}^k \oplus \mathbb{R}^{m-k}.$$

Thus, we can represent Φ'_{nx} in a "coordinate form":

$$\Phi'_{nx}(u,v) = (A_{nx}u + a_{nx}(u,v), B_{nx}v + b_{nx}(u,v)) \tag{3.14}$$

where $u \subset \mathbb{R}^k$, $v \in \mathbb{R}^{m-k}$,

$$\left.\begin{aligned} A_{nx} &= \tau'_{\Phi^{n+1}x} \circ (d\Phi_{\Phi^n x}|_{E_{1\Phi^n x}}) \circ (\tau'_{\Phi^n x})^{-1}, \\ B_{nx} &= \tau'_{\Phi^{n+1}x} \circ (d\Phi_{\Phi^n x}|_{E_{2\Phi^n x}}) \circ (\tau'_{\Phi^n x})^{-1}, \end{aligned}\right\}$$

a_{nx} and b_{nx} are C^P mappings $a_{nx}(0,0) = 0$, $b_{bx}(0,0) = 0$, $da_{nx}(0,0) = 0$, $db_{nx}(0,0) = 0$.

Since the maps τ'_y, $y \in \Lambda$, are isometries, it follows from (2.2.3) that

$$\left.\begin{aligned} &\|A_{nx}\| < \lambda'(x) \\ &(\|B^{-1}_{nx}\|)^{-1} > \mu'(x). \end{aligned}\right\} \tag{3.15}$$

Let $t_{nx} = (a_{nx}, b_{nx})$.

The following theorem plays the key role in the subsequent procedure of the reduction of our situation to Pesin's abstract invariant manifold theorem

THEOREM 3.1. *There exists a constant* $H > 0$ *such that for all* $x \in \Lambda_k \cap \Omega^t_{\alpha,\gamma}$, $n \geq 0$, $z_1, z_2 \in B_n(x)$ *one has*

$$\|dt_{nx}(z_1) - dt_{nx}(z_2)\| < \frac{HA(x)}{\gamma^b \nu(x)[\alpha^b \nu(x)]^n} \|z_1 - z_2\|,$$

where the constant b *and the functions* $A(x)$ *and* $\nu(x)$ *come from* (1.3) *Theorem* 2.2, *and formulas* (2.8), *respectively*.

Remark. Here for the first time the fact that Φ has higher smoothness appears in a crucial way.

Proof. As already stated,

$$\Lambda_n(x) = \{w \in T_{\Phi^n x}N;\ \|w\|_{\Phi^n x} < (Q(x)\alpha^{2b|n|})^2$$

and consequently, $\Lambda_n(x)$ is a convex neighborhood of the origin in $T_{\Phi^n x}N$. The map Φ_{nx} is defined on $\Lambda_n(x)$ and transforms this set into $T_{\Phi^{n+1}x}N$. We have for $z_1, z_2 \in B_n(x)$

$$\left.\begin{aligned} &\|dt_{nx}(z_1) - dt_{nx}(z_2)\| = \|(d\Phi'_{nx}(z_1) - d\Phi'_{nx}(0)) + \\ &- (d\Phi'_{nx}(z_2) - d\Phi'_{nx}(0))\| = \|d\Phi'_{nx}(z_1) - d\Phi'_{nx}(z_2)\| = \\ &= \|d\Phi_{nx}(w_1) - d\Phi_{nx}(w_2)\|', \end{aligned}\right\} \tag{3.16}$$

where $w_i = (\tau'_{\Phi^n x})^{-1}(z_i) \in T_{\Phi^n x}N$, $i = 1,2$, and $\|\psi\|'$ means the operator norm for the linear operator

$$\psi: (T_{\Phi^n x}N, \|\cdot\|'_{\Phi^n x}) \to (T_{\Phi^{n+1}x}N, \|\cdot\|'_{\Phi^{n+1}x}).$$

Now we can use (2.2.5) and (2.2.1) to relate the norms $\|\cdot\|'_x$ with the norms $\|\cdot\|_x$ generated by the Riemannian metric ρ.

From (2.2.5) and (2.2.1), it is easy to see that

$$\|\psi\|' \le \sqrt{2}\, A(x)\exp(\,(n+1)\varepsilon(x))\|\psi\|$$

Consequently

$$\left.\begin{aligned} &\|d\Phi_{nx}(w_1) - d\Phi_{nx}(w_2)\|' \le \sqrt{2}\, A(x)\exp(\,(n+1)\varepsilon(x))\cdot \\ &\cdot\|d\Phi_{nx}(w_1) - d\Phi_{nx}(w_2)\| \le \sqrt{2}\, A(x)\exp(\,(n+1)\varepsilon(x))\cdot \\ &\cdot\|w_1 - w_2\|_{\Phi^n x}\cdot\Big(\max_{\substack{w=\lambda w_1+(1-\lambda)w_2\\ 0\le\lambda\le 1}}\|d^2\Phi_{nx}(w)\|\Big). \end{aligned}\right\} \tag{3.17}$$

Since $w_1, w_2 \in \Lambda_n(x)$ and $\Lambda_n(x)$ is convex, the last factor in the right-hand part of (3.17) is less than or equal to

$$\sup_{w\in\Lambda_n(x)} \|d^2\Phi_{nx}(w)\|.$$

Combining (3.16) (3.17) and using (2.2.5) and (3.3.3) we obtain

$$\left.\begin{aligned} &\|dt_{nx}(z_1) - dt_{nx}(z_2)\| < \\ &< 2A(x)\exp(\,(n+1)\varepsilon(x)) \sup_{y\in\Lambda_n(x)} \|d^2\Phi_{nx}(w)\|\cdot\|z_1 - z_2\| \le \\ &\le 2A(x)\exp(\,(n+1)\varepsilon(x))\,\frac{c_6}{\gamma^b\alpha^{bn}}\,\|z_1 - z_2\|. \end{aligned}\right\}$$

Now, as $\nu(x) = \exp(-\varepsilon(x))$, we obtain

$$\|dt_{nx}(z_1) - dt_{nx}(z_2)\| \le \frac{2c_6 A(x)}{\gamma^b \nu(x)\,[\alpha^b\nu(x)]^n}\,\|z_1 - z_2\|.$$

Thus, the statement of the theorem is true with $H = 2c_6$. □

3.4 Let us consider now for $\ell \ge 2$, the measurable set $I(\ell,k,r,s,\alpha,\gamma) \overset{\text{def}}{=} \Lambda^{\ell}_{k,r,s} \cap \Omega_{\alpha,\gamma} \cap \Lambda_0^{\ell,\delta_\alpha}$. A priori this set may not be compact. By the Lusin theorem one can find a compact subset $\Lambda^{\ell}_{k,r,s,\alpha,\gamma} \subset I(\ell,k,r,s,\alpha,\gamma)$ such that $\mu(\Lambda^{\ell}_{k,r,s,\alpha,\gamma}) \ge (\ell - \frac{1}{\ell})\cdot \mu(I(\ell,k,r,s,\alpha,\gamma))$ and the subspaces E_{1x} and E_{2x} depend continuously on x for $x \in \Lambda^{\ell}_{k,r,s,\alpha,\gamma}$. Moreover one can choose these sets in such a way that $\Lambda^{\ell}_{k,r,s,\alpha,\gamma} \subset \Lambda^{\ell+1}_{k,r,s,\alpha,\gamma}$ for $\ell = 1,2,\ldots$. Let us fix these sets once and for all.

It is clear that $\lim_{\substack{\ell\to\infty \\ \gamma\to 0}} \mu(\Lambda^{\ell}_{k,r,s,\alpha,\gamma}) = \mu(\Lambda_{k,r,s})$ (cf. (2.10)).

4. THE PROOF OF LEMMA 3.3 AND SOME RELATED TOPICS

4.1 We will now prove the Lemma 3.3.

(3.3.1) Let $y \in W_n(x)$. We want to estimate the distance between $\Phi(y)$ and $\Phi^{n+1}(x)$. For this purpose let us consider the shortest geodesic Γ joining $\Phi^n(x)$ and y; such a geodesic lies entirely in the neighborhood $W_n(x)$.

Let us suppose that $\Phi(\Gamma)$ is not entirely contained in $\{y;\ \rho(y,\Phi^{n+1}(x)) \le \frac{R_V(\Phi^{n+1}(x))}{2}\}$ (cf. Condition (Ba)). Thus by continuity arguments there exists the unique point $w \in \Gamma$ such that

$$\rho(\Phi(w),\Phi^{n+1}(x)) = \frac{R_V(\Phi^{n+1}(x))}{2} \tag{4.1}$$

and that for all $u \in I$, $u \neq w$, where by I we denote the interval of Γ between $\Phi^n(x)$ and w, one has $\rho(\Phi(u),\Phi^{n+1}(x)) < \frac{R_V(\Phi^{n+1}(x))}{2}$. Thus in particular, for all $u \in I$, $\Phi_{0\Phi^n(x)}(\tilde{u}) = \Phi_{nx}(\tilde{u})$ is well defined where $\tilde{u} = \exp^{-1}_{\Phi^n(x)}(u)$

From condition (Bb) it follows immediately that for every $u \in I$ one has $\|d\Phi_u\| \le q^2\|d\Phi_{nx}(\tilde{u})\|$ and

$$\rho(\Phi(w),\Phi^{n+1}(x)) \le \ell(\Gamma)\cdot\max_{u\in I}\|d\Phi_u\| \le q^2\ell(\Gamma)\cdot\max_{u\in I}\|d\Phi_{nx}(\tilde{u})\| \tag{4.2}$$

where $\ell(\Gamma)$ denotes the length of Γ. Now, from the Mean Value Theorem one has

$$\|d\Phi_{nx}(\tilde{u})\| \le \|d\Phi_{nx}(0)\| + \ell(\Gamma)\cdot\max_{z\in I}\|d^2\Phi_{nx}(\tilde{z})\|, \tag{4.3}$$

and consequently from (4.2)

$$\rho(\Phi(w),\Phi^{n+1}(x)) \le q^2\ell(\Gamma)\cdot(\|d\Phi_{nx}(0)\| + \ell(\Gamma)\cdot\max_{z\in I}\|d^2\Phi_{nx}(\tilde{z})\|). \tag{4.4}$$

Since $x \in \Omega_{\alpha,\gamma}$, from (3.10) we have for $z \in I \subset W_n(x)$

$$\rho(z,A) \ge \rho(\Phi^n(x),A) - \rho(\Phi^n(x),z) \ge$$

$$\geq \gamma\alpha^{|n|} - \rho(\Phi^n(x),z) \geq \gamma\alpha^{|n|} - \frac{1}{4}\gamma\alpha^{|n|} = \frac{3}{4}\gamma\alpha^{|n|}.$$

Moreover, as $W_n(x) \subset \{y;\, \rho(y,\Phi^n(x)) < R_N(x)\}$, from the condition (1.3), one obtains that for any $z \in I$,

$$\|d^2\Phi_{nx}(\tilde{z})\| \leq \frac{c_2(\frac{4}{3})^b}{\gamma^b\alpha^{|n|b}}. \tag{4.5}$$

On the other hand, as $x \in \Omega^t_{\alpha,\gamma}$ from (3.7), (3.10), (4.4) and (4.5) one obtains that

$$\left.\begin{aligned}&\rho(\Phi(w),\Phi^{n+1}(x)) \leq q^2 Q(x)\alpha^{2b|n|}\left(\frac{t}{\alpha^{b|n|}} + Q(x)\alpha^{2b|n|}\frac{c_2(\frac{4}{3})^b}{\gamma^b\alpha^{b|n|}}\right) < \\ &< 2tq^2Q(x)\alpha^{b|n|}.\end{aligned}\right\} \tag{4.6}$$

Indeed, from (3.11) one has $Q(x) \leq \frac{\gamma^b}{c_2}(\frac{3}{4})^b$. Now, from (3.11) one has $Q(x) \leq \frac{CR\gamma\alpha^g}{6tq^2} < \frac{R}{5}$ and $Q(x) \leq \frac{CR\gamma\alpha^g}{6tq^2} \leq \frac{C\gamma\alpha^g}{6tq^2}$. Thus as $b > 2g$ (see Sec. 1.1) one obtains that

$$\left.\begin{aligned}&2tq^2Q(x)\alpha^{b|n|} \leq \tfrac{1}{3}\min(R,C(\gamma\alpha^{|n|+1})^g) \leq \tfrac{1}{3}\min(R,C[\rho(\Phi^{n+1}(x),A)]^g) \leq \\ &\leq \frac{R_N(\Phi^{n+1}(x))}{3}.\end{aligned}\right\}$$

This, together with (4.6) implies that

$$\rho(\Phi(w),\Phi^{n+1}(x)) \leq \frac{R_N(\Phi^{n+1}(x))}{3}.$$

Since $R_N(y) \leq R_V(y)$, then the last inequality contradicts (4.1). This proves (3.3.1).

(3.3.2) Let $h \in M_n(x)$, i.e. $h = \exp^{-1}_{\Phi^n(x)} y$, where $y \in W_n(x)$. Let Γ be the shortest geodesic joining $\Phi^n(x)$ and y; $\Gamma \subset W_n(x)$. From (4.3) one obtains immediately (3.3.2) with $c_5 = 2$.

(3.3.3) follows immediately from (4.5); $c_6 = c_2(\frac{4}{3})^b$.

(3.3.4) The proof of (3.3.4) is based on the following equality (see formula (5.6.2) from [Car] applied to $n = 1$)

$$f(h+u) = f(h) + df(h)u + \int_0^1 (1-t)d^2f(h+tu)(u,u)dt, \tag{4.7}$$

where f is a twofold continuously differentiable mapping from an open, convex subset V of an Euclidean space E into another Euclidean space, $h, u \in E$, $h, h+u \in V$. In particular this implies that

$$\|f(h+u) - f(h)\| \geq \|f'(h)u\| - (\sup_{z\in V} \|d^2f(z)\|)\|u\|^2. \tag{4.8}$$

Now, applying (4.8) to $f = \Phi_{nx}$, $h = 0 \in T_{\Phi^n x}N$ and to $u \in M_n(x)$, as $x \in \Omega^t_{\alpha,\gamma}$, from (3.7) and (3.3.3) one obtains that if $u \in \partial M_n(x) \subset T_{\Phi^n x}N$, i.e., if $\|u\| = Q(x)\alpha^{2b|n|}$, then

$$\left.\begin{aligned} &\|\Phi_{nx}(u)\| \geq \frac{\alpha^{b|n|}}{t} Q(x)\alpha^{2b|n|} - \frac{c_6}{\gamma^b\alpha^{b|n|}} Q^2(x)\alpha^{4b|n|} > \\ &> Q(x)\alpha^{2b|n|}\left(\frac{1}{t} - \frac{c_6 Q(x)}{\gamma^b}\right) > \frac{Q(x)\alpha^{2b|n|}}{2t} \geq (Q(x)\alpha^{2b|n|})^2 \end{aligned}\right\}$$

because from (3.11) one has $Q(x) < \frac{\gamma^b}{2tc_2(\frac{4}{3})^b}$. Thus Φ_{nx} maps origin into origin and the image $\Phi_{nx}(\partial M_n(x))$ lies outside $\Lambda_{n+1}(x)$. This implies that the image of the ball $M_n(x)$ contains $\Lambda_{n+1}(x)$.

(3.3.5) The prooof of (3.3.5) is based on the following easy inequality. Let f be a C^2 mapping from an open convex subset V of 0 of some Euclidean space E into another Euclidean space. Then for every $u \in V$ and every $h \in E$ one has

$$\|df(u)(h)\| \geq \|df(0)(h)\| - (\sup_{0\leq s\leq 1} \|d^2f(su)\|)\|u\|\|h\|. \tag{4.9}$$

Indeed, (4.9) follows from the obvious identity

$$df(u)(h) = df(0)(h) + \int_0^1 \frac{d}{ds}(df(su)(h))ds$$

and from $\frac{d}{ds}(df(su)(h)) = d^2f(su)(u,h)$.

From (3.3.4) one has $\Phi^{-1}_{nx}(\Lambda_{n+1}(x)) \subset M_n(x)$. Thus any number c_7 such that $\|d\Phi^{-1}_{nx}(w)(g)\| \leq c_7\|g\|$ for any $w \in \Lambda_{n+1}(x)$ and

$g \in T_{\Phi^{n+1}(x)} N$ satisfy $\|d\Phi_{nx}(y)(h)\| \geq \frac{1}{c_7}\|h\|$ for any $y \in M_n(x)$ and $h \in T_{\Phi^n(x)} N$ and vice versa. Indeed $w = \Phi_{nx}(y)$ for some $y \in M_n(x)$.

Applying (4.9) to $f = \Phi_{nx}$ one obtains that for every $y \in M_n(x)$ and $h \in T_{\Phi^n x} N$

$$\|d\Phi_{nx}(y)(h)\| \geq \|d\Phi_{nx}(0)(h)\| - (\sup_{h \in M_n(x)} \|d^2\Phi_{nx}(y)\|)\|y\|\|h\|.$$

Now, using (3.7) and (3.3.3) one obtains that

$$\|d\Phi_{nx}(y)(h)\| \geq \left(\frac{\alpha^{b|n|}}{t} - \frac{c_6}{\gamma^b \alpha^{b|n|}} Q(x)\alpha^{2b|n|}\right)\|h\| \geq \frac{\alpha^{b|n|}}{2t}\|h\|. \qquad (4.12)$$

because from (3.11), $Q(x) \leq \frac{\gamma^b}{2tc_6} = \frac{\gamma^b}{2tc_2(\frac{4}{3})^b}$. Thus one can take $c_7=2t$.

(3.3.6) Differentiating two times the following identity

$$\Phi_{nx} \circ \Phi_{nx}^{-1}(y) = y$$

where $y \in \Lambda_{n+1}(x)$, one obtains the inequality

$$\|d^2(\Phi_{nx}^{-1})(h)\| \leq \|d^2(\Phi_{nx})(\Phi_{nx}^{-1}h)\| \cdot \|d\Phi_{nx}^{-1}(h)\|^3.$$

Now, (3.3.6) follows from (3.3.3) and (3.3.5). We leave the details to the reader. □

4.2 We will now sketch quickly the arguments implying that if the mappings $\Phi_0, \ldots, \Phi_{k-1}$ satisfies assumptions (1.1)-(1.3), then for the mapping $\Phi = \Phi_{k-1} \circ \ldots \circ \Phi_0$ the exact counterpart of Lemma 3.3 holds. As for the mapping Φ the Condition (1.2) is satisfied, than all results of Parts I-III can be applied to Φ.

Let M, V and μ be as in Sec. 1.1. For every j, $0 \leq j \leq k-1$, let us consider an open subset N_j of V and a mapping $\Phi_j : N_j \to V$ satisfying all our assumptions. Let us denote $A_j = M \setminus N_j$ for $0 \leq j \leq k-1$.

Let us consider now, k disjoint copies of M noted M_0, $M_1, \ldots, M_{k-1}$, $M_k = M_0$. Let us denote $\overline{M} = \bigcup_{i=0}^{k-1} M_i$ and let us define on $\overline{M}$ the metric $\tilde{\rho}$ as follows

$$\overline{\rho}(x,y) = \begin{cases} \rho(x,y) & \text{if } x,y \text{ belong to some } M_i,\ 0 \le i \le k-1 \\ \sum_{i=0}^{k-1} \operatorname{diam}(M_i) + 1 & \text{if this is not the case.} \end{cases}$$

Let us define the mapping $\overline{\Phi}:\overline{M} \to \overline{M}$ as follows: $\overline{\Phi}(M_i) = M_{i+1}$ and $\overline{\Phi}$ restricted to M_i coincides with Φ_i, $0 \le i \le k-1$. As the measure $\overline{\mu}$ on $\overline{M}$ we take the normalized measure which coincides with μ on every M_i. Clearly, $\overline{\mu}$ is $\overline{\Phi}$ invariant.

Let us define the subsets $\overline{N}_i \subset M_i$, $0 \le i \le k-1$:

$$\overline{N}_0 = N_0 \cap \Phi_0^{-1}(N_1) \cap (\Phi_1 \circ \Phi_0)^{-1}(N_2) \cap \dots \cap (\Phi_{k-2} \circ \dots \circ \Phi_0)^{-1}(N_{k-1})$$

$$\overline{N}_1 = N_1 \cap \Phi_1^{-1}(N_2) \cap (\Phi_2 \circ \Phi_1)^{-1}(N_3) \cap \dots \cap (\Phi_{k-1} \circ \dots \circ \Phi_1)^{-1}(N_0)$$

— — — — — — — — — — — — — — —

$$\overline{N}_{k-1} = N_{k-1} \cap \Phi_{k-1}^{-1}(N_0) \cap (\Phi_0 \circ \Phi_{k-1})^{-1}(N_1) \cap \dots \cap (\Phi_{k-3} \circ \dots \circ \Phi_0 \circ \Phi_{k-1})^{-1}(N_{k-1}).$$

Let us note that Φ is well defined on each $\overline{N}_i$. As $\mu(N_i) = 1$ for all i, then one easily sees that $\mu(\tilde{N}_i) = 1$, $0 \le i \le k-1$. In particular, any $\overline{N}_i$ is an open non-empty submanifold of V. Let us define $\overline{N} = \bigcup_{i=0}^{k-1} \overline{N}_i ; \overline{\mu}(\overline{N}) = 1$. Thus if $x \in \overline{N}$, $\overline{\Phi}(x)$ is well defined and $\overline{\Phi}(x) \in \overline{V} = \bigcup_{i=0}^{k-1} V_i$, where $V_i \subset M_i$ is a copy of V.

Now, like in the case of one mapping let us define the set $\tilde{\overline{N}} = \bigcap_{n \in \mathbb{Z}} \overline{\Phi}^n(\overline{N})$ on which all iterates of $\overline{\Phi}$ are well defined, $\overline{\mu}(\tilde{\overline{N}}) = 1$, and the set $\overline{\Omega}_{\alpha,\gamma} = \{x \in \tilde{\overline{N}};\ \rho(\overline{\Phi}^n(x), A_{j(n)}) \ge \gamma\alpha^{|n|}$ for all integers $n\}$, where $0 < \alpha < 1$, $\gamma > 0$ and where $j(n)$ is defined by

$$j(n) = \begin{cases} i & \text{if } n = pk + i \text{ for some } p \ge 0,\ 0 \le i \le k-1 \\ k-i & \text{if } n = -(pk+i) \text{ for some } p \ge 0,\ 0 \le i \le k-1. \end{cases}$$

Like for one map, we define the sets $\overline{\Omega}^t_{\alpha,\gamma}$. The exact counterpart of the Lemma 3.3 remains valid for any $x \in \overline{\Omega}^t_{\alpha,\gamma}$ and the mapping $\overline{\Phi}$. As for $x \in N_0$, $\Phi(x) = \overline{\Phi}^k(x)$, then our assertion concerning Φ follows from the fact that it is true for $\overline{\Phi}$. We leave the details to the reader.

5. THE FORMULATION OF PESIN'S ABSTRACT INVARIANT MANIFOLD THEOREM

We formulate now Theorem 2.1.1 from $[\mathrm{Pes}]_1$ in slightly less general form than in Pesin's paper. This theorem deals with the existence of stable manifolds for a sequence of smooth maps defined near the origin of $\mathbb{R}^m$ and satisfying some non-uniformly hyperbolic conditions.

Let U_1, U_2 be open neighborhoods of the origin in the Euclidean spaces $\mathbb{R}^k$ and $\mathbb{R}^{m-k}$, respectively. Then $U = U_1 \times U_2$ is a neighborhood of the origin in $\mathbb{R}^m$.

Suppose that $\{f_n\}$, $n = 0,1,2\ldots$, is a sequence of $C^p (p \geq 2)$ maps $f_n\colon U \to \mathbb{R}^m$ such that each f_n preserves the origin of $\mathbb{R}^m$ and the derivative $df_n(0)$ preserves the decomposition $\mathbb{R}^m = \mathbb{R}^k \times \mathbb{R}^{m-k}$. In other words, f_n can be represented in a form similar to (3.8). Namely, for $u \in U_1$, $v \in U_2$,

$$f_n(u,v) = (A_n u + a_n(u,v),\ B_n v + b_n(u,v)), \tag{5.1}$$

where A_n, B_n are linear maps, $a_n(0,0) = 0$, $b_n(0,0) = 0$, $da_n(0,0) = 0$, $db_n(0,0) = 0$. Note that a_n and b_n are also of class C^p.

Let us denote $f_n \circ \ldots \circ f_0$ by F_n. The map F_n is defined in some neighborhood of the origin which generally depends on n.

Let us also denote the ball

$$\{u \in \mathbb{R}^k;\ \|u\| \leq r\} \text{ by } B_r^k,$$

where $\|\cdot\|$ denotes the standard Euclidean norm

THEOREM 5.1. (cf. $[\mathrm{Pes}]_1$, Theorem 2.1.1). *Suppose that the sequence $\{f_n\}_{n\geq 0}$ has form (5.1) and in addition the following is true:*

(i) *All mappings B_n are invertible and there exist real numbers λ and μ, $0 < \lambda < \min(1,\mu)$ such that for all positive integers n,*

$$\|A_n\| < \lambda, \quad \|B_n^{-1}\|^{-1} \geq \mu.$$

(ii) *Let $t_n(u,v) = (a_n(u,v), b_n(u,v))$. There exist real numbers K and ν such that $\lambda < \nu < 1$, $K > 0$, and for every $z_1, z_2 \in U$ and and every non-negative integer n,*

$$\|dt_n(z_1) - dt_n(z_2)\| \le K\nu^{-n}\|z_1 - z_2\|. \tag{5.2}$$

Then for every real number $\kappa, \lambda < \kappa < \min(\mu,\nu)$, _there exist numbers_ $C = C(\kappa) > 0$, $r_0 = r_0(\kappa) > 0$ _and a_ C^{p-1} _map_ ϕ _such that_

$B^k_{r_0} \subset U_1$, $\phi: B^k_{r_0} \to \mathbb{R}^{m-k}$ _and_

(5.1.1) $\phi(0) = 0$, $d\phi(0) = 0$ _and for all_ $u_1, u_2 \in B^k_{r_0}$

$$\|d\phi(u_1) - d\phi(u_2)\| \le C\|u_1 - u_2\|.$$

(5.1.2) _For every_ $u \in B^k_{r_0}$ _and any non-negative integer_ n, _the map_ F_n _is defined on the graph of_ ϕ,

$F_n(u,\phi(u)) \in U$ _and_

$$\|F_n(u,\phi(u))\| \le 200\kappa^{n+1}\|(u,\phi(u))\|.$$

(5.1.3) _Let_ $(u,v) \in U$, $u \in B^k_{r_0}$, _and let_ $F_n(u,v)$ _be correctly defined for all non-negative integers_ n. _If there exists_ $\tilde{C} > 0$ _such that_

$$F_n(u,v) \in U \quad \text{and} \quad \|F_n(u,v)\| < \tilde{C}\kappa^n$$

for all non-negative integers n, _then_ $v = \phi(u)$.

(5.1.4) _There exist continuous positive functions_ $\psi_i = \psi_i(\lambda,\mu,\nu,\kappa)$ $i = 1,2$, _defined for all_ $(\lambda,\mu,\nu,\kappa) \in \mathbb{R}^4$ _satisfying the inequalities_

$$0 < \lambda < \kappa < \min(\mu,\nu), \ \nu < 1,$$

such that

$$r_0 = \frac{1}{K}\psi_1(\lambda,\mu,\nu,\kappa) \quad \text{and} \quad C = K\psi_2(\lambda,\mu,\nu,\kappa).$$

As follows from (5.1.2) and (5.1.3) the function ϕ essentially does not depend on κ: in other words ϕ's defined for different values of κ coincide in their common domain.

6. INVARIANT MANIFOLDS FOR MAPS SATISFYING CONDITIONS (1.1)-(1.3)

6.1 Now we are prepared to formulate and prove the main result of

Part I - the counterpart of Theorem 2.2.1 from $[\mathrm{Pes}]_1$.

Let us define the sets $\Lambda_{k,r,s,\alpha} = \bigcup_{\substack{\ell>2\\ \gamma>0}} \Lambda^{\ell}_{k,r,s,\alpha,\gamma}$ (cf. Sec. 3.4) and $\Lambda_{r,s,\alpha} = \bigcup_{k=1}^{m} \Lambda_{k,r,s,\alpha}$. All those sets are Φ invariant. Let us denote also $\Lambda^{\ell}_{r,s,\alpha,\gamma} = \bigcup_{k=1}^{m} \Lambda^{\ell}_{k,r,s,\alpha,\gamma}$.

For fixed α and $x \in \Lambda_{r,s,\alpha}$ let us define

$$\gamma(x) = \sup\{r \le R;\ \rho(\Phi^n(x),A) \ge r\alpha^{|n|} \text{ for all integers } n\}$$

and

$$\ell(x) = \inf\{\ell \ge 2;\ x \in \Lambda^{\ell}_{k,r,s,\alpha,\gamma(x)}\}.$$

As $x \in \Lambda^{\ell(x)}_{r,s,\alpha,\gamma(x)}$ then (cf. (3.2.1), Theorem 2.1 and (3.7))

$$\gamma(\Phi(x)) \ge \alpha\gamma(x) \quad \text{and} \quad \ell(\Phi(x)) \le \ell(x)\alpha^{-b}\exp(\varepsilon_{r,s}). \tag{6.1}$$

In the future, when writing $x \in \Lambda^{\ell}_{k,r,s,\alpha,\gamma}$ or $x \in \Lambda^{\ell}_{r,s,\alpha,\gamma}$ by γ and ℓ we understand respectively $\gamma(x)$ and $\ell(x)$ defined above.

(6.2) For $x \in \Lambda_{r,s,\alpha}$ and $r > 0$ let us denote the set

$$B_r^{k(x)} \times B_r^{m-k(x)} \text{ in } \mathbb{R}^m \text{ by } B(x,r)$$

and the set

$$\exp_x(\tau'_x)^{-1}B(x,r) \text{ by } U(x,r).$$

Furthermore, for $x \in \Lambda_{r,s,\alpha}$ and $n \ge 0$ we set

$$U_n(x) = \exp_{\Phi^n x}(\tau'_{\Phi^n x})^{-1}B\left(x, \frac{(Q(x)\alpha^{2bn})^2}{2}\right).$$

It follows from (2.2.5) that $U_n(x) \subset V_n(x)$, where $V_n(x)$ is defined by (3.9).

We will construct the local stable manifold $V(x)$ of a point $x \in \Lambda_{r,s,\alpha}$ in the form

$$V(x) = \exp_x(\tau'_x)^{-1}(\text{graph } \phi_x), \tag{6.2}$$

where

$$\phi_x : B^{k(x)}_{\delta(x)} \to B^{m-k(x)}_{\delta(x)}, \tag{6.3}$$

and the radius $\delta(x)$ is sufficiently small but bounded away from zero on any set of the form $\Lambda^{\ell}_{k,r,s,\alpha,\gamma}$.

Naturally, we expect that $x \in V(x)$ and $T_x V(x) = E_{1x}$, which is equivalent to the following conditions

$$\phi_x(0) = 0, \; d\phi_x(0) = 0. \tag{6.4}$$

Subsequent properties of the manifolds $V(x)$ are listed below in the formulation of Theorem 6.1 and in Section 7.

THEOREM 6.1. Let $\Phi: N \to V$ be a map of class C^P $(p \geq 2)$ satisfying conditions (1.1)-(1.3). Let $\alpha = \alpha(r,s)$ be a number such that

$$0 < \alpha < 1, \; r \cdot \exp(100\varepsilon_{r,s}) < \alpha^{10b} \text{(*)} \tag{6.5}$$

(b comes from (1.3) and $\varepsilon_{r,s}$ is defined by (2.5)).

Moreover, let $\kappa(x)$ be a Φ-invariant function defined on the set $\Lambda_{r,s}$ (cf. (2.10)) such that

$$\lambda'(x) < \kappa(x) < \kappa(x)\exp(2\varepsilon_{r,s}) < \min(\mu'(x), \alpha^b \nu(x)). \tag{6.6}$$

(For example, as follows from (6.5), (2.5) and (2.8)

$$\kappa(x) = \lambda(x)\exp(4\varepsilon_{r,s})$$

satisfies these inequalities.)

Then there exists a positive measurable function

$$\delta(x) < \frac{Q^2(x)}{2} \tag{6.7}$$

defined on the set $\Lambda_{k,r,s,\alpha}$ and a family of C^{p-1} (**) maps

$$\phi_x, x \in \Lambda_{k,r,s,\alpha}$$

of the form (6.3) satisfying (6.4) such that the following additional properties hold.

(*) Let us underline, that this is the first line, where a relation between α and r and s appears.

(**) For C^P smoothness see $[\text{Rue}]_{2,3}$.

(6.1.1) *For every non-negative integer* n, *the map*

$$F_{nx} = \Phi'_{nx} \circ \dots \circ \Phi'_{ox}$$

(*where* Φ'_{ix} *is defined by* (3.13)) *is defined on the graph of* ϕ_x *and*

$$F_{nx}(\text{graph } \phi_x) \subset B(x, Q^2(x)\alpha^{4b(n+1)}).$$

Moreover, for $y \in V(x)$, *where* $V(x)$ *is defined by* (6.2),

$$\rho(\Phi^n x, \Phi^n y) \le 200\sqrt{2}\, A(x)(\kappa(x))^n \rho(x,y).$$

(6.1.2) *If for some* $y \in U(x,\delta(x))$ *and for some constant* C, *one has for all non-negative integers* n

$$\Phi^n y \in U_n(x) \quad \text{and}$$

$$\rho(\Phi^n(x), \Phi^n(y)) \le C(\kappa(x))^n,$$

then $y \in V(x)$.

(6.1.3) $\delta(\Phi^n x) \ge \delta(x)\alpha^{2bn}\exp(-2\varepsilon_{r,s} n)$

and

$$\inf_{x \in \Lambda^{\ell}_{r,s,\alpha,\gamma}} \delta(x) = \delta^{\ell}_{r,s,\alpha,\gamma} > 0.$$

(6.1.4) $\Phi(V(x) \cap U(x,\delta(x)\exp(-2\varepsilon_{r,s})) \cap U(\Phi(x),\delta(\Phi(x))) \subset V(\Phi(x))$

(6.1.5) *There exists a measurable function* G *defined on the set* $\Lambda_{r,s,\alpha}$ *such that*

$$G(x) \ge \alpha^b G(\Phi(x)), \quad \sup_{x \in \Lambda_{r,s,\alpha,\gamma}} G(x) = G_{r,s,\alpha,\gamma} < \infty$$

and for $y \in V(x)$ *one has*

$$\Gamma(T_x V(x), (d\,\exp_x^{-1})(T_y V(x))) \le G(x)A^2(x)\rho(x,y)$$

where the aperture $\Gamma(\cdot,\cdot)$ *is defined in Sec.* 2.1 *and the function* $A(x)$ *is defined in Theorem* 2.2.

To underline that the constructed local invariant manifold corresponds to the Lyapunov Characteristic Exponents bounded away by a member $T \le 0$, we will call it local T-stable invariant manifold (TLSM) of x. In the literature one considers primarily the case $T = 0$ and in this case one speaks simply about local stable invariant manifolds. Nevertheless in the recent work [Led]$_3$, TLSM for $T < 0$ play a key role.

6.3 Proof We will deduce Theorem 6.1 from Theorem 5.1 .

Let us fix a point $x \in \Lambda^{\ell}_{k,r,s,\alpha,\gamma}$, $1 \le k \le m$ and consider the following sequence of maps $\{f_n\}_{n\ge 0}$

$$f_n(z) = \alpha^{-4b(n+1)}\Phi'_{nx}(\alpha^{4bn}x) \tag{6.8}$$

where the maps Φ'_{nx} are defined by (3.13). Any of such map is defined in the some neighborhood of the origin which a priori may depend on n. All these maps are of class C^P.

Lemmas 3.3.1 and 2.2.5 imply that all the maps f_n are defined in the neighborhood $B(x,\frac{Q^2(x)}{2}) = D$ of the origin in $\mathbb{R}^m$.

Thus, we can use (3.14) and write down the representation for f_n in D in the form similar to (5.1). Namely

$$\left.\begin{aligned} &f_n(u,v) = (A_n u + a_n(u,v),\ B_n v + b_n(u,v)) = \\ &(\alpha^{-4b}A_{nx}u + \alpha^{-4b(n+1)}a_{nx}(\alpha^{4bn}(u,v)), \\ &\alpha^{-4b}B_{nx}v + \alpha^{-4b(n+1)}b_{nx}(\alpha^{4bn}(u,v))). \end{aligned}\right\} \tag{6.9}$$

PROPOSITION 6.1 If $\alpha = \alpha(r,s)$ satisfies (6.5), then the assumptions of Theorem 5.1 are true for the sequence of maps $\{f_n\}_{n\ge 0}$ defined by (6.8), with

$$\lambda = \frac{\lambda'(x)}{\alpha^{4b}},\ \mu = \frac{\mu'(x)}{\alpha^{4b}},\ \nu = \nu(x)\alpha^b$$

and

$$K = K(x) = \frac{HA(x)}{\gamma^b\nu(x)\alpha^{4b}} .$$

Here b comes from (1.3), H from Theorem 3.1 and $\nu(x)$ is defined by (2.8). As $H \ge 2$ and $A(x) \ge \frac{1}{\sqrt{2}}$, $K(x) > 1$.

Proof. (i) The inequality

$$\lambda < \min(1,\mu)$$

follows immediately from (6.5) and (2.9). The inequalities for $\|A_n\|$ and $\|B_n^{-1}\|^{-1}$ follow immediately from (3.15).

(ii) The inequality $\lambda < \nu < 1$ follows from (6.5). Inequality (5.2) is an easy consequence of Theorem 3.1 (cf. (6.9)). □

Remark. From (6.6) it follows immediately that the number $\kappa = \frac{\kappa(x)}{\alpha^{4b}}$ satisfies the iequalities

$$\lambda < \kappa < \min(\mu,\nu).$$

Now we can return to the proof of Theorem 6.1. In virtue of Proposition 6.1 we can apply Theorem 5.1 to the sequence of maps $\{f_n\}_{n\geq 0}$ defined by (6.8), with λ,μ,ν,K defined in that proposition and $\kappa = \frac{\kappa(x)}{\alpha^{4b}}$. The application of Theorem 5.1 gives the numbers $r_0 = r_0(x)$, $C = C(x)$, and the map $\phi = \phi_x$. As $K = K(x) > 1$, then without any loss of generality we can admit that $C = C(x) > 1$. Indeed, for this aim it suffices to define in (5.1.4) a new C by $C = K(1 + \psi_2(\lambda,\mu,\nu,\kappa))$.

Let us set

$$\delta(x) = \frac{r_0(x)}{C(x)} \tag{6.10}$$

Since

$$r_0 < \frac{Q^2(x)}{2} < 1$$

condition (6.3) follows from (5.1.1) and the definition of $\delta(x)$; (6.4) also follows from (5.1.1).

From now on the proof follows exactly along the lines of the proof of Theorem 2.2.1 from $[\text{Pes}]_1$. Since the proof in $[\text{Pes}]_1$ is somewhat sketchy we will, for the sake of completeness, give the proof in our case in full detail.

(6.1.1) Suppose that $z \in B(x,\delta(x))$ and that $(f_n \circ \dots \circ f_0)(z)$ is defined. Then

$$\left.\begin{aligned}(f_n \circ \dots \circ f_0)(z) &= \alpha^{-4b(n+1)} F_{nx}(z) = \\ &= \alpha^{-4b(n+1)} (\tau'_{\Phi^{n+1}x} \circ \exp^{-1}_{\Phi^{n+1}x} \circ \Phi^{n+1} \circ \exp_x \circ (\tau'_x)^{-1})(z).\end{aligned}\right\}$$

Let now $y \in V(x)$. Then by (6.2),

$$\tau_x'(\exp_x)^{-1}(y) = (u,\phi_x(u))$$

for some $u \in B_{\delta(x)}^{k(x)}$. From (5.1.2) we have for all $n \geq 0$,

$$F_{nx}(u,\phi_x(u)) \subset B(x,Q(x)\alpha^{4b(n+1)}).$$

Moreover, (5.1.2) implies that,

$$\|(\exp^{-1}_{\Phi^n x} \circ \Phi^n)(y)\|'_{\Phi^n x} \leq 200(\kappa(x))^n \|(u,\phi(u))\|'_x.$$

Thus by (2.2.5), one obtains

$$\rho(\Phi^n x,\Phi^n y) = \|(\exp^{-1}_{\Phi^n x} \circ \Phi^n)(y)\|_{\Phi^n x} \leq \sqrt{2}\|(\exp^{-1}_{\Phi^n x} \circ \Phi^n)(y)\|'_{\Phi^n x} \leq$$

$$\leq 200\sqrt{2}(\kappa(x))^n \|(u,\phi(u)\|'_x \leq 200\sqrt{2}A(x)(\kappa(x))^n \|(u,\phi(u))\|_x =$$

$$= 200\sqrt{2}A(x)(\kappa(x))^n \rho(x,y).$$

(6.1.2) Suppose that $y \in U(x,\delta(x)) \subset U_0(x)$. Then

$$(\tau_x' \circ \exp_x^{-1})(y) = (u,v), \text{ where } u \in B_{\delta(x)}^{k(x)}.$$

Let us assume that $\Phi^n(y) \in U_n(x)$ for all non-negative integers n. Then $(f_{n-1} \circ \dots \circ f_o)(u,v)$ is well defined. Following the same method as in the proof of (6.1.1), using (2.2.1), one deduces from the inequality

$$\rho(\Phi^n x,\Phi^n y) \leq C(\kappa(x))^n$$

that

$$\|(f_{n-1} \circ \dots \circ f_0)(u,v)\| \leq A(\Phi^n x)(\kappa(x)\alpha^{-4b})^n \rho(\Phi^n x,\Phi^n y) \leq$$

$$\leq CA(x)(\alpha^{-4b}\kappa(x)\exp(\varepsilon_{r,s}))^n.$$

Let us denote the number

$\alpha^{-4b}\kappa(x)\exp(\varepsilon_{r,s})$ by κ_1.

It follows from (2.5) and (6.5) that

$$\lambda < \kappa_1 < \min(\mu,\nu).$$

Therefore we can apply (5.1.3) and obtain that $v = \phi(u)$.

(6.1.3) By (3.2.1) if $x \in \Omega_{\alpha,\gamma}$ then for every $n \geq 0$, $\Phi^n(x) \in \Omega_{\alpha,\gamma\alpha^n}$.

On the other hand, as stated in Proposition 6.1,

$$K(x) = \frac{HA(x)}{\gamma^b \nu(x) \alpha^{4b}}. \tag{6.11}$$

Thus, for every $n \geq 0$, in virtue of (2.2.1) one has

$$K(\Phi^n x) = \frac{HA(\Phi^n x)}{\gamma^b \alpha^{nb} \nu(x) \alpha^{4b}} \leq \frac{HA(x) \exp(2\varepsilon_{r,s} n)}{\gamma^b \alpha^{nb} \nu(x) \alpha^{4b}} \tag{6.12}$$

Now, we pass to the proof of (6.1.3). From (5.1.4) and (6.10)-(6.12) we have

$$\frac{\delta(\Phi^n x)}{\delta(x)} = \frac{r_0(\Phi^n x)}{C(\Phi^n x)} \cdot \frac{C(x)}{r_0(x)} = \frac{K^2(x)}{K^2(\Phi^n x)} \geq \alpha^{2bn} \exp(-2\varepsilon_{r,s} n).$$

The last assertion of (6.1.3) follows immediately from (5.1.4), (2.2.2) and the definition of $K(x)$.

(6.1.4) Let us begin with the following remark. Let instead of original TLSM $V(x)$ corresponding to the parameters $\lambda, \mu, \nu, \kappa, K$ and $\frac{Q^2(x)}{2}$, we consider a constructed TLSM $V'(x)$ corresponding to the same parameters but instead of $\frac{Q^2(x)}{2}$ we take $\frac{Q^2(x)\omega}{2}$ where $0 < \omega < 1$. Then as follows from Theorem 5.1 and from the construction of TLSM

$$V'(x) = V(x) \cap U(x, \delta(x)\omega).$$

Now, let $y \in \Phi(V(x) \cap U(x, \delta(x)\exp(-\varepsilon_{r,s})))$. Then $y = \Phi(z)$, when $z \in V(x) \cap U(x, \delta(x)\exp(-\varepsilon_{r,s}))$. By the above remark applied to $\omega = \exp(-\varepsilon_{r,s})$ one sees that $z \in V'(x)$. Therefore by (6.1.1) for all $n \geq 0$ one has

$$\Phi^n(y) = \Phi^{n+1}(z) \in U'_{n+1}(x),$$

where $U'_{n+1}(x) = \exp_{\Phi^{n+1}x}(\tau'_{\Phi^{n+1}x})^{-1} B(\Phi^{n+1}(x), \frac{Q^2(x)}{2\exp(2\varepsilon_{r,s})} \alpha^{4b(n+1)})$.

From the definition of $Q(x)$ and (6.1) it follows easily that for every $n \geq 0$ one has

$$U'_{n+1}(x) \subset U_n(\Phi(x)).$$

Consequently

$$\Phi^n(y) = \Phi^{n+1}(z) \in U_n(\Phi(x)).$$

On the other hand, as $z \in V(x)$, by (6.1.1), for all $n \geq 0$

$$\Phi^n(y) = \Phi^{n+1}(z) \in U_{n+1}(x)$$

and

$$\rho(\Phi^n(x),\Phi^n(z)) = \rho(\Phi^{n-1}(\Phi(x)),\Phi^{n-1}(y)) \leq NA(x)(\kappa(x))^n\rho(x,z)$$

$$= NA(x)\kappa(x)(\kappa(x))^{n-1}\rho(x,z) = C(\kappa(x))^{n-1},$$

where, $N = 200\sqrt{2}$ and

$$C = NA(x)\kappa(x)\rho(x,z).$$

Since $y \in U(\Phi(x),\delta(\Phi(x)))$ then (6.1.2) implies that $y \in V(\Phi(x))$.

(6.1.5) First of all note that from (5.1.1), (5.1.4) and Proposition 6.1, it follows that for any $v \in B(x)$, one has

$$\|d\phi_x(v)\| \leq C(x)\|v\|, \tag{6.13}$$

where $C(x) = K(1 + \psi_2(\lambda,\mu,\nu,\kappa))$ and $K = \dfrac{HA(x)}{\gamma^\ell \nu(x)\alpha^{4b}}$. Consequently,

$$C(x) = \frac{A(x)}{\gamma^b} L(x),$$

where L is a Φ-invariant function constant on the set $\Lambda_{r,s,\alpha}$.

Let

$$y \in V(x),\ (\tau'_x \circ \exp_x^{-1})(y) = (u,\phi_x(u)).$$

Let us consider the linear manifold

$$F = (d\exp_x^{-1})(T_yV(x)) \subset T_xN.$$

We want to estimate $\Gamma(T_xV(x),F')$ where F' is the linear subspace parallel to F. Let $\|\cdot\|$ and $\|\cdot\|'$ be two norms on the space H such that for some constants $a > 0$, $b > 0$ and for $0 \neq v \in H$

$$a\|v\|' \le \|v\| \le b\|v\|'.$$

Then it is easy to see that for any two linear manifolds $E, F \subset H$, $\dim E = \dim F$, one has

$$\frac{a}{b}\,\Gamma(E,F) \le \Gamma'(E,F) \le \frac{b}{a}\,\Gamma(E,F),$$

where $\Gamma(E,F), \Gamma'(E,F)$ denote the apertures computed with respect to the norm $\|\cdot\|$ and $\|\ \|'$ respectively.

Using this remark and (2.2.3) one sees that in our situation

$$\Gamma(T_xV(x),F') \le \sqrt{2}\,A(x)\Gamma'(T_xV(x),F').$$

It is easy to see that

$$\Gamma'(T_xV(x),F') \le \|d\phi_x(u)\|.$$

Therefore using (6.13) one obtains

$$\Gamma'(T_xV(x),F') \le C(x)\|u\| \le C(x)(\|u\|+\|\phi_x(u)\|)$$

$$\le \sqrt{2}\,C(x)\|(u,\phi_x(u))\| \le \sqrt{2}\,C(x)A(x)\|\exp_x^{-1}(y)\|_x =$$

$$= \sqrt{2}\,C(x)A(x)\rho(x,y).$$

Thus, we finally obtain

$$\Gamma(T_xV(x),F') \le G(x)A^2(x)\rho(x,y),$$

where $G(x) = 2C(x)$. As for $x \in \Omega_{\alpha,\gamma}$, $\Phi(x) \in \Omega_{\alpha,\gamma\alpha}$, then one easily deduces that

$$G(x) \ge \alpha^{b}G(\Phi(x)).$$

Moreover, as L is constant on the set $\Lambda_{r,s,\alpha}$ from (2.2.2) one has

$$\sup_{x\in\Lambda_{r,s,\alpha}} G(x) < +\infty. \qquad \square$$

6.4 The size $\delta(x)$ of this manifold $V(x) = V(x,\kappa)$ depends on κ. Nevertheless, if κ_1 is a Φ-invariant function defined on the set $\Lambda_{r,s}$ such that $\kappa(x) < \kappa_1(x)$ for $x \in \Lambda_{r,s}$ and κ_1 satisfies inequalities (6.6) then as follows from (6.1.2)

$$V(x,\kappa_1) \cap U(x,\delta(x)) = V(x,\kappa).$$

More generally, if one chooses λ,μ,ν,κ and K in an arbitrary manner but such that Theorem 5.1 is applicable to the sequence $\{f_n\}_{n\geq 0}$ defined by (6.8), one has the corresponding TLSM. The size of the TLSM depends on the choice of parameters λ,μ,ν,κ,K. Nevertheless, as follows from Theorem 6.1.2 the germ of the TLSM is unique. From now on, for

$$x \in \Lambda^{\ell}_{r,s,\alpha,\gamma},$$

we will consider exclusively the TLSM $V(x)$, corresponding to the parameters

$$\lambda = \frac{r\exp(3\varepsilon_{r,s})}{\alpha^{4b}}, \quad \mu = \frac{s\exp(-3\varepsilon_{r,s})}{\alpha^{4b}}, \quad \nu = \alpha^{b}\exp(-\varepsilon_{r,s}),$$

$$\kappa = r\exp(4\varepsilon_{r,s}) \text{ and } K = \frac{HA^{\ell}_{r,s}}{\gamma^{b}\nu\alpha^{4b}}.$$

For these parameters by Proposition 6.1, the assumptions of Theorem 5.1 are satisfied for the sequence of mappings $\{f_n\}_{n\geq 0}$ defined by (6.8). All statements of Theorem 6.1 remains true for them.

7. SOME ADDITIONAL PROPERTIES OF LOCAL STABLE MANIFOLDS

7.1 We formulate this section some results similar to the results of Section 2.3 $[\text{Pes}]_1$.

THEOREM 7.1.

(7.1.1) *There exists* $L = L(\ell,r,s,\alpha,\gamma)$ *such that if* $x,y \in \Lambda^{\ell}_{r,s,\alpha,\gamma}, y \in U(x,L)$ *and* $y \notin V(x)$, *then* $V(x) \cap V(y) \cap U(x,L) = \emptyset$.

(7.1.2) *There exists* $M = M(\ell,r,s,\alpha,\gamma)$ *such that if*

$$x,y \in \Lambda^{\ell}_{r,s,\alpha,\gamma} \text{ and } y \in V(x),$$

then

$$V(y) \cap U(x,M) \subset V(x).$$

(7.1.3) *If* $x \in \Lambda^{\ell}_{k,r,s,\alpha,\gamma}$, $x_i \in \Lambda^{\ell}_{k,r,s,\alpha,\gamma}, i = 1,2,\ldots, x_i \to x$, *then*

<u>for</u> <u>every</u> $q : 0 < q < \delta^{\ell}_{r,s,\alpha,\gamma}$,

$$V(x_1) \cap U(x,q) \to V(x) \cap U(x,q)$$

<u>in</u> C^1 <u>topology</u>.

<u>Proof</u>. (7.1.1) We will prove that if

$$0 < L < \frac{(Q(x))^2}{1600(A^{\ell}_{r,s})^3}$$

then $V(x) \cap V(y) \cap U(x,L) = \emptyset$. Let us suppose that this is not true. Then one can find a point $z \in V(x) \cap V(y) \cap U(x,L)$.

From (6.1.1) applied to $z \in V(x)$ and $z \in V(y)$ respectively one obtain that for all $n \geq 0$

$$\left.\begin{aligned} &\rho(\phi^n(x),\phi^n(y)) \leq \rho(\phi^n(x),\phi^n(z)) + \rho(\phi^n(z),\phi^n(y)) \leq \\ &\leq 200\sqrt{2}\, A^{\ell}_{r,s}(\kappa_{r,s})^n[\rho(x,z) + \rho(y,z)] \leq \\ &\leq 800(A^{\ell}_{r,s})^2(\kappa_{r,s})^n L = C(\kappa_{r,s})^n. \end{aligned}\right\} \tag{7.1}$$

The last inequality follows from the fact that the Riemannian diameter $\operatorname{diam} U(x,L) \leq 2A^{\ell}_{r,s}L$.

Thus, we have verified the second assumption of (6.1.2). We have from (6.5)

$$\kappa_{r,s} = r \exp 4\varepsilon_{r,s} < \nu\alpha^{4b}. \tag{7.2}$$

We want to check that for all $n \geq 0$

$$\Phi^n y \in U_n(x). \tag{7.3}$$

This together with (7.1) will assure that $y \in V(x)$ which is a contradiction.

We want to check that for all $n \geq 0$

$$\Phi^n y \in U_n(x).$$

From the definition of $U_n(x)$ and from Theorem 2.2 it follows that

$$U_n(x) \supset \{w \in M ; \rho(w,\Phi^n(x)) \leq \frac{Q^2(x)}{2A^{\ell}_{r,s}} \alpha^{4bn} e^{-n\varepsilon_{r,s}}\}.$$

Thus to finish the proof it is sufficient to see that for every $n \geq 0$

$$800(A^{\ell}_{r,s})^2 L(\kappa_{r,s})^n \le \frac{Q^2(x)}{2A^{\ell}_{r,s}} \alpha^{4bn} e^{-n\varepsilon_{r,s}}.$$

But this is so in virtue of (7.2) and of our choice of L.

(7.1.2) The proof goes along the same lines as the proof of (7.1.1). We leave the details to the reader.

(7.1.3) By definition of the sets $\Lambda^{\ell}_{r,s,\alpha,\gamma}$ (cf. Sec. 3.4), $E_{1x_i} \to E_{1x}$. Thus, by Theorem 6.1, for every q, $0 < q < \delta^{\ell}_{r,s,\alpha,\gamma}$, and for i big enough, the manifold

$$V_i(x) \cap U(x,q)$$

is of the form

$$\{\exp_x(\tau'_x)^{-1}(u,\chi_i(u)); \|u\| \le q\},$$

where $\chi_i : B^k_q \to \mathbb{R}^{m-k}$, $k = k(x) = k(x_i)$ for all i under consideration.

It follows from (5.1.1), (5.1.4) and Proposition 6.1 that there exists such a constant $C^{\ell}_{r,s,\alpha,\gamma}$ that for i big enough and for every $u_1,u_2 \in B^k_q$ one has

$$\|d\chi_i(u_1) - d\chi_i(u_2)\| \le C^{\ell}_{r,s,\alpha,\gamma}\|u_1 - u_2\|.$$

This implies that the family $\{\chi_i\}$ is compact in C^1 topology. Let ψ be a limit point of the family $\{\chi_i\}$, i.e. for some sequence $i_p \to \infty$ one has $\chi_{i_p} \to \psi$ in C^1 topology. Since $\chi_{i_p}(0) \to 0$ and $d\chi_{i_p}(0) \to 0$ as $p \to \infty$, $\psi(0) = 0$ and $d\psi(0) = 0$.

Let $u \in B^k_q$. Let us define

$$\left.\begin{aligned} z_{i_p} &= \exp_x(\tau'_x)^{-1}(u,\chi_{i_p}(u)) \\ z &= \exp_x(\tau'_x)^{-1}(u,\psi(u)). \end{aligned}\right\}$$

Let us fix $n > 0$. Because $x_i \to x$ and $z_{i_p} \to z$, for every $\varepsilon > 0$, for p big enough, one has

$$\left.\begin{aligned} &\rho(z_{i_p},z) \le \varepsilon,\ \rho(x_{i_p},x) \le \varepsilon, \\ &\rho(\Phi^n(z_{i_p}),\Phi^n(z)) \le \varepsilon \\ &\rho(\Phi^n(x_{i_p}),\Phi^n(x)) \le \varepsilon \end{aligned}\right\} \tag{7.4}$$

But $z_{i_p} \in V(x_{i_p})$, so that from (6.1.1) and (2.2.2) one has

$$\rho(\Phi^n(z_{i_p}),\Phi^n(x_{i_p})) \le NA^{\ell}_{r,s}(\kappa_{r,s})^n \rho(z_{i_p},x_{i_p})$$

and consequently from (7.4)

$$\rho(\Phi^n(z),\Phi^n(x)) \le NA^{\ell}_{r,s}(\rho(x,z) + 2\varepsilon). \tag{7.5}$$

As $\varepsilon > 0$ is arbitrary, (7.5) implies that for all $n \ge 0$, one has

$$\rho(\Phi^n(z),\Phi^n(x)) \le \tilde{C}(\kappa_{r,s})^n \rho(x,z), \tag{7.6}$$

where $\tilde{C} = NA^{\ell}_{r,s}$. Moreover

$$z \in U(x,\delta^{\ell}_{r,s,\alpha,\gamma}), \tag{7.7}$$

because $u \in B^k_q$ and $0 < q < \delta^{\ell}_{r,s,\alpha,\gamma}$.

Thus, to finish the proof it remains to be shown that for all $n \ge 0$

$$\Phi^n(z) \in U_n(x). \tag{7.8}$$

This suffices because then (7.6)-(7.8) and (6.1.2) will imply that $z \in V(x)$.

Let us fix $n \ge 0$. From (6.1.1) one has that for any $i \ge 0$, $\Phi^n(z_i) \in U_n(x_i)$ because $z_i \in V(x_i)$. Now, from the definition of the functions $\gamma(x)$ and $\ell(x)$ one sees easily that

$$\gamma(x) \ge \limsup_{x_i \to x} \gamma(x_i) \quad \text{and} \quad \ell(x) \le \liminf_{x_i \to x} \ell(x_i)$$

Thus, as $z_i \to z$ and $x_i \to x$, from the definition of the sets U_n, one obtains that

$\Phi^n(z) \in U_n(x)$. □

The following statement is an immediate consequence of the proof of (7.1.3).

COROLLARY 7.1. *Numbers* $\delta^{\ell}_{r,s,\alpha,\gamma}$ *can be chosen so small that for any*

$$x, y \in \Lambda^{\ell}_{r,s,\alpha,\gamma},\ y \in U(x, \tfrac{1}{2}\delta^{\ell}_{r,s,\alpha,\gamma})$$

one has

$$V(y) \cap U(x, \delta^{\ell}_{r,s,\alpha,\gamma}) = \{\exp_x(\tau'_x)^{-1}(u, \phi_y(u)); u \in B^{k(x)}_{\delta^{\ell}_{r,s,\alpha,\gamma}}\}$$

where

$$\phi_y: B^{k(x)}_{\delta^{\ell}_{r,s,\alpha,\gamma}} \to B^{m-k(x)}_{\delta^{\ell}_{r,s,\alpha,\gamma}}$$

is a map of class C^{p-1} *such that*

$$\max_{y \in \Lambda^{\ell}_{r,s,\alpha,\gamma} \cap U(x, \frac{1}{2}\delta^{\ell}_{r,s,\alpha,\gamma})} \ \max_{u \in B^{k(x)}_{\delta^{\ell}_{r,s,\alpha,\gamma}}} (\|\phi_y(u)\| + \|d\phi_y(u)\|) \le 1.$$

Let $\delta^{\ell}_{r,s,\alpha,\gamma}$ be chosen according to Corollary 7.1. Let

$$x \in \Lambda^{\ell}_{k,r,s,\alpha,\gamma}.$$

By the family of local T-stable manifolds $S^{\ell}_{k,r,s,\alpha,\gamma}(x)$ we will mean the set of all local T-stable manifolds passing through all the points

$$y \in \Lambda^{\ell}_{k,r,s,\alpha,\gamma} \cap U(x, \tfrac{1}{2}\delta^{\ell}_{r,s,\alpha,\gamma}).$$

Finally let us formulate the following theorem, which is an exact analogue of Propositions 2.3.1.1 and 2.3.1.2 of $[\text{Pes}]_1$. Let $x \in \Lambda_{r,s,\alpha}$ and $y \in V(x)$. Since the points $\Phi^n(x)$ and $\Phi^n(y)$ converge with exponential speed, the linearized equation for Φ along the trajectory of y may be considered as an equation obtained from the linearized equation for Φ along the trajectory of x

by a rapidly decreasing perturbation. From this remark and from perturbation theorems (see Theorems 15.2.1 and 17.1.1 of [Byl], see also $[\mathrm{Rue}]_{2,3}$ and [Cam]) one obtains

THEOREM 7.2.

(7.2.1) *Let* $x \in \Lambda^{\ell}_{r,s,\alpha,\gamma}$ *and let* $y \in V(x)$. *Then the point* y *is regular in the future* (see Sec. 2.2) and

$$s(x) = s(y), \chi_i(x) = \chi_i(y), \ k_i(x) = k_i(y), \ 1 \le i \le s(x).$$

(7.2.2) *Let* $x \in \Lambda^{\ell}_{r,s,\alpha,\gamma}$ *and let* $y \in \Lambda_{r,s} \cap V(x)$. *There exists* $K = K(\ell,r,s,\alpha,\gamma)$ *such that for every* $n \ge 0$ *one has*

$$\|d\Phi^n v\|_{\Phi^n(y)} \le K\, r^n \exp(\varepsilon_{r,s} n)\|v\|_y \quad \underline{\text{for}} \quad v \in E_{1y} = T_y V(x),$$

$$\|d\Phi^n v\|_{\Phi^n(y)} \ge K^{-1} s^n \exp(-\varepsilon_{r,s} n)\|v\|_y \quad \underline{\text{for}} \quad v \in E_{2y}.$$

(7.2.3) *Let* $x \in \Lambda_{r,s,\alpha}$ *and* $y \in V(x)$, *then for every vector* $v \in T_y V(x)$

$$\limsup_{n\to\infty} \frac{1}{n} \log\|d\Phi^n_y(v)\| < T$$

(*cf. Sec.* 2.3).

PART II

ABSOLUTE CONTINUITY

A. Katok and J.-M. Strelcyn

1. INTRODUCTION

The present part contains a full and very detailed proof of so called absolute continuity of families of invariant manifolds corresponding to the Lyapunov Characteristic Numbers not bigger than fixed $T < 0$ (TLSM) for differentiable maps with singularities described in Sec. 1 of Part I.

In the case of diffeomorphisms of class $C^{1+\varepsilon}$ these proofs are given by Ya.B. Pesin in Sec. 3 of his fundamental paper $[Pes]_1$. However, the proof in $[Pes]_1$ contain several gaps in many places he does not give enough details of his arguments. All this makes necessary to present the proof of absolute continuity in clear and complete form. After this is done, the generalization from the diffeomorphism case to more general case of maps with singularities does not lead to any extra complications.

Our proof of absolute continuity follows very closely the general plan of Pesin's proof with modifications due to errors and incomplete arguments in his presentation. Probably this is the reason why our proof is too lengthy; it is conceivable that more fundamental revision of the proof would shorten it substantially.

Sections 2 - 7 are preliminary; the proof itself is contained in Sections 8 - 10.

A very rough sketch of the idea of the proof of Absolute Continuity Theorem is given just after its formulation in Sec. 4.2. Sec. 11 is devoted to one consequence of our main results, very important to the application in ergodic theory. Sec. 12 is devoted to the absolute continuity in infinite dimensional Hilbert Space setting. Moreover, an extension of finite dimensional framework described in Sec. 1_I to somewhat larger class of mapping is also indicated. Sec. 13 contains the formulation of some important result from $[Pes]_2$ and $[Kat]_2$ which remain true in our framework.

The notations used in the present part are very complicated. They are described mostly in Secs. 2.2 - 2.6. The reader may be advised not to read Section 2 carefully at the beginning, but rather consult with it when a need appears.

Let us note that C. Robinson and L.S. Young see [Rob] constructed an example of a C^1 Anosov diffeomorphism of two-dimensional torus for which the absolute continuity fails for stable and unstable foliations. Unfortunately, in their example, there is no finite invariant measure equivalent to Lebesgue measure. However, there is a little doubt that an area-preserving example of that sort also exist.

2. PRELIMINARIES AND NOTATIONS

2.1. The basic notations used here without special notification are the same as in Part I. Nevertheless, many symbols used in this part have a different meaning than in Part I. But it never leads to ambiguities.

The complexity of notations is partially a consequence of the necessity of the simultaneous use of both Riemannian and Lyapunov norms. As far as the references to the formulas, theorems, etc. from Part I are concerned, we will write simply "formula 3.10_I" instead of "formula 3.10 from Part I", etc.

2.2. The proof of Absolute Continuity requires the following reinforcement of Condition (Bb) from Sec. 2_I.

Condition Bc. For every $\varepsilon > 0$ there exists $r_\varepsilon > 0$ such that for every $x,y \in V$ satisfying $\rho(x,y) < \min(r_\varepsilon, R_V(x)) \overset{\text{def}}{=} R_\varepsilon(x)$ one has $\|d\exp_x(w)\| \le 1 + \varepsilon$ where $w = \exp_x^{-1}(y)$ and $\|d(\exp_x^{-1})(y)\| \le$ $\le 1 + \varepsilon$.

Obviously Condition (Bc) is always satisfied when M is a smooth compact Riemannian manifold and V is a smooth open submanifold of M.

2.3. Starting from this moment we will always suppose that for the mapping $\Phi: N \to V$ preserving the measure μ, all conditions from Sec. 1_I are satisfied.

2.4. Let us assume that the numbers ℓ, k, r, s are fixed. Let us fix $\varepsilon_{r,s}$ so small and $\alpha, 0 < \alpha < 1$, such that the conditions

$$r\exp(100\,\varepsilon_{r,s}) < 1 \quad\text{and}\quad r\exp(100\,\varepsilon_{r,s}) < s \qquad (2.5)_I$$

$$r\exp(50\,\varepsilon_{r,s}) < \alpha^{10b} \qquad (6.5)_I$$

are satisfied. Then we define numbers λ, μ, κ and ν as follows

$$\begin{cases} \lambda = \lambda_{r,s} = r\exp(3\varepsilon_{r,s}),\ \mu = \mu_{r,s} = s\exp(-3\varepsilon_{r,s}) \\ \chi = \chi_{r,s} = r\exp(4\varepsilon_{r,s}),\ \nu = \nu_{r,s} = \exp(-\varepsilon_{r,s}) \end{cases} \qquad (2.1)$$

Let us note that these quantities are different from the ones defined in Proposition 6.1_I and at the end of Sec. 6_I and denoted by the same symbols.

Let us define also

$$\left.\begin{aligned}\xi = \xi_{r,s} = r\exp(7\varepsilon_{r,s}),\ \xi' = \frac{\xi}{\nu} = r\exp(8\varepsilon_{r,s}),\\ \hat{\xi} = r\exp(12\varepsilon_{r,s})\end{aligned}\right\}\tag{2.2}$$

$$\eta = \eta_{r,s} = \exp(-7\varepsilon_{r,s})\tag{2.3}$$

There are no particular reasons for choosing the constants in definition (2.1) - (2.3). What we actually need is the following inequalities which easily follow from our choice of the constants and from $(2.5)_I$ and $(6.5)_I$.

$$\chi < \xi < \xi' < \hat{\xi} < \mu\tag{2.4}$$

$$\frac{\lambda}{\mu} < \eta\tag{2.5}$$

$$\xi < \xi' < \eta\tag{2.6}$$

$$\frac{\xi}{\nu} < \eta\tag{2.7}$$

2.5. We assume that the number α satisfies $(2.5)_I$, $(6.5)_I$ and the following inequality:

$$\exp(-\varepsilon_{r,s}) < \alpha^{2bp}\tag{2.8}$$

where $p \overset{\text{def}}{=\!=} m - k \geq 1$ and b comes from condition 1.3_I.

$\|\cdot\|$ denotes the standard Euclidean norm in $\mathbb{R}^m$.

Let $w \in \Lambda$ (of. Sec. 2.2_I). By $\|\cdot\|_w$ we denote the Riemannian norm in T_wN and by $\|\cdot\|'_w$ the Lyapunov norm (See Sec. 2_I)

Let $w \in \Lambda^{\ell}_{r,s}$ (cf. Sec. 2.3_I). From the inequalities $(2.2.1)_I$ and $(2.2.5)_I$ one sees that for every integer n and $v \in T_{\Phi^n(w)}N$,

$$\frac{1}{\sqrt{2}}\|v\|_{\Phi^n(w)} \leq \|v\|'_{\Phi^n(w)} \leq A^{\ell}_{r,s}\exp(n\varepsilon_{r,s})\|v\|_{\Phi^n(w)}.\tag{2.9}$$

Sometimes, when $w \in \Lambda$ is fixed, we will write τ'_n instead of $\tau'_{\Phi^n(w)}$ and for $v \in \mathbb{R}^m$, $\|v\|_n$ instead of $\|(\tau'_{\Phi^n(w)})^{-1}v\|_{\Phi^n(w)}$.

We will frequently consider objects related to the Riemannian metric and similar objects related to Lyapunov metric. Usually we will use the same symbols for those objects adding $\sim$ above the symbol in the Lyapunov metric case. In particular, when $w \in \Lambda_{k,r,s}$ is fixed, we will denote

$$\begin{cases} B_n^i(q) = \{v \in E_i; \|v\|_n \le q\}, \\ \widetilde{B}^i(q) = \{v \in E_i; \|v\| \le q\}, \\ B_n(q) = B_n^1(q) \times B_n^2(q), \\ \widetilde{B}(q) = \widetilde{B}^1(q) \times \widetilde{B}^2(q), \end{cases} \tag{2.10}$$

where $E_1 = \mathbb{R}^k$ and $E_2 = \mathbb{R}^p$; $p = m - k$.

Furthermore, for $v \in E_i$, $i = 1,2$, and $r > 0$.

$$\begin{cases} B_n^i(v,r) = \{z \in E_i;\ \|z - v\|_n \le r\} \\ B^i(v,r) = B_0^i(v,r) \\ \widetilde{B}^i(v,r) = \{z \in E_i;\ \|z - v\| \le r\} \end{cases} \tag{2.11}$$

and for $z = (u,v)$, $u \in E_1$, $v \in E_2$.

$$\begin{cases} B_n(z,r) = B_n^1(u,r) \times B_n^2(v,r) \\ \widetilde{B}(z,r) = \widetilde{B}^1(u,r) \times \widetilde{B}^2(v,r) \end{cases} \tag{2.12}$$

$$\widetilde{U}(w,q) = \exp_w(\tau_w')^{-1}(\widetilde{B}(q)) \tag{2.13}$$

2.6. As in Part I, for $n \ge 0$, and $w \in \Lambda^{\ell}_{r,s,\alpha,\gamma}$ we will use the notations

$$\Phi_{nw} = \exp^{-1}_{\Phi^{n+1}(w)} \circ \Phi \circ \exp_{\Phi^n(w)} \qquad (3.2)_I$$

$$\Phi'_{nw} = \tau'_{\Phi^{n+1}(w)} \circ \exp^{-1}_{\Phi^{n+1}(w)} \circ \Phi \circ \exp_{\Phi^n(w)} \circ (\tau_n')^{-1}. \qquad (3.13)_I$$

Sometimes, when $w \in \Lambda$ is fixed, we will write Φ_n instead of Φ_{nw} and Φ_n' instead of Φ'_{nw}.

$$V_n(w) = \{y \in M;\ \rho(y,\Phi^n(w)) < (Q(w)\alpha^{2bn})^2 \qquad (3.9)_I$$

and

$$U_n(w) = \exp_{\Phi^n w}(\tau'_{\Phi^n w})^{-1}\widetilde{B}\left(\frac{Q^2(w)}{2}\alpha^{2bn}\right). \tag{2.14}$$

By Lemma 3.3_I, Φ_{nw} is a diffeomorphic empedding of $\Lambda_n(w) \stackrel{\text{def}}{=} \exp^{-1}_{\Phi^n(w)}(V_n(w))$ into $T_{\Phi^{n+1}(w)}N$, while Φ'_{nw} is a diffeomorphic embedding of $B_n(w) = \tau'_n(\Lambda_n(w))$ into $\mathbb{R}^m$.

Since the map Φ^{-1} is well defined on $V_n(w)$ (see Lemma 3.3_I) then for $n \geq 0$

$$\Phi^{-1}_{nw} = \exp^{-1}_{\Phi^n(w)} \circ \Phi^{-1} \circ \exp_{\Phi^{n+1}(w)}.$$

Consequently for $1 \leq p \leq n+1$ we have

$$\Phi^{-p}_{nw} \stackrel{\text{def}}{=} \exp^{-1}_{\Phi^{n-p+1}(w)} \circ \Phi^{-p} \circ \exp_{\Phi^{n+1}(w)}$$

$$= \Phi^{-1}_{(n-p+1)w} \circ \Phi^{-1}_{(n-p+2)w} \circ \ldots\ldots\ldots \circ \Phi^{-1}_{nw}.$$

In particular

$$\Phi^{-n}_{(n-1)w} = \Phi^{-1}_{0w} \circ \Phi^{-1}_{1w} \circ \ldots\ldots\ldots \circ \Phi^{-1}_{(n-1)w}.$$

2.7. Let E,F be two linear manifolds in an Euclidean space H. We denote by $\gamma(E,F)$ the minimal angle between E and F, i.e.

$$\gamma(E,F) = \min_{\substack{0 \neq e \in E' \\ 0 \neq f \in F'}} \gamma(e,f), \tag{2.15}$$

where $\gamma(e,f)$ denote the angle between e and f and where E', F' are linear subspaces of H parallel to E and F respectively.

It is easy to see that always $2\Gamma(E,F) \geq \gamma(E,F)$, where the aperture $\Gamma(E,F)$ was defined in Sec. 2.1_I (cf. also Sec. 3.4).

If we replace the Euclidean structure in H by another one generating the norm $\|\cdot\|'$ such that for every v

$$a\|v\|' \leq \|v\| \leq b\|v\|'$$

then the new angle $\gamma'(E,F)$ computed with respect to the norm $\|\cdot\|'$ satisfies the inequalities

$$\frac{a}{2b}\gamma(E,F) \leq \gamma'(E,F) \leq \frac{2b}{a}\gamma(E,F). \tag{2.16}$$

2.8. As usual by $\overline{Y}$ we denote the closure of the set Y. When $Y \subset N$, Int Y denotes the interior of Y. diam Y denotes the diameter of Y.

The cardinality of set Y will be denoted by $\#Y$.

As usual $\emptyset$ denotes the empty set.

If $f: X \to Y$ is a mapping and $\overline{X}_1 \subset X$, then by $f|_{X_1}$ we denote the restriction of f to X_1.

Let $Y \subset \mathbb{R}^m$ be an open set. Although this does not conform to the standard notations, we will write $\Psi \in C^p(\overline{Y}, \mathbb{R}^k)$ if $\Psi: \overline{Y} \to \mathbb{R}^k$ is a continuous mapping whose restriction to Y is of class C^p.

3. SOME FACTS FROM MEASURE THEORY AND FROM LINEAR ALGEBRA

3.1. We recall here some well known facts related to the general definition of Jacobian of absolutely continuous mappings between measure spaces.

Let $(X, \mathfrak{A}, \mu)$ and $(Y, \mathfrak{B}, \nu)$ be two measure spaces with probability measures μ and ν defined on σ - algebras $\mathfrak{A}$ and $\mathfrak{B}$ respectively.

Let $p: X \to Y$ be a measurable mapping, i.e. for every $B \in \mathfrak{B}$, $p^{-1}(B) \in \mathfrak{A}$. Let us assume that p is bimeasurable, i.e. p is bijective and p and p^{-1} are measurable.

The bimeasurable mapping p is called absolutely continuous if for each set $A \in \mathfrak{A}$ such that $\mu(A) = 0$ one has $\nu(p(A)) = 0$. Let us assume that p is absolutely continuous.

Let us define on $\mathfrak{A}$ a new probability measure μ_p by the formula, $\mu_p(A) = \nu(p(A))$. The measure μ_p is absolutely continuous with respect to the measure μ. Thus by Radon - Nikodym Theorem one can consider the measurable function $J(p) = \frac{d\mu_p}{d\mu}$ defined on X, which is called the Jacobian of the mapping p.

It is easy to see that when p and p^{-1} are absolutely continuous, then $J(p) = \frac{1}{J(p^{-1}) \circ p}$ for μ almost all points X (we admit $\frac{1}{0} = +\infty$ and $\frac{1}{+\infty} = 0$).

Let us suppose that X is a compact subset of a smooth submanifold W of $\mathbb{R}^m$. $\mathfrak{A}$ is the σ - algebra of all Borel subsets of X and μ is the restriction to X of some absolutely continuous measure on W.

Let $x \in X$. A sequence $\{A_r\}_{r \geq 1}$ of open balls of X will be called an x - sequence if $x \in A_r$ for any $r \geq 1$ and $\lim_{r \to \infty} \operatorname{diam}(A_r) = 0$.

From the Lebesgue - Vitali Theorem on differentiation (see for example Chapt. 10 of $[\text{Shi}]_1$) it follows that

$$J(p)(x) = \lim_{r\to\infty} \frac{\nu(p(A_r))}{\mu(A_r)} \tag{3.1}$$

for almost all points $x \in X$ and for all x - sequences $\{A_r\}_{r\geq 1}$.

3.2. If H is a finite dimensional vector space with a scalar product $(\cdot,\cdot)_H$, then by Vol_H we will denote the corresponding volume on H.

Let E and F be two real vector spaces of the same finite dimension, provided with scalar products $(\cdot,\cdot)_E$ and $(\cdot,\cdot)_F$ respectively. Let $E_1 \subset E$ be a linear subspace of E and $A: E \to F$ be a linear mapping. Let us define

$$\left| A|_{E_1} \right| = \frac{\text{Vol}_{F_1}(A(U))}{\text{Vol}_{E_1}(U)} ,$$

where U is an arbitrary open and bounded subset of E_1 and where F_1 is an arbitrary linear subspace of F of the same dimension as E_1 and $A(U) \subset F_1$. We will denote $\left|A|_E\right|$ by $|\det A|$.

Let X and Y be two Riemannian manifolds of the same finite dimension and $T: X \to Y$ be a C^1 diffeomorphism. Riemannian metrics on X and Y define volume elements and consequently measures on X and Y which we denote by ν_X and ν_Y respectively.

PROPOSITION 3.1.

(3.1.1) For any $z \in X$ one has $\frac{d(\nu_Y \circ T)}{d\nu_X}(z) = |\det dT(z)|$

(3.1.2) For any function $f \in L^1(Y,\nu_Y)$ one has

$$\int_X (f\circ T) \cdot |\det dT| d\nu_X = \int_Y f \, d\nu_Y .$$

This is essentially a standard change of variables formula.

3.3. For the rest of this section we will deal with a slightly more general situation than usual. Namely, let H be a finite or infinite dimensional Hilbert space.

Let ν_p denote the p dimensional Hausdorff measure in $\mathbb{R}^p \oplus H$ (cf. [Rog] for Hausdorff measures). When H is finite dimensional

ν_p restricted to a p dimensional submanifold of $\mathbb{R}^p \oplus H$ coincides with the p dimensional volume (Lebesgue measure) on this submanifold (cf [Rog]). For $U \subset \mathbb{R}^p$, by $\nu_p(U)$ we will denote its p dimensional Lebesgue measure.

PROPOSITION 3.2. *Let $U \subset \mathbb{R}^p$ be an open bounded set and let $f\colon U \to H$ be a C^1 mapping. If $\sup_{v \in U} \|df(v)\| \le d$, then*

$$\nu_p(U) \le \nu_p(\text{graph } f) \le (1+d^2)^{\frac{p}{2}} \nu_p(U). \tag{3.2}$$

Proof. If $S \subset U$ and if $\operatorname{diam}(S) = r$, then

$$r \le \operatorname{diam}\{(v,f(v)); v \in S\} \le \sqrt{1+d^2}\, r$$

where the diameter is taken with respect to the Riemannian metric induced on the graph of f. Thus every cover of U by balls determines a cover of the graph by balls which are $\sqrt{1+d^2}$ times larger. Now the assertion follows from the definition of the Hausdorff measure. ■

3.4. If E,F are two closed linear manifolds in a Hilbert space H, then by $\Gamma(E,F)$ we denote the aperture between E and F i.e.

$$\Gamma(E,F) = \sup_{\substack{e \in E' \\ \|e\|=1}} \left(\inf_{v \in F'} \|e-v\| \right)$$

where E', F' are linear subspaces of H parallel to E, F respectively. If we replace the Euclidean structure in H by another one generating the norm $\|\cdot\|'$ such that for every v

$$a\|v\|' \le \|v\| \le b\|v\|'$$

then the new aperture $\Gamma'(E,F)$ satisfies the inequalities

$$\frac{a}{b}\Gamma(E,F) \le \Gamma'(E,F) \le \frac{b}{a}\Gamma(E,F). \tag{3.4}$$

Starting from now we will consider only the case when E and F are linear subspaces of H. If K is a linear subspace of H, then P_K denotes the orthogonal projection of H onto K.

Let us consider the following metric θ in the Grassmanian manifold of all p - dimensional linear subspaces of H

$$\theta(E,F) = \|P_E - P_F\|.$$

The following lemma (sec Sec. 39 of [Akh]) explains the geometric meaning of this metric.

LEMMA 3.1. $\theta(E,F) = \max\{\Gamma(E,F), \Gamma(F,E)\}$

3.5. The main goal of the present subsection is the proof of the following lemma which may also be of interest in itself.

LEMMA 3.2. *For every* $p \in \mathbb{N}$ *there exists a number* $C_1 = C_1(p) > 0$ *such that for every two Hilbert Spaces* H_1 *and* H_2, *for any* $a \geq 1$ *and any two linear operators* $A,B: H_1 \to H_2$ *with* $\|A\| \leq a$, $\|B\| \leq a$ *and any two linear subspaces* $E_1, E_2 \subset H_1$ *of dimension* p *one has*

$$\left| \left|A\big|_{E_1}\right| - \left|B\big|_{E_2}\right| \right| \leq C_1 a^p (\|A-B\| + \Gamma(E_1,E_2))$$

Proof. Replacing if necessary H_1 by the space spanned by E_1 and E_2 we can assume that $\dim H_1 \leq 2p$. Indeed, the aperture of two subspaces is independent of the choice of space containing them. Let us fix a linear subspace $\widetilde{H}_2$ of H_2 such that $A(H_1) \subset \widetilde{H}_2$, $B(H_1) \subset \widetilde{H}_2$ and $\dim \widetilde{H}_2 = 2p$. Now let us denote by $\widetilde{H}_1$ a $2p$ dimensional Hilbert space containing H_1 as its subspace. Let us denote by $H_1^{\perp}$ the orthogonal complement of H_1 in $\widetilde{H}_1$. Let us extend the operators $A,B: H_1 \to \widetilde{H}_2$ to the operators from $\widetilde{H}_1$ to H_2 by taking $A\big|_{H_1^{\perp}} = B\big|_{H_1^{\perp}} = 0$. Replacing if necessary A by UA and B by UB where $U: \widetilde{H}_2 \to \widetilde{H}_1$ is an isometry, we can reduce the situation to the case when $H_1 = H_2 = E$, $\dim E = 2p$. We can also assume that $a = 1$, because otherwise we can consider $A_1 = a^{-1}A$ and $B_1 = a^{-1}B$ instead of A and B.

Let us pass now to some introductory remarks.

Let E be a real Euclidean space of finite dimension m. By $\overset{p}{\oplus} E$, $\overset{p}{\otimes} E$ and $\overset{p}{\Lambda} E$ we denote its p - fold direct sum, p - fold tensor power and p - fold exterior power respectively. Let us recall that $\overset{p}{\Lambda} E \subset \overset{p}{\otimes} E$. All the scalar products and all the Euclidean norms will be denoted up to the end of this subsection by $(\cdot,\cdot)$ and $\|\cdot\|$ respectively.

PROPOSITION 3.3. *For any* $p \geq 1$ *there exists a number* $C(p) > 0$

<u>such that for all vectors</u> $\{e_i\}_{1\le i\le p}$ <u>and</u> $\{f_i\}_{1\le i\le p}$, e_i, $f_i \in E$, $\|e_i\| \le 1$, $\|f_i\| \le 1$, $1 \le i \le p$, <u>one has</u>

$$\left\| \bigotimes_{i=1}^{p} e_i - \bigotimes_{i=1}^{p} f_i \right\| \le C(p) \sum_{i=1}^{p} \left\| e_i - f_i \right\|.$$

<u>Proof</u>. Let us consider a mapping $T\colon \overset{p}{\oplus} E \to \overset{p}{\otimes} E$ given by the formula $T(h_1,\ldots\ldots,h_p) = \bigotimes_{i=1}^{p} h_i$, where $h_i \in E$, $1 \le i \le p$. This mapping is a C^∞ mapping and thus Proposition 3.1. follows from the Mean Value Theorem. ∎

Let us denote by π_p the canonical linear projection of $\overset{p}{\otimes} E$ onto $\overset{p}{\Lambda} E$ given by the operator of antisymmetrization. It is well known that $\left\| \bigwedge_{i=1}^{p} e_i \right\|$ is equal to the p-dimensional volume in E of the parallelepiped spanned by $e_1,\ldots\ldots,e_p$. Thus if $C\colon E \to F$ is a linear map, $E_1 \subset E$ is a linear subspace with a basis $e_1,\ldots\ldots,e_p$ then

$$\left| C\big|_{E_1} \right| = \frac{\left\| \bigwedge_{i=1}^{p} Ce_i \right\|}{\left\| \bigwedge_{i=1}^{p} e_i \right\|}. \tag{3.5}$$

Obviously the statement of Lemma 3.2. follows from two statements (a) and (b):

$$\left| \left| A\big|_{E_1} \right| - \left| B\big|_{E_1} \right| \right| \le C'\|A - B\|, \tag{a}$$

$$\left| \left| B\big|_{E_1} \right| - \left| B\big|_{E_2} \right| \right| \le C''\Gamma(E_1,E_2), \tag{b}$$

where $C' = C'(p) > 0$ and $C'' = C''(p) > 0$.

<u>Proof of (a)</u>. Let $e_1,\ldots\ldots,e_p$ be an orthonormal basis of E_1. Thus by (3.5) one has $\left| A\big|_{E_1} \right| = \left\| \bigwedge_{i=1}^{p} Ae_i \right\|$ and $\left| B\big|_{E_1} \right| = \left\| \bigwedge_{i=1}^{p} Be_i \right\|$. Consequently, using Proposition 3.3. one obtains that

$$\left| \left| A|_{E_1} \right| - \left| B|_{E_1} \right| \right| = \left| \left\| \bigwedge_{i=1}^{p} Ae_i \right\| - \left\| \bigwedge_{i=1}^{p} Be_i \right\| \right| \le$$

$$\le \left\| \bigwedge_{i=1}^{p} Ae_i - \bigwedge_{i=1}^{p} Be_i \right\| = \left\| \pi_p \left(\bigotimes_{i=1}^{p} Ae_i - \bigotimes_{i=1}^{p} Be_i \right) \right\| \le$$

$$\le \left\| \pi_p \right\| \cdot \left\| \bigotimes_{i=1}^{p} Ae_i - \bigotimes_{i=1}^{p} Be_i \right\| \le \left\| \pi_p \right\| C(p) \sum_{i=1}^{p} \|Ae_i - Be_i\| \le C' \|A - B\|$$

where $C' = p\|\pi_p\|C(p)$.

Proof of (b). For every pair of subspaces E_1 and E_2 we will find a neighborhood U_{E_1,E_2} of $(E_1,E_2) \in G_{m,p} \times G_{m,p}$ and a number $d_U > 0$ such that for all $(F_1,F_2) \in U_{E_1,E_2}$

$$\left| \left| B|_{F_1} \right| - \left| B|_{F_2} \right| \right| \le d_U \Gamma(F_1,F_2). \tag{3.6}$$

Obviously (b) follows from (3.6) by standard compactness arguments, because $\Gamma(\cdot,\cdot)$ is a continuous function on the compact product space $G_{2p,p} \cdot G_{2p,p}$, where $G_{m,p}$ denotes the Grassmannian manifold of all p-dimensional subspaces of $\mathbb{R}^m$.

First, let us remark that if $E_1 \neq E_2$ and consequently $\Gamma(E_1,E_2) > 0$ we can obviously find $d > 0$ such that

$$\left| \left| B|_{E_1} \right| - \left| B|_{E_2} \right| \right| < d\Gamma(E_1,E_2).$$

Since expressions at both sides of that inequality depend on E_1 and E_2 continuously, the same inequality holds for every F_1, F_2 from some neighborhood of (E_1,E_2).

Now let us consider the case when $E_1 = E_2$. Let

$$U_{E_1} = \{(F_1,F_2) \in G_{2p,p} \times G_{2p,p};\ \Gamma(F_i,E_1) < \frac{1}{100p!} \quad i = 1, 2\}.$$

Let us estimate $\Gamma(F_1,F_2)$ for $(F_1,F_2) \in U_{E_1}$. Since $\dim F_1 = \dim F_2 = \dim E_1 = p$ and no vector from F_1 or F_2 is orthogonal to E_1, then the operator P_{E_1} is surjective on both F_1 and F_2.

Let $v \in F_1$, $\|v\| = 1$; we have by definition

$$\|v - P_{E_1} v\| \le \Gamma(F_1,E_1) < \frac{1}{100p!}.$$

By the surjectivity of P_{E_1} on F_2 there exists $w \in F_2$ such that $P_{E_1}w = P_{E_1}v$. Furthermore, as $\left\|P_{E_1}v\right\| \le 1 + \frac{1}{100p!}$,

$$\|w\| \le \frac{\left\|P_{E_1}w\right\|}{\sqrt{1-\Gamma(F_2,E_1)^2}} \le 1 + \frac{1}{50p!} < 2. \tag{3.7}$$

Consequently

$$\|w - P_{E_1}w\| \le \|w\| \cdot \Gamma(F_2,E_1) < \frac{1}{50p!}$$

and finally

$$\|v - P_{F_2}v\| \le \|v - w\| \le \|v - P_{E_1}v\| + \|w - P_{E_1}w\| < \frac{1}{25p!} .$$

Since $v \in F_1$, $\|v\| = 1$, is arbitrary we obtain

$$\Gamma(F_1,F_2) < \frac{1}{25p!} . \tag{3.8}$$

For $(F_1,F_2) \in U_{E_1}$ let us fix an orthonormal basis $e_1,\ldots\ldots,e_p$ in F_1 and denote $f_i = P_{F_2}e_i$, $i = 1,\ldots\ldots\ldots,p$. Of course $f_1,\ldots\ldots\ldots,f_p$ is not an orthonormal basis in F_2 (except only the case when $F_1 = F_2$) but as we will see soon, it is a basis which is not very far from the orthonormal one.

The volume of the parallelipiped generated by $f_1,\ldots\ldots,f_p$ is equal to the absolute value of the Gramm determinant

$$J_1 = \det[(f_i,f_j)]_{i,j=1,\ldots\ldots,p}.$$

Since $e_i = f_i + h_i$ where $\|h_i\| \le \Gamma(F_1,F_2)$ then by (3.7) and (3.8) one can easily see that for $i = 1,\ldots\ldots,p$

$$\left|(f_i,f_i) - 1\right| < 5\Gamma(F_1,F_2) < \frac{1}{5p!} \tag{3.9}$$

and for $i \ne j$, $i, j = 1,\ldots\ldots,p$

$$|(f_i,f_j)| \le 5\Gamma(F_1,F_2) < \frac{1}{5p!} . \tag{3.10}$$

From the definition of the determinant one obtains

$$\begin{cases} |J_1 - 1| \le 2^p \cdot 5\Gamma(F_1,F_2) + p!(1 + 5\ (F_1,F_2)) \cdot 5\Gamma(F_1,F_2) < \\ < (5 \cdot 2^p + 2p!)\Gamma(F_1,F_2) < \frac{1}{2} . \end{cases} \tag{3.11}$$

In particular, $J_1 \neq 0$ and the vectors $f_1,\ldots\ldots,f_p$ form a basis in F_2.

Now we can proceed to the estimation of $\left| |B|_{F_1}| - |B|_{F_2}| \right|$.

By (3.5) we can express this quantity through Gramm determinants, namely

$$\left| B \right|_{F_1} \Big| = |J_0| = \left| \det\left[(Be_i, Be_j)\right]_{i,j=1,\ldots,p} \right|, \quad |B|_{F_2} = \frac{|J_2|}{|J_1|},$$

where

$$J_2 = \det[(Bf_i, Bf_j)]_{i,j=1,\ldots,p}.$$

Thus, we have, using (3.11)

$$\begin{cases} \left| |B|_{F_1}| - |B|_{F_2}| \right| = \left| \dfrac{|J_0|\cdot|J_1| - |J_2|}{|J_1|} \right| \le \dfrac{|J_0 J_1 - J_2|}{|J_1|} \le \\ \le 2|J_0 J_1 - J_2| \le 2|J_1| \cdot |J_0 - J_2| + 2|J_2| \cdot |J_1 - 1| \le \\ \le 3|J_0 - J_2| + 2(|J_0| + |J_0 - J_2|) \cdot |J_1 - 1| . \end{cases} \tag{3.12}$$

Since $\|B\| \le 1$ then $|J_0| \le C(p)$ so it is left to estimate $|J_0 - J_2|$. We do that similarly to the estimation of $|J_1 - 1|$ above. Namely, we have $Be_i = Bf_i + Bh_i$ and since $\|B\| \le 1$,

$$\|Bh_i\| \le \|B\| \cdot \|h_i\| \le \Gamma(F_1,F_2).$$

Therefore

$$|(Bf_i,Bf_j) - (Be_i,Be_j)| \le 5\Gamma(F_1,F_2). \tag{3.13}$$

We will use the following inequality: let $X = (x_{i,j})_{i,j=1,\ldots,p}$ $Y = (y_{i,j})_{i,j=1,\ldots,p}$ be two $p \times p$ matrices and $|x_{i,j} - y_{i,j}| < \alpha$. Then

$$|\det Y - \det X| \le p!2^p(\max(\|X\|,1))^p \cdot \alpha. \tag{3.14}$$

The norm of the matrix $[(Be_i, Be_j)]_{i,j=1,\ldots,p}$ is bounded from above by a constant $C_1(p)$. Hence, we can apply (3.12) and using (3.14) we obtain

$$|J_0 - J_2| \leq D(p)\Gamma(F_1, F_2) \tag{3.15}$$

with another constant $D(p)$. Finally by (3.12), (3.11) and (3.15) we have for $(F_1, F_2) \in U_{E_1}$

$$\left| \left| B\Big|_{F_1} \right| - \left| B\Big|_{F_2} \right| \right| \leq d(p) \cdot \Gamma(E_1, E_2)$$

where $d(p) = 3D(p) + 2(C(p) + D(p)) \cdot (5 \cdot 2^p + 2p!)$. ∎

Let us remark that Lemma 3.1. and Lemma 3.2. imply the following elegant inequality.

<u>COROLLARY 3.1</u>. <u>For every</u> $p \in \mathbb{N}$, <u>there exists a constant</u> $C_1(p)$ <u>such that for every two linear operators</u> A, B <u>in the Hilbert space</u> H, $\|A\| \leq 1$, $\|B\| \leq 1$ <u>and every two supspaces</u> $E, F \subset H$ <u>of the same finite dimension</u> p <u>one has</u>

$$\left| \left| A\Big|_E \right| - \left| B\Big|_F \right| \right| \leq C_1(p)(\|A - B\| + \|P_E - P_F\|).$$

3.6. For a linear operator $C: E \to F$ let us denote the graph of C by G_C: $G_C = \{(x, Cx) \in E \times F;\ x \in E\}$.

<u>PROPOSITION 3.4</u>. <u>For any two linear operators</u> $A, B: E \to F$

$$\Gamma(G_A, G_B) \leq 2(\|A\| + \|B\|).$$

<u>Proof</u>. One can see from (3.3) that without less of generality we can assume in the definition of the aperture Γ that $\|v\| \leq 1$. Thus

$$\Gamma(G_A, G_B) = \sup_{\{x \in E;\, \|x\|^2 + \|Ax\|^2 = 1\}} \left(\inf_{\{y \in E;\, \|y\|^2 + \|By\|^2 \leq 1\}} \|(x, Ax) - (y, By)\| \right). \tag{3.16}$$

But $\|(Ax, x) - (By, y)\| \leq \|(Ax, x) - (0, y)\| + \|(By, y) - (0, y)\| = \|(Ax, x - y)\| + \|By\| \leq \|Ax\| + \|x - y\| + \|B\|$ because $\|y\| < 1$. Thus, from (3.16) one deduces that

$$\Gamma(G_A,G_B) \le \sup_{\{x\in E;\|x\|\le 1\}} \left(\inf_{\{y\in E;\|y\|^2+\|By\|^2\le 1\}} \|Ax\| + \|x-y\| + \|B\| \right) \le$$

$$\le \|A\| + \|B\| + \sup_{\{x\in E;\|x\|\le 1\}} \left(\inf_{\{y\in E;\|y\|^2+\|By\|^2\le 1\}} \|x-y\| \right) \le$$

$$\le \|A\| + \|B\| + \sup_{\{x\in E;\|x\|\le 1\}} \left(\inf_{\{y\in E;\|y\|\le \frac{1}{1+\|B\|}\}} \|x-y\| \right) \le$$

$$\le \|A\| + \|B\| + 1 - \frac{1}{1+\|B\|} \le \|A\| + 2\|B\| .$$

The last inequality follows from the inequality $1 - \frac{1}{1+x} \le x$ valid for $x \ge 0$. ∎

4. FORMULATION AND A SKETCH OF THE PROOF OF ABSOLUTE CONTINUITY

4.1. We will use the notations from Part I (especially from sections 6,7) along with some extra notations described in Section 2 without special notification.

Thus, we begin with the family $S^{\ell}_{k,r,s,\alpha,\gamma}(x)$ of TLSM (cf. Sec. 7_I). Let us define for any number q, $0 < q \le \frac{1}{2}\delta^{\ell}_{r,s,\alpha,\gamma}$

$$\hat{\Lambda}^{\ell}_{k,r,s,\alpha,\gamma}(x,q) = \bigcup_{y\in\Lambda^{\ell}_{k,r,s,\alpha,\gamma}\cap\tilde{U}(x,q)} V(y) \cap \tilde{U}(x,q), \tag{4.1}$$

where $\tilde{U}(x,q)$ is defined in (2.13).

As $\Lambda^{\ell}_{k,r,s,\alpha,\gamma}$ is a closed set (cf. Sec. 3.4_I), from Lemma $7.1.3_I$ one sees that $\hat{\Lambda}^{\ell}_{k,r,s,\alpha,\gamma}(x,q)$ is a closed subset of $\tilde{U}(x,q)$.

Consider a smooth submanifold (with boundary) W of $\tilde{U}(x,q)$ such that

$$\begin{cases} (\tau'_x \circ \exp_x^{-1}(W) \text{ is the graph of a smooth mapping } \psi \text{ such that} \\ \\ \psi \in C^1(\tilde{B}^2(q),E_1) \text{ (cf. (2.10))} \end{cases} \tag{4.2}$$

and such that W intersects any TLSM $V(y)$, $y \in \Lambda^{\ell}_{k,r,s,\alpha,\gamma} \cap \tilde{U}(x,q)$ at most at one point and that this intersection is transversal.

Any submanifold W as above will be called a submanifold transversal to the family $S^{\ell}_{k,r,s,\alpha,\gamma}(x)$ in the neighborhood $\tilde{U}(x,q)$.

Let us denote

$$|W| = \max_{v\in\tilde{B}^2(q)} \|\psi(v)\| + \max_{v\in\tilde{B}^2(q)} \|d\psi(v)\|, \tag{4.3}$$

where $\|d\Psi(v)\|$ denotes the operator norm induced by the Euclidean norms on E_1 and E_2. From Corollary 7.1_I it is easy to see that for every submanifold W of $\tilde{U}(x,q)$ satisfying (4.2), if $|W| \le \frac{1}{2}$ then W is transversal to the family $S^{\ell}_{k,r,s,\alpha,\gamma}(x)$ in $\tilde{U}(x,q)$.

Let us consider two submanifolds W^1 and W^2 transversal to the family $S^{\ell}_{k,r,s,\alpha,\gamma}(x)$ in $\tilde{U}(x,q)$. Of course, a TLSM which intersects W^1 may not intersect W^2 or vice versa.

By Theorem $7.1.3_I$ there exist two open submanifolds $\tilde{W}^1$ and $\tilde{W}^2$ of W^1 and W^2 respectively, such that the so-called Poincaré map $p = p_{\tilde{W}^1,\tilde{W}^2}$

$$p: \tilde{W}^1 \cap \hat{\Lambda}^{\ell}_{k,r,s,\alpha,\gamma}(x,q) \to \tilde{W}^2 \cap \hat{\Lambda}^{\ell}_{k,r,s,\alpha,\gamma}(x,q)$$

is well defined and is a homeomorphism between $\tilde{W}^1 \cap \hat{\Lambda}^{\ell}_{k,r,s,\alpha,\gamma}(x,q)$ and $\tilde{W}^2 \cap \hat{\Lambda}^{\ell}_{k,r,s,\alpha,\gamma}(x,q)$. This map (see Fig. 1) is defined as follows: if $w \in \tilde{U}(x,q) \cap \Lambda^{\ell}_{k,r,s,\alpha,\gamma}$ and if $y = \tilde{W}^1 \cap V(w)$, then

$$p(y) = \tilde{W}^2 \cap V(w). \tag{4.4}$$

Let us note that $\left(p_{\tilde{W}^1,\tilde{W}^2}\right)^{-1} = p_{\tilde{W}^2,\tilde{W}^1}$.

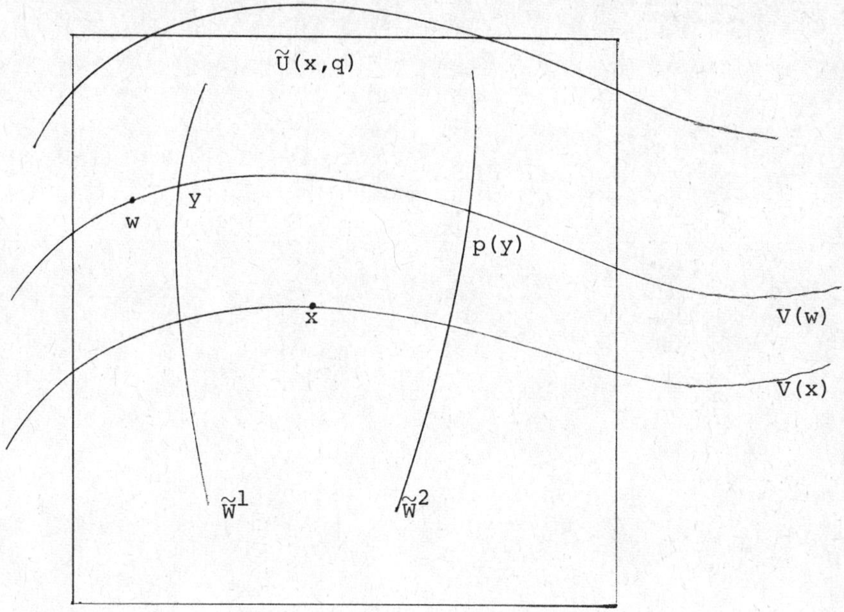

Fig. 1. Here $\widetilde{U}(x,q)$ is symbolically represented as a square centered at x. $V(x)$ always intersects $\widetilde{W}^1$ and $\widetilde{W}^2$. Let us note that the intersection of TLSM with $\widetilde{U}(x,q)$ is not necessarily connected.

Let ν be the measure on N induced by the Riemannian metric ρ on N. We do not assume that $\nu(N) < \infty$ neither we assume any relation between ν and the Φ-invariant measure μ. In fact, in this part, we are not going to use the measure μ explicitely anymore. This measure was used in Part I only to prove that the sets $\Lambda^{\ell}_{k,r,s}$, $\Lambda^{t,\delta}_{0}$ and $\Omega_{\alpha,\gamma}$ are non-empty and henceforth it is present in our considerations.

The family $S^{\ell}_{k,r,s,\alpha,\gamma}(x)$ restricted to $\widetilde{U}(x,q)$ is called absolutely continuous if for any two submanifolds W^1, W^2 transversal to the family $S^{\ell}_{k,r,s,\alpha,\gamma}(x)$ in $\widetilde{U}(x,q)$, the Poincaré map $p_{\widetilde{W}^1,\widetilde{W}^2}$ is absolutely continuous.

This means that for any two submanifolds W^1 and W^2 as above and for any measurable et $A \subset \widetilde{W}^1$ such that $\nu_{W^1}(A) = 0$ one has $\nu_{W^2}(p(A)) = 0$. Here, ν_W denotes the measure induced on the smooth submanifold W of N by the Riemannian metric p.

Let us note by $V^{\ell}_{k,r,s,\alpha,\gamma}$ the set of all points $x \in \Lambda^{\ell}_{k,r,s,\alpha,\gamma}$ such that for infinitely many $n \in \mathbb{N}$, $\Phi^n(x) \in \Lambda^{\ell}_{k,r,s,\alpha,\gamma}$. From Poincaré Recurrence Theorem, $\mu(V^{\ell}_{k,r,s,\alpha,\gamma}) = \mu(\Lambda^{\ell}_{k,r,s,\alpha,\gamma})$.

Let us denote by $R^{\ell}_{k,r,s,\alpha,\gamma}(x)$ the collection of all elements of the family $S^{\ell}_{k,r,s,\alpha,\gamma}(x)$ which intersect the set

$$V^{\ell}_{k,r,s,\alpha,\gamma} \cap \tilde{U}\left(x, \frac{\delta^{\ell}_{r,s,\alpha,\gamma}}{2}\right).$$

We define the absolute continuity of the family $R^{\ell}_{k,r,s,\alpha,\gamma}(x)$ restricted to $\tilde{U}(x,q)$ in the same way as for the family $S^{\ell}_{k,r,s,\alpha,\gamma}(x)$.

Theorem 4.1 formulated below and proved in sections 5 - 10 is the main result of this part. It is a counterpart of Pesin's absolute continuity theorem (Theorem 3.2.1. of $[\text{Pes}]_1$) for diffeomorphisms.

<u>THEOREM 4.1</u>. <u>Let us fix</u> ℓ,k,r,s,α <u>and</u> γ <u>satisfying</u> $(2.5)_I$, $(6.5)_I$ <u>and</u> (2.8).

(4.1.1) <u>For every</u> $x \in \Lambda^{\ell}_{k,r,s,\alpha,\gamma}$, <u>the family</u> $R^{\ell}_{k,r,s,\alpha,\gamma}(x)$ <u>restricted to</u> $\tilde{U}(x,\delta^{\ell}_{r,s,\alpha,\gamma})$ <u>is absolutely continuous</u>.

(4.1.2) <u>There exists a number</u> $q^{\ell}_{r,s,\alpha,\gamma}$, $0 < q^{\ell}_{r,s,\alpha,\gamma} < \frac{\delta^{\ell}_{r,s,\alpha,\gamma}}{2}$ <u>such that for every</u> $x \in \Lambda^{\ell}_{k,r,s,\alpha,\gamma}$ <u>and every two submanifolds</u> W^1 <u>and</u> W^2 <u>transversal to the family</u> $S^{\ell}_{k,r,s,\alpha,\gamma}(x)$ <u>restricted to</u> $\tilde{U}(x,q^{\ell}_{r,s,\alpha,\gamma})$ <u>and such that</u> $|W^1| \le \frac{1}{2}$ and $|W^2| \le \frac{1}{2}$, <u>the mapping</u> $p_{\tilde{W}^1,\tilde{W}^2}$ <u>is absolutely continuous</u>.

(4.1.3) <u>Moreover, if</u> $\nu(\Lambda^{\ell}_{k,r,s,\alpha,\gamma}) > 0$ <u>and</u> $x \in \Lambda^{1}_{k,r,s,\alpha,\gamma}$ <u>is a density point of</u> $\Lambda^{\ell}_{k,r,s,\alpha,\gamma}$, <u>then for every</u> $B > 0$ <u>there exist numbers</u> $q^{\ell}_{r,s,\alpha,\gamma}(B)$, $0 < q^{\ell}_{r,s,\alpha,\gamma}(B) < \frac{\delta^{\ell}_{r,s,\alpha,\gamma}}{2}$ <u>and</u> $\varepsilon^{\ell}_{r,s,\alpha,\gamma}(B) > 0$, <u>such that for every two submanifolds</u> W^1 <u>and</u> W^2 <u>transversal to the family</u> $S^{\ell}_{k,r,s,\alpha,\gamma}(x)$ <u>restricted to</u> $\tilde{U}(x,q^{\ell}_{r,s,\alpha,\gamma}(B))$ <u>such that</u> $|W^i| \le \varepsilon^{\ell}_{r,s,\alpha,\gamma}(B)$ <u>one has</u> $\nu_{W^i}(\tilde{W}^i \cap \hat{\Lambda}^{\ell}_{k,r,s,\alpha,\gamma}(x,q^{\ell}_{r,s,\alpha,\gamma}(B)) >$ > 0, $i = 1, 2$. <u>Furthermore, for</u> ν_{W^1} <u>almost all points</u>

$y \in \tilde{W}^1 \cap \hat{\Lambda}^{\ell}_{k,r,s,\alpha,\gamma}(x, q^{\ell}_{r,s,\alpha,\gamma}(B))$ <u>the</u> <u>Jacobian</u> $J(p)(y)$ (<u>See</u> <u>Sec</u>. 3.1.) <u>satisfies</u> <u>the</u> <u>inequality</u>

$$|J(p)(y) - 1| \leq B. \tag{4.5}$$

<u>Remarks</u>.

1) It will follow from the proof, that there exists a constant $m^{\ell}_{r,s,\alpha,\gamma} > 0$ such that for every B, $0 < B < 1$, one can take

$$q^{\ell}_{r,s,\alpha,\gamma}(B) = \varepsilon^{\ell}_{r,s,\alpha,\gamma}(B) = \frac{B}{m^{\ell}_{r,s,\alpha,\gamma}} .$$

2) If $\nu(\Lambda_{k,r,s}) > 0$ and if α, $0 < \alpha < 1$ is fixed, then for all ℓ big enough and all γ small enough, $\nu(\Lambda^{\ell}_{k,r,s,\alpha,\gamma}) > 0$.

Let us emphasize that the statement (4.1.3) is fully sufficient to the proof of criteria of K-property given in $[\text{Pes}]_2$ and reported in Sec. 13.2.

4.2. Let us describe the general structure of the proof.

First of all let us note that all the difficulties are concentrated in the proof of statement (4.1.2). Statement (4.1.3) easily follows from estimates obtained in the proof of statement (4.1.2). The proof of statement (4.1.1) also follows from the proof of (4.1.2). Indeed, roughly speaking, for some n big enough, one can divide the manifolds $\Phi^n(\tilde{W}^i)$, $i = 1,2$, into corresponding parts which satisfy the assumptions of (4.1.2) with respect to the family $\Phi^n(R^{\ell}_{k,r,s,\alpha,\gamma}(x))$.

Moreover, as in the proof of the existence of TLSM in Part I, instead of working directly with the mapping Φ, we work almost all the time (Secs. 5 - 9) with the local mappings $\Phi'_{0w} = \tau'_{\Phi(w)} \circ \circ \exp^{-1}_{\Phi(w)} \circ \Phi \circ \exp_w \circ (\tau'_w)^{-1}$ which are defined in some neighborhoods of 0 in $\mathbb{R}^m$.

A characteristic feature of the proof is the necessity of frequent verifications that various composition of mappings, etc., are in fact well defined. This problem already occurs in the Pesin paper as well as in Part I, but here it is one of the landmarks of the first half of the proof.

To prove Theorem 4.1, we want to compare $\nu_{W^1}(D)$ with $\nu_{W^2}(p(D))$, where D is a measurable subset of $\tilde{W}^1 \cap \hat{\Lambda}^{\ell}_{k,r,s,\alpha,\gamma}$. The main idea of the proof which goes back to Ya.G. Sinai (see $[\text{Ano}]_2$) is to deduce the comparison between $\nu_{W^1}(D)$ and $\nu_{W^2}(p(D))$ from the comparison between $\nu_{\Phi^n(W^1)}(\Phi^n(D))$ and $\nu_{\Phi^n(W^2)}(\Phi^n(p(D)))$ for n big enough.

Indeed, as the submanifolds W^1 and W^2 are transversal to the family of TLSM $S^{\ell}_{k,r,s,\alpha,\gamma}(x)$, then one hopes that $\Phi^n(\tilde{W}^1 \cap \hat{\Lambda}^{\ell}_{k,r,s,\alpha,\gamma}(x,q))$ and $\Phi^n(\tilde{W}^2 \cap \hat{\Lambda}^{\ell}_{k,r,s,\alpha,\gamma}(x,q))$ become closer and closer as $n \to \infty$. And really it is so. After using the formula of the change of variables in the integral, one deduces the desired comparison between $\nu_{W^1}(D)$ and $\nu_{W^2}(p(D))$. Unfortunately, the realization of this idea is not easy. First of all for n big enough the set $\Phi^n(\tilde{W}^1 \cap \hat{\Lambda}^{\ell}_{k,r,s,\alpha,\gamma}(x,q))$ in general does not belong to one local chart. Thus, to avoid this difficulty, we cover $\Phi^n(\tilde{W}^1 \cap \hat{\Lambda}^{\ell}_{k,r,s,\alpha,\gamma}(x,q))$ by sufficiently small pieces $\{W^1_{in}\}_{1\le i\le p(n)}$, such that for every i, $1 \le i \le p(n)$, W^1_{in} belongs to one local chart. Moreover, the next step consisting in comparison between $\nu_{\Phi^n(W^1)}(\Phi^n(D \cap W^1_{in}))$ and $\nu_{\Phi^n(W^2)}(\Phi^n(p(D \cap W^1_{in})))$ is also rather complicated.

The realization of the above idea is contained in Secs. 5 - 10 and is divided between eight Lemmas I - VIII as follows.

The proof is sharply divided in two parts. The first part formed by Secs. 5 - 7, contains all necessary tools and preliminaries to the proof properly speaking, which is given in Secs. 8 - 10.

After some important preliminaries described in Sec. 5, which contains among others Lemma I, we prove in Sec. 6 the Lemma II, which can be considered as the Main Lemma. Roughly speaking, this Lemma describes the asymptotic behavior of $\Phi^n(A)$, for large n, where A is a small piece of $\tilde{W}^1$ whose size depends on n.

Sec. 7 is mainly devoted to the careful description of the construction which leads to the small submanifolds $\{W^1_{in}\}_{1\le i\le p(n)}$ mentioned above.

In order to be able to estimate $\nu_{\Phi^n(W^2)}(\Phi^n(p(D)))$, we need some special coverings of $\Phi^n(D)$. We devote Sec. 8 to the construction of these coverings and to the proof of their good geometrical properties. In Sec. 8 we prove Lemmas III - V. Lemma III can be considered as a second main lemma and Lemmas IV and V are devoted to the construction of desired covering. Unlike all other sections, Sec. 8 has a true geometrical character.

Sec. 9 is mainly devoted to the proof of third main lemma Lemma VI giving the estimation which allows to compare $\nu_{W^1}(A)$ and $\nu_{W^2}(p(A))$ for $A \subset \widetilde{W}^1$ as above, if only one knows the measures $\nu_{\Phi^n(W^1)}(\Phi^n(A))$ and $\nu_{\Phi^n(W^2)}(\Phi^n(p(A)))$. A more global Lemma VII is an easy corollary of Lemma VI.

Finally, in Sec. 10 we prove Theorem 4.1. The proof uses some special coverings whose construction is based on coverings constructed in Sec. 8. The proof uses Lemma VIII which is the adaptation of Lemma VII to the elements of just constructed coverings. Let us note that the exact values of constants from Theorem 4.1 appear only at the end of Sec. 10.

Sec. 11 contains a very important consequence of Absolute Continuity Theorem 4.1. Namely we show that the conditional measures (see Sec. 11.1), induced on TLSM

$$\left\{V(y) \cap \widetilde{U}(x,q^{\ell}_{k,r,s,\alpha,\gamma}\right\}_{y \in \widetilde{U}(x,q^{\ell}_{k,r,s,\alpha,\gamma}) \cap \Lambda^{\ell}_{k,r,s,\alpha,\gamma}}$$

by an arbitrary measure which is absolutely continuous with respect to the Riemannian measure ν, are absolutely continuous with respect to the measures induced on these submanifolds by the Riemannian metric ρ. The idea of the proof goes back to Ja.G. Sinai (see $[\text{Ano}]_2$).

The role of Lyapunov metric in the formulation of Theorem 4.1 is completely illusory. In fact this metric intervenes only through the sets $\widetilde{U}(x,q)$ and by the quantities $|W^1|$ and $|W^2|$. But as on the set $\Lambda^{\ell}_{r,s,\alpha,\gamma}$, the Lyapunov and the Riemannian metrics are uniformly equivalent, nothing will change if instead of Lyapunov metric we use to formulate the Theorem 4.1. the Riemannian metric.

As far as the proof of Theorem 4.1 is concerned, the Lyapunov metric plays a substantial role. In particular this is true for the Main Lemma II, but also for many other places. As it was explained in Part I, the use of Lyapunov metric allows to reduce the study of mapping $\Phi|_{\Lambda^{\ell}_{r,s}}$ to the study of mapping satisfying the uniform hyperbolic conditions (see $(2.2.3)_I$).

Finally, let us underline that Absolute Continuity Theorem is not a local result despite the fact that it is concerned with a small neighborhood of a point x. Indeed, the definition of the family $S^{\ell}_{k,r,s,\alpha,\gamma}(x)$ involves the mapping Φ on the whole N.

5. START OF THE PROOF - I

5.1. Starting from this moment up to the end of the proof of Theorem 4.1, $x \in \Lambda^{\ell}_{k,r,s,\alpha,\gamma}$ will be fixed once and for all.

Let us consider a point $w \in \Lambda^{\ell}_{k,r,s,\alpha,\gamma} \cap \tilde{U}\left(x, \frac{\delta^{\ell}_{r,s,\alpha,\gamma}}{2}\right)$. Let us note $\hat{V}(w) = (\tau'_w \circ \exp_w^{-1})(V(w))$.

Let us choose a point $y \in V(w) \cap \tilde{U}\left(w, \frac{\delta^{\ell}_{r,s,\alpha,\gamma}}{2}\right)$. Let

$$P_0 = (u_0, v_0) = (\tau'_w \circ \exp_w^{-1})(y) \in \hat{V}(w) \cap \tilde{B}\left(\frac{\delta^{\ell}_{r,s,\alpha,\gamma}}{2}\right) \text{(see (2.10))}. \tag{5.1}$$

As follows from $(6.1.1)_I$ and from (2.9) for every $n \geq 1$ one has

$$\left\{\begin{array}{l} P_n = (u_n, v_n) \overset{\text{def}}{=\!=} \left(\prod_{i=0}^{n-1} \Phi'_i\right)(P_0) = \left(\tau'_{\Phi^n(w)} \circ \exp^{-1}_{\Phi^n(w)}\right)(\Phi^n(y)) \in \\ \in B_n(w), \end{array}\right. \tag{5.2)}$$

where $B_n(w)$ is defined in Sec. 2.6.

Moreover, the proof of $(6.1.1)_I$ together with (2.9) gives

$$\left\{\begin{array}{l} \|(u_n, v_n)\| = \left\|\left(\prod_{i=0}^{n-1} \Phi'_i\right)(P_0)\right\| \leq 200(\kappa_{r,s})^n \|(u_0, v_0)\| < \\ \qquad < 100(\kappa_{s,r})^n \delta^{\ell}_{r,s,\alpha,\gamma} . \end{array}\right. \tag{5.3}$$

In the future, when the point $(u_0, v_0) \in \hat{V}(w) \cap \tilde{U}\left(w, \frac{\delta^{\ell}_{r,s,\alpha,\gamma}}{2}\right)$ will be fixed, we will use the notation u_n and v_n exclusively in the sense defined above, without additional explanation.

For a fixed $q_0 > 0$ and $\delta_0 > 0$ let us note for $n \geq 0$

$$q_n = q_0(\alpha^{8b}\exp(-15\varepsilon_{r,s}))^n \tag{5.4}$$

$$\delta'_n = \delta_0(\xi')^n. \tag{5.5}$$

where $\xi' = r\exp(8\varepsilon_{r,s}) < 1$.

LEMMA 5.1 (I)

(5.1.1). *Let* $C > 0$ *and* $0 < q_0 < 1$. *Then there exists a number* $n_0 = n_0(\ell,r,s,\alpha,\gamma,C,q_0)$ *such that for every* δ_0, $0 < \delta_0 < 1$, *every point* $(u_0,v_0) \in \hat{V}(w)$ *and every* $n \geq n_0$ *one has*

$$\tilde{B}^1(u_n,\delta_n') \subset B_n^1(q_n) \quad \text{and} \quad \tilde{B}^2(v_n,\delta_n') \subset B_n^2(q_n) \tag{5.6}$$

(*cf*. (2.10) *and* (2.11)) *and*

$$\begin{cases} B_n^1(q_n) \times B_n^2(q_n) \subset \left\{ h \in \mathbb{R}^m;\ \|h\| < \frac{1}{\sqrt{2}} (Q(\ell,\alpha,\gamma)\alpha^{2bn})^2 \right\} \subset \\ \qquad \subset B_n(w) \stackrel{\text{def}}{=\!=} \tau_n' \circ \exp_n^{-1}(V_n(w)). \end{cases} \tag{5.7}$$

(cf. $(3.9)_I$, (2.9) *and Sec*. 2.6)).

(5.1.2). *Let* $q_0 = \frac{\delta^\ell_{r,s,\alpha,\gamma}}{2}$. *Let* $(u_0,v_0) \in \hat{V}(w)$. *If* $\|(u_0,v_0)\| \leq \leq \frac{q_0}{400\sqrt{2}}$ *and if* $0 < \delta_0 < \frac{q_0}{4}$, *then for every* $n \geq 0$ *the inclusions* (5.6) *and* (5.7) *hold*.

Proof (5.1.1). We will prove only the first inclusion of (5.6) and (5.7). The proof of the second inclusion of (5.6) as well as the proof of (5.1.2) are completely similar.

Let $u \in \tilde{B}^1(u_n,\delta_n')$ for some $n \geq 0$. Thus $\|u-u_n\| \leq \delta_n' = \delta_0(\xi')^n$. Now, from (5.3) and from the assumptions of the Lemma it follows that

$$\left.\begin{aligned} \|u\| &\leq \|u_n\| + \delta_n' \leq \|(u_n,v_n)\| + \delta_n' \leq 200(\kappa_{r,s})^n\|(u_0,v_0)\| + \delta_n' \leq \\ &\leq 100\ \delta^\ell_{r,s,\alpha,\gamma}(\kappa_{r,s})^n + \delta_0(\xi')^n < \\ &< 100((\kappa_{r,s})^n + (\xi')^n) < 200(\xi')^n, \end{aligned}\right\}$$

because $\delta^\ell_{r,s,\alpha,\gamma} < Q(\ell,\alpha,\gamma) < 1$ (see $(6.7)_I$ and because $\kappa < \xi'$ (see (2.4)).

From (2.9) one has that $\tilde{B}^1(\frac{q_n}{\sqrt{2}}) \subset B_n^1(q_n)$. Thus to finish the proof of the first inclusion of (5.6), it is sufficient to verify that there exist a number $n_{01} = n_{01}(\ell,r,s,\alpha,\gamma,C)$ such that for

$n \geq n_{01}$ one has $200(\xi')^n < q_n = q_0[\alpha^{8b}\exp(-15\varepsilon_{r,s})]^n$. But this is so, because $(6.5)_I$ implies that $\xi' = r\exp(8\varepsilon_{r,s}) < \alpha^{8b}\exp(-15\varepsilon_{r,s})$.

The proof of (5.7) is evident, because for $n \geq n_{02}(\ell,s,r,\alpha,\gamma)$ one has

$$[\alpha^{8b}\exp(-15\varepsilon_{r,s})]^n < \frac{1}{\sqrt{2}}\,[Q(\ell,\alpha,\gamma)\alpha^{2bn}]^2.$$ ■

Let us note that in fact this proof implies that for $n \geq n_0$ (under assumptions of Lemma 5.1.1) and for $n \geq 0$ (under assumptions of Lemma 5.1.2) one has $\tilde{B}^2(v_n,\delta_n') \subset \operatorname{Int} B_n^2(q_n) = \{v \in \mathbb{R}^p;\ \|v\| < q_n\}$ as shown in Fig. 2.

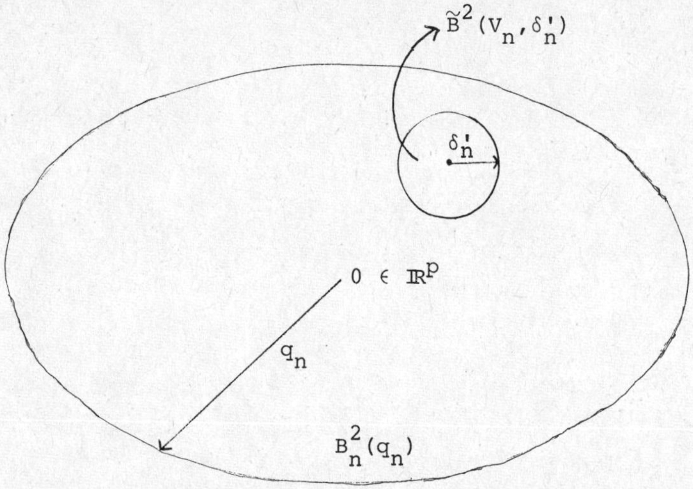

Fig. 2. Considered with respect to the standard Euclidean norm in $\mathbb{R}^p$, $B_n^2(q_n)$ is, generally speaking, an ellipsoid.

5.2. Lemma 5.1.1 will play in the future only a technical role. It permits us to consider for any $n \geq n_0$, $\Phi_n'(u,v)$ for $u \in \tilde{B}^1(q_n)$ and $v \in \tilde{B}^2(v_n,\delta_n')$ because for $n \geq n_0$, Φ_n' is well defined on $B_n(q_n) = B_n^1(q_n) \times B_n^2(q_n) \subset B_n(w)$.

The role of (5.1.2) will be completely analogous. It permits to consider for every $n \geq 0$ the mapping $\Phi_n'(u,v)$ for $u \in \tilde{B}^1(q_n)$ and $v \in \tilde{B}^2(v_u,\delta_n')$ because for every $n \geq 0$, Φ_n' is well defined on $B_n(q_n) = B_n^1(q_n) \times B_n^2(q_n) \subset B_n(w)$. The last inclusion follows from the fact that for every $n \geq 0$ has $\sqrt{2}\, q_n =$ $=\sqrt{2} q_0 [\alpha^{8b} \exp(-15\varepsilon_{r,s})]^n < [Q(\ell,\alpha,\gamma)\alpha^{2bn}]^2$. Indeed, this follows from $(6.7)_I$ because $q_0 = \dfrac{\delta^\ell_{r,s,\alpha,\gamma}}{2} < (Q(\ell,\alpha,\gamma))^2$.

6. THE FIRST MAIN LEMMA

6.1. Let $C > 0$ be some number. Let us define $q(1,C) =$ $=q(\ell,r,s,\alpha,\gamma,C)$ by

$$q(1,C) = \min\left(\frac{\delta^\ell_{r,s,\alpha,\gamma}}{8q(A^\ell_{r,s}+1)}, \frac{\mu-\hat{\xi}}{\Gamma}, \frac{\mu\eta-\lambda}{\Gamma(\frac{1}{C}+\eta)}\right) \tag{6.1}$$

(cf. (2.1) - (2.3)), where q comes from the condition (Bb) from Sec. 1.1_I, $\Gamma = \Gamma(\ell,r,s) = 2\dfrac{HA^\ell_{r,s}}{\gamma^b \nu}$, H comes from the Theorem 3.1_I and b comes from condition $(1.3)_I$. From (2.4), $\mu > \hat{\xi}$ and from (2.5), $\mu\eta > \lambda$. Thus $q(1,C) > 0$. The definition of $q(1,C)$ is the first step towards the definition of numbers $q^\ell_{r,s,\alpha,\gamma}(B)$ from Theorem 4.1.2.

6.2. The following First Main Lemma consists of two parts. The first part describes the behaviour of $\Phi^n(w)$ for n big enough, where W is a submanifold close to the subspace E_{2W}, whose size in the Lyapunov metric is about δ'_n. The second part deals with the case when W is very close to the subspace E_{2W}. Then we are able to describe the behavior of $\Phi^n(W)$ for all $n \geq 1$. This lemma can be considered in our framework as a counterpart of so called λ-lemma (see for example [New] or [Pal]). Like Lemma 5.1, the first part is used in the proof of (4.1.1), the second in the proof of (4.1.2).

<u>LEMMA 6.1 (II)</u>

(6.1.1). <u>Let</u> $x \in \Lambda^\ell_{k,r,s,\alpha,\gamma}$ <u>and let</u> W <u>be a submanifold transversal to the family</u> $S^\ell_{k,r,s,\alpha,\gamma}(x)$ <u>in</u> $\tilde{U}(x,\delta^\ell_{r,s,\alpha,\gamma})$.

<u>Let us consider a point</u> $w \in \Lambda^\ell_{k,r,s,\alpha,\gamma} \cap \operatorname{Int} \tilde{U}\left(x, \dfrac{\delta^\ell_{r,s,\alpha,\gamma}}{2}\right)$ <u>such that TLSM</u> $V(w)$ <u>intersect</u> W <u>at the point</u> z_0. <u>Let us note by</u>

$(u_0,v_0) \in \mathbb{R}^m = \mathbb{R}^k \oplus \mathbb{R}^p$, the point $(u_0,v_0) = \tau_w' \circ \exp_w^{-1}(z_0)$. Thus for $\Delta_0 > 0$ small enough, there exists the unique C^1 mapping $\Psi_0\colon \tilde{B}^2(v_0,\Delta_0) \to \mathbb{R}^k$ such that $\Psi_0(v_0) = u_0$ and that $\exp_w \circ (\tau_w')^{-1}\{(\Psi_0(v),v);\ v \in \tilde{B}^2(v_0,\Delta_0)\}$ is a submanifold of W.

Then there exists a number $n_1 = n_1(\ell,k,r,s,\alpha,\gamma,W) \ge n_0$ (where n_0 comes from Lemma 5.1), a number $\tilde{\delta}_{n_1} = \tilde{\delta}_{n_1}(\ell,k,r,s,\alpha,\gamma,w,W) > 0$ such that for every $n \ge n_1$ there exist the unique mapping Ψ_n of class C^1, $\Psi_n\colon \tilde{B}^2(v_n,\tilde{\delta}_n') \to \mathbb{R}^k$, where $\tilde{\delta}_n' = \tilde{\delta}_{n_1}(\xi')^n$ such that

$$\exp_{\Phi^{n_1}(w)} \circ (\tau'_{\Phi^{n_1}(w)})^{-1}\{(\Psi_{n_1}(v),v);\ v \in \tilde{B}^2(v_{n_1},\tilde{\delta}'_{n_1})\} \subset \Phi^{n_1}(W)$$

and that for every $n \ge n_1$ one has $\Psi_n(v_n) = u_n$ and

$$\{(\Psi_n(v),v);\ v \in \tilde{B}^2(v_n,\tilde{\delta}_n')\} \subset B_n(w) = \tau_n' \circ \exp_n^{-1}(V_n(w)). \qquad (6.2)_n$$

Thus the mapping Φ_n' is well defined on the graph of the mapping Ψ_n.

For every δ_{n_1}, $0 < \delta_{n_1} < \tilde{\delta}_{n_1}$, one has

$$\{(\Psi_{n+1}(v),v);\ v \in \tilde{B}^2(v_{n+1},\delta'_{n+1})\} \subset \{\Phi_n'(\Psi_n(v),v);\ v \in \tilde{B}^2(v_n,\delta_n')\}, \qquad (6.3)_n$$

where $\delta_n' = \delta_{n_1}(\xi')^n$.

$$a_n = \max_{v \in \tilde{B}^2(v_n,\delta_n')} \|\Psi_n(v)\| \le \hat{\xi}^n \qquad (6.4)_n$$

$$b_n = \max_{v \in \tilde{B}^2(v_n,\tilde{\delta}_n')} \|d\Psi_n(v)\| \le \eta^{n-n_1} \qquad (6.5)_n$$

(6.1.2). Let $w \in \Lambda^{\ell}_{k,r,s,\alpha,\gamma}$. Let $0 < C \le 1$. Let $0 < q_0 \le q(1,C)$, where $q(1,C)$ is defined by (6.1). Let $0 < \delta_0 \le \frac{q_0}{4}$. Let $(u_0,v_0) \in \hat{V}(w)$ and let $\|(u_0,v_0)\| \le \frac{q_0}{800\sqrt{2}}$. Let $\delta_n' = \delta_0(\xi')^n$.

Let $\Psi_0\colon \tilde{B}^2(v_0,\delta_0) \to \mathbb{R}^k$ be a mapping of class C^1 such that $\Psi_0(v_0) = u_0$ and

$$\max_{v\in\widetilde{B}^2(v_0,\delta_0)} \|\Psi_0(v)\| \le \frac{1}{4} q_0 \tag{6.6}$$

$$\max_{v\in\widetilde{B}^2(v_0,\delta_0)} \|d\Psi_0(v)\| \le C. \tag{6.7}$$

<u>Then there exists a unique sequence</u> $\{\Psi_n\}_{n\ge 1}$ <u>of mappings of class</u> C^1, $\Psi_n\colon \widetilde{B}^2(v_n,\delta_n') \to \mathbb{R}^k$, <u>such that for every</u> $n \ge 0$ <u>one has</u>

$$\{(\Psi_n(v),v);\ v \in \widetilde{B}^2(v_n,\delta_n')\} \subset B_n(w) = \tau_n' \circ \exp_n^{-1}(V_n(w)) \qquad (6.8)_n$$

<u>and thus the mapping</u> Φ_n' <u>is well defined on the graph of the mapping</u> Ψ_n,

$$\left\{\begin{array}{l} \{(\Psi_{n+1}(v),v);\ v \in \widetilde{B}^2(v_{n+1},\delta_{n+1}')\} \subset \\ \subset \{\Phi_n'(\Psi_n(v),v);\ v \in \widetilde{B}^2(v_n,\delta_n')\}, \end{array}\right. \qquad (6.9)_n$$

$$a_n = \max_{v\in\widetilde{B}^2(v_n,\delta_n')} \|\Psi_n(v)\| \le (\tfrac{1}{4}+C)q_0\xi^n \qquad (6.10)_n$$

$$b_n = \max_{v\in\widetilde{B}^2(v_n,\delta_n')} \|d\Psi_n(v)\| \le C\eta^n. \qquad (6.11)_n$$

Let us note that $(6.3)_n$ and $(6.9)_n$ define Ψ_{n+1} from Ψ_n.

<u>6.3. Proof</u>. We will first prove in all details (6.1.1). The proof of (6.1.2) follows the same lines, but is somewhat simpler. We will quickly scetch this proof after the proof of (6.1.1). For the convenience of the reader the proof of (6.1.1) is divided on four steps. Proof goes by induction.

<u>Proof of (6.1.1)</u>

<u>Step 1</u>. In this step we will determine n_1 and $\widetilde{\delta}_{n_1}$ and we will check $(6.2)_{n_1}$, $(6.4)_{n_1}$ and $(6.5)_{n_1}$.

Let us note by Γ_0 the graph of the mapping Ψ_0, i.e., $\Gamma_0 = \{(\Psi_0(v),v) \in \mathbb{R}^m: v \in \tilde{B}^2(v_0,\Delta_0)\}$. As Ψ_0 is a map of class C^1, then no vector from the tangent space to Γ_0 at (u_0,v_0) is parallel to $\mathbb{R}^k$ in $\mathbb{R}^m = \mathbb{R}^k \oplus \mathbb{R}^p$. For every $n \geq 0$, the diffeomorphism $H'_n = \Phi'_n \circ \dots \circ \Phi'_0$ is well defined in some sufficiently small neighborhood X_n of (u_0,v_0) (cf. Theorem 6.1_I). Let us consider the p - dimensional submanifold $\Gamma_n = H'_n(\Gamma_0 \cap X_n)$ of $\mathbb{R}^m$. As the derivatives $dH'_n(u_0,v_0) = d\Phi'_n \circ \dots \circ d\Phi'_0(u_0,v_0)$ preserve the splitting $\mathbb{R}^m = \mathbb{R}^k \oplus \mathbb{R}^p$, then the tangent space, denoted L_n, to Γ_n at the point $(u_n,v_n) = H'_n(u_0,v_0)$ is not parallel to $\mathbb{R}^k$. Consequently, in some sufficiently small neighborhood Y_n of the point (u_n,v_n), Γ_n is the graph of some (uniquely defined) smooth C^1 mapping $\Psi_n: \tilde{B}^2(v_n,\delta_n) \to \mathbb{R}^k$, where $\delta_n > 0$ is a sufficiently small number. In particular $\Psi_n(v_n) = u_n$ and

$$\max_{v \in \tilde{B}^2(v_n,\delta_n)} \|d\Psi_n(v)\| < +\infty.$$

Now, we will estimate $\|d\Psi_n(v_n)\|$.

Let us recall that

$$\Phi'_n = \tau'_{n+1} \circ \exp^{-1}_{n+1} \circ \Phi \circ \exp_n \circ (\tau'_n)^{-1}, \quad n \geq 0 \qquad (3.13)_I$$

$$\Phi'_n(u,v) = (A_n u + a_n(u,v), B_n v + b_n(u,v)), u \in \mathbb{R}^k, \ v \in \mathbb{R}^p. \qquad (3.14)_I$$

Let us denote also $t_n = (a_n,b_n)$. From $(2.2.3)_I$ and from the definition of λ and μ one has

$$\begin{cases} \|A_n(u)\| \leq \lambda\|u\| & \text{for } u \in \mathbb{R}^k \\ \\ \|B_n(v)\| \geq \mu\|v\| & \text{for } v \in \mathbb{R}^p. \end{cases} \qquad (6.12)$$

PROPOSITION 6.1. <u>There exists</u> $n_{11} = n_{11}(\ell,r,s,\alpha,\gamma)$ <u>such that for every</u> $n \geq n_{11}$ <u>one has</u>

$$\|dt_n(u_n,v_n)\| < \exp(-10n\varepsilon_{r,s}).$$

<u>Proof</u>. As $(u_0,v_0) \in \hat{V}(w)$, then from (5.2) one knows that for every $n \geq 1$ one has $(u_n,v_n) \in B_n(w)$. Thus for every $n \geq 1$ we can apply Theorem 3.1_I. From this theorem, from (5.3), as $dt_n(0,0) = 0$ one obtains

$$\begin{cases} \|dt_n(u_n,v_n)\| = \|dt_n(u_n,v_n) - dt_n(0,0)\| \leq \\ \leq \dfrac{200HA^{\ell}_{r,s}}{\gamma^b\nu[\alpha^b\nu]^n}(\kappa_{r,s})^n\|(u_0,v_0)\| < \dfrac{200HA^{\ell}_{r,s}\delta^{\ell}_{r,s,\alpha,\gamma}}{\gamma^b\nu} \cdot \left(\dfrac{r\exp(5\varepsilon_{r,s})}{\alpha^b}\right)^n. \end{cases}$$

Now, from $(6.5)_I$, one deduces that there exists $n_{11} = n_{11}(\ell,r,s,\alpha,\gamma) \geq$ $\geq n_0$ such that for $n \geq n_{11}$ one has

$$\|dt_n(u_n,v_n)\| \leq C(\ell,r,s,\alpha,\gamma)\left(\frac{r\exp(5\varepsilon_{r,s})}{\alpha^b}\right)^n \leq \exp(-10n\varepsilon_{r,s}). \qquad \blacksquare$$

Now, as $\hat{\xi} = r\exp(12\varepsilon_{r,s}) < \mu$ (cf. (2.4)), as $\frac{\lambda}{\mu} < \eta < 1$ (cf. (2.5)) and as $\kappa_{r,s} < \xi$ (cf. (2.4)), then there exists $n_{12} = n_{12}(\ell,r,s,\alpha,\gamma) \geq n_{11}$ such that for all $n \geq n_{12}$ one has

$$\begin{cases} \dfrac{\lambda+\exp(-10n\varepsilon_{r,s})}{\mu-\exp(-10n\varepsilon_{r,s})} < \eta, \\ \\ 400\,\delta^{\ell}_{r,s,\alpha,\gamma}(\kappa_{r,s})^n < \xi^n. \end{cases} \tag{6.13}$$

<u>PROPOSITION 6.2</u>. $\lim_{n\to\infty}\|d\Psi_n(u_n,v_n)\| = 0$. <u>More precisely there exists a constant</u> $Q = Q(\ell,k,r,s,\alpha,\gamma,W) > 0$, <u>such that for all</u> $n \geq n_{12}$ <u>and all</u> $t \geq 0$,

$$\|d\Psi_{n+t}(u_{n+t},v_{n+t})\| \leq Q\eta^t.$$

<u>Proof</u>. As all the linear mappings

$$d\Phi'_n(u_n,v_n) = (A_n + da_n(u_n,v_n), B_n + dB_n(u_n,v_n)),\ n \geq 0, \tag{6.14}$$

preserves the splitting $\mathbb{R}^m = \mathbb{R}^k \oplus \mathbb{R}^p$, then the linear mapping $dH_n'(u_0,v_0)$ also preserves it.

Let us consider $n \geq n_{12}$. So that (6.13) is satisfied. The tangent space L_n to the graph of Ψ_n at the point (u_n,v_n) is the graph $L_n = \{(K_n v,v);\ v \in \mathbb{R}^p\}$ of the linear mapping $K_n = d\Psi_n(v_n)\colon \mathbb{R}^p \to \mathbb{R}^k$. We have $L_n = dH_n'(u_0,v_0)L_0$.

As $L_{n+1} = d\Phi_n'(u_n,v_n)L_n$, then from (6.14) one has

$$\begin{cases} L_{n+1} = \{((A_n + da_n(u_n,v_n) \circ K_n v, (B_n + db_n(u_n,v_n))v); v \in \mathbb{R}^p\} = \\ \quad = \{((A_n + da_n(u_n,v_n)) \circ K_n \circ (B_n + db_n(u_n,v_n))^{-1}v, v); \\ \qquad\qquad v \in \mathbb{R}^p\} = \\ \quad = \{(K_{n+1}v,v);\ v \in \mathbb{R}^p\}. \end{cases} \tag{6.15}$$

From (6.12) and from Proposition 6.1 one obtains that for $n \geq n_{12}$

$$\|A_n+da_n(u_n,v_n)\| \leq \|A_n\| + \|da_n(u_n,v_n)\| \leq \lambda + \exp(-10_n\varepsilon_{r,s}). \tag{6.16}$$

Moreover, by the same arguments, one obtains that for $v \in \mathbb{R}^p$,

$$\begin{cases} \|B_n v+db_n(u_n,v_n)v\| \geq \|B_n v\| - \|db_n(u_n,v_n)v\| \geq \\ \qquad \geq (\mu - \exp(-10n\varepsilon_{r,s})\|v\|, \end{cases}$$

i.e.

$$\|[B_n + db_n(u_n,v_n)]^{-1}\| \leq \frac{1}{\mu-\exp(-10n\varepsilon_{r,s})}. \tag{6.17}$$

Now, (6.15)-(6.17) together with (6.13) implies that for all $n \geq n_{12}$

$$\|K_{n+1}\| \leq \|K_n\|\eta.$$

Consequently for all $n \geq n_{12}$ and $t > 0$

$$\|K_{n+t}\| \leq \|K_n\|\eta^t.$$

Thus, to finish the proof it is enough only to prove that

$$Q(\ell,k,r,s,\alpha,\gamma,W) = \sup\|d\Psi_{n_{12}}(v_{n_{12}})\| < +\infty,$$

where sup is taken over all parts $z_0 \in W \cap V(w)$, where $w \in \Lambda^{\ell}_{k,r,s,\alpha,\gamma} \cap \text{Int}\, \tilde{U}(x, \frac{\delta^{\ell}_{r,s,\alpha,\gamma}}{2})$. But this is so in virtue of the fact that for any fixed $q > 0$, $\Phi^q(\Lambda^{\ell}_{k,r,s,\alpha,\gamma})$ is a closed set and the decomposition $T_wN = E_{1w} \oplus E_{2w}$ is continuous on that set. Indeed, one can apply the above remarks to $q = n_{12}$. ∎

Beginning from now, for all $n \geq n_{12}$ we fix the numbers $\delta_n > 0$ so small that

$$\max_{v \in \tilde{B}^2(v_n, \delta_n)} \| dt_n(\Psi_n(v), v) \| \leq \exp(-10n\varepsilon_{r,s}), \tag{6.18}$$

$$\max_{v \in \tilde{B}^2(v_n, \delta_n)} \| d\Psi_n(v) \| \leq 2Q\eta^{n-n_{12}} \tag{6.19}$$

and that

$$\max_{v \in \tilde{B}^2(v_n, \tilde{\delta}_n)} \| \Psi_n(v) \| < \hat{\xi}^n. \tag{6.20}$$

This is possible by Proposition 6.1 and 6.2 and (5.3) together with the second of (6.13).

Let us fix the number $n_1 = n_1(\ell, r, s, \alpha, \gamma, W) \geq n_{12}$ such that for all $n \geq n_1$ one has

$$\frac{1200\, HA^{\ell}_{r,s}}{\gamma^{b_\nu}} (\xi')^n < \exp(-10n\varepsilon_{r,s}) \tag{6.21}$$

$$\hat{\xi} < \mu - \exp(-10n\varepsilon_{r,s}) \tag{6.22}$$

$$301(\xi')^n < \hat{\xi}^n \tag{6.23}$$

and $2Q\eta^{n_1} < \eta^{n_{12}}$.

This last condition, together with (6.19) implies that for $n \geq n_1$ one has

$$\max_{v \in \tilde{B}^2(v_n, \delta_n)} \| d\Psi_n(v) \| \leq \eta^{n-n_1} \leq 1. \tag{6.24}$$

In particular, from (6.24) one has that $\|\Psi_{n_1}(v) - \Psi_{n_1}(v_{n_1})\| \le \|v-v_{n_1}\|$ and thus that $\{\Psi_{n_1}(v);\ v \in \tilde{B}^2(v_{n_1},\tilde{\delta}_{n_1})\} \subset \tilde{B}^1(v_{n_1},\tilde{\delta}_{n_1})$. Now, from (5.1.1), as $n_1 \ge n_0$, one obtains that $(6.2)_{n_1}$ is true. Clearly, (6.24) imply also $(6.5)_{n_1}$. $(6.4)_{n_1}$ follows from (6.20).

Step 2. Beginning from this moment we forget completely the meaning of Ψ_n and of $\tilde{\delta}_n$ introduced above, for $n \ne n_1$. In what follows these symbols will have a different meaning.

Now, for every $n \ge n_1$ we will prove that $(6.2)_n$ and $(6.5)_n$ allows to construct the mapping Ψ_{n+1} satisfying $(6.3)_n$, $(6.2)_{n+1}$, $(6.4)_{n+1}$ and $(6.5)_{n+1}$. This, together with $(6.2)_{n_1}$, $(6.4)_{n_1}$ and $(6.5)_{n_1}$ proves Lemma 6.1 by induction.

Thus let us suppose that for some $n \ge n_1$, $(6.2)_n$, $(6.4)_n$ and $(6.5)_n$ are true. In particular, from $(6.2)_n$ one knows that $t_n(\Psi_n(v),v)$ is well defined for $v \in \tilde{B}^2(v_n,\tilde{\delta}'_n)$.

PROPOSITION 6.3. Let $n \ge n_1$ and let $v^1,v^2 \in \tilde{B}^2(v_n,\tilde{\delta}'_n)$. Then

$$\|t_n(\Psi_n(v^1),v^1) - t_n(\Psi_n(v^2),v^2)\| \le e^{-10n\varepsilon_{r,s}}\|v^1-v^2\|.$$

Proof. By the Mean Value Theorem one obtains that

$$\left\{\begin{aligned} &\|t_n(\Psi_n(v^1),v^1) - t_n(\Psi_n(v^2),v^2)\| \le \\ &\qquad \le \sup_{z\in\Gamma_n}\|dt_n(z)\|(\|\Psi_n(v^1) - \Psi_n(v^2)\| + \|v^1-v^2\|) \end{aligned}\right. \tag{6.25}$$

where $\Gamma_n = \{(\Psi_n(v),v);\ v \in \tilde{B}^2(v_n,\tilde{\delta}'_n)\}$.

Let us estimate $\sup_{z\in\Gamma_n}\|dt_n(z)\|$. From the proof of Proposition 6.1 and from Theorem 3.1_I one gets

$$\begin{cases} \|dt_n(z)\| \le \|dt_n(u_n,v_n)\| + \|dt_n(z) - dt_n(u_n,v_n)\| \le \\ \le \dfrac{200\, HA^{\ell}_{r,s}\delta^{\ell}_{r,s,\alpha,\gamma}}{\gamma^{b}\nu} \left[\left(\dfrac{r\exp(5\varepsilon_{r,s})}{\alpha^{b}} \right)^{n} + \sqrt{2}\,\tilde{\delta}'_n \right] \le \\ < \dfrac{600\, HA^{\ell}_{r,s}\delta^{\ell}_{r,s,\alpha,\gamma}}{\gamma^{b}\nu} (r\exp(8\varepsilon_{r,s}))^{n}. \end{cases} \tag{6.26}$$

Indeed, from $(6.4)_n$, $z \in \Gamma_n \subset \tilde{B}^1(u_n,\tilde{\delta}'_n) \times \tilde{B}^2(v_n,\tilde{\delta}'_n)$ and thus $\|z-(u_n,v_n)\| \le \sqrt{2}\,\tilde{\delta}'_n$. Moreover, from (2.8), $\dfrac{r\exp(5\varepsilon_{r,s})}{\alpha^b} < \xi' = r\exp(8\varepsilon_{r,s})$.

Again from the Mean Value Theorem and from $(6.5)_n$ one obtains that

$$\|\Psi_n(v^1) - \Psi_n(v^2)\| \le \|v^1-v^2\|. \tag{6.27}$$

Thus, (6.25)-(6.27) together with (6.21) imply

$$\begin{cases} \|t_n(\Psi_n(v^1),v^1) - t_n(\Psi_n(v^2),v^2)\| \le \\ \le \dfrac{1200\, HA^{\ell}_{r,s}\delta^{\ell}_{r,s,\alpha,\gamma}}{\gamma^{b}\nu} (r\exp(8\varepsilon_{r,s}))^{n}\|v^1-v^2\| \le \\ \le \exp(-10n\varepsilon_{r,s})\|v^1-v^2\|. \end{cases}$$

■

Step 3. Let us define the mapping $T_n : \tilde{B}^2(v_n,\tilde{\delta}'_n) \to \mathbb{R}^p$

$$T_n v = B_n v + b_n(\Psi_n(v),v).$$

PROPOSITION 6.4. *For every* $v^1,v^2 \in \tilde{B}^2(v_n,\tilde{\delta}'_n)$ *one has*

$$\|T_n v^1 - T_n v^2\| \ge \hat{\xi}\|v^1-v^2\|,$$

where $\hat{\xi} = r\exp(12\varepsilon_{r,s})$.

Proof. As $t_n = (a_n,b_n)$, then from Proposition 6.3 and from (6.22) one obtains that

$$\begin{cases} \|T_n v^1 - T_n v^2\| \geq \|B_n(v^1-v^2)\| - \|b_n(\Psi_n(v^1),v^1) - b_n(\Psi_n(v^2),v^2)\| \geq \\ \geq \mu\|v^1 - v^2\| - \exp(-10n\varepsilon_{r,s})\|v^1 - v^2\| = \\ = (\mu-\exp(-10n\varepsilon_{r,s}))\|v^1 - v^2\| \geq \hat{\xi}\|v^1 - v^2\|. \end{cases}$$

■

PROPOSITION 6.5. *For every* $0 < \delta_{n_1} < \tilde{\delta}_{n_1}$,

$$L = \{\Phi'_n(\Psi_n(v),v); v \in \tilde{B}^2(v_n,\delta'_n)\} \cap \{\mathbb{R}^k \times \tilde{B}^2(v_{n+1},\delta'_{n+1})\}$$

is the graph of a mapping $\Psi_{n+1}\colon \tilde{B}^2(v_{n+1},\delta'_{n+1}) \to \mathbb{R}^k$ *of class* C^1.

Proof. By Proposition 6.4., T_n is a C^1 injective immersion whose image contains the ball of radius $\hat{\xi}\delta'_n > \xi'\delta'_n = \delta'_{n+1}$ around the point $T_n v_n = v_{n+1}$. Thus, in particular, $T^{-1}_{n_1}$ is defined and C^1 on $\tilde{B}^2(v_{n+1},\delta'_{n+1})$. This allows us to express Ψ_{n+1} as $\Psi_{n+1} = \Phi'_n \circ (\Psi_n \times \mathrm{Id}) \circ T^{-1}_n$, where Id denotes the identity mapping in $\mathbb{R}^p$.

■

Thus $(6.3)_{n+1}$ is fully proved. Let us note that from the construction of Ψ_{n+1}, it follows that $\Psi_{n+1}(v_{n+1}) = u_{n+1}$.

Step 4. We will now prove $(6.5)_{n+1}$.
In virtue of $(3.14)_I$ and $(6.4)_n$ one can write

$$\begin{cases} \Phi'_n(\Psi_n(v),v) \overset{\text{def}}{=\!=} (\bar{u},\bar{v}) = \\ = (A_n\Psi_n(v) + a_n(\Psi_n(v),v), B_n v + b_n(\Psi_n(v),v). \end{cases} \tag{6.28}$$

Let $v, v + \tau \in \tilde{B}^2(v_{n+1},\delta'_{n+1})$ and let $\tilde{v} = T^{-1}_n v, \tilde{v} + \tilde{\tau} = T^{-1}_n(v+\tau)$. By the definition of T_n we have

$$\tau = B_n\tilde{\tau} + b_n(\Psi_n(\tilde{v}+\tilde{\tau}),\tilde{v} + \tilde{\tau}) - b_n(\Psi_n(\tilde{v}),\tilde{v}). \tag{6.29}$$

Now, since $\Phi'_n(\Psi_n(\tilde{v}),\tilde{v}) = (\Psi_{n+1}(v),v)$, one has from $(3.14)_I$

$$\begin{cases} \Psi_{n+1}(v) = A_n\Psi_n(\tilde{v}) + a_n(\Psi_n(\tilde{v}),\tilde{v}), \\ \Psi_{n+1}(v+\tau) = A_n\Psi_n(\tilde{v}+\tilde{\tau}) + a_n(\Psi_n(\tilde{v}+\tilde{\tau}),\tilde{v}+\tilde{\tau}). \end{cases} \tag{6.30}$$

It is clear that

$$b_{n+1} \le \limsup_{\|\tau\|\to 0} \frac{\|\Psi_{n+1}(v+\tau)-\Psi_{n+1}(v)\|}{\|\tau\|} .$$

From $(6.31)_{n+1}$, (6.30), from Proposition 6.3 with $v^1 = \tilde{v}$ and $v^2 = \tilde{v} + \tilde{\tau}$, from the estimations for A_n and B_n given by (6.12), from $(6.5)_n$ and from (6.22), as $n \ge n_1 \ge n_{12}$, one obtains

$$\begin{cases} \dfrac{\|\Psi_{n+1}(v+\tau) - \Psi_{n+1}(v)\|}{\|\tau\|} = \\ = \dfrac{\|A_n[\Psi_n(\tilde{v}+\tilde{\tau})-\Psi_n(\tilde{v})]+[a_n(\Psi_n(\tilde{v}+\tilde{\tau}),\tilde{v}+\tilde{\tau})-a_n(\Psi_n(\tilde{v}),\tilde{v})]\|}{\|B_n\tilde{\tau}+b_n(\Psi_n(\tilde{v}+\tilde{\tau}),\tilde{v}+\tilde{\tau})-b_n(\Psi_n(\tilde{v}),\tilde{v})\|} \le \\ \le \dfrac{\|A_n\|\cdot\|\Psi_n(\tilde{v}+\tilde{\tau})-\Psi_n(\tilde{v})\|+\|a_n(\Psi_n(\tilde{v}+\tilde{\tau}),\tilde{v}+\tilde{\tau})-a_n(\Psi_n(\tilde{v}),\tilde{v})\|}{\|B_n\tilde{\tau}\|-\|b_n(\Psi_n(\tilde{v}+\tilde{\tau}),\tilde{v}+\tilde{\tau})-b_n(\Psi_n(\tilde{v}),\tilde{v})\|} \le \\ \le \dfrac{\lambda\dfrac{\|\Psi_n(\tilde{v}+\tilde{\tau})-\Psi_n(\tilde{v})\|}{\|\tau\|}+\exp(-10n\varepsilon_{r,s})}{\mu-\exp(-10n\varepsilon_{r,s})} \le \\ \le \dfrac{\lambda\eta^{n-n_1}+\exp(-10n\varepsilon_{r,s})}{\mu-\exp(-10n\varepsilon_{r,s})} = \eta^{n-n_1}\dfrac{\lambda+\exp(-3n\varepsilon_{r,s})\exp(-7n_1\varepsilon_{r,s})}{\mu-\exp(-10n\varepsilon_{r,s})} \le \\ \le \eta^{n-n_1}\dfrac{\lambda+\exp(-10n_1\varepsilon_{r,s})}{\mu-\exp(-10n\varepsilon_{r,s})} \le \eta^{n-n_1}\dfrac{\lambda+\exp(-10n_1\varepsilon_{r,s})}{\mu-\exp(-10n_1\varepsilon_{r,s})} \le \eta^{n+1-n_1}. \end{cases}$$

Thus $b_{n+1} \le \eta^{n+1-n_1}$, i.e., $(6.5)_{n+1}$ is true.

We pass now to the proof of $(6.4)_{n+1}$.

Let us estimate $a_{n+1} = \max_{v\in\tilde{B}^2(\tilde{v}_{n+1},\delta'_{n+2})} \|d\Psi_{n+1}(v)\|$. First let let us note that for $(u_0,v_0) \in \hat{V}(w)$ one has $\|(u_0,v_0)\| \le \sqrt{2}\,\delta^{\ell}_{r,s,\alpha,\gamma}$.

We have

$$\Psi_{n+1}(v) = \Psi_{n+1}(v_{n+1}) + [\Psi_{n+1}(v) - \Psi_{n+1}(v_{n+1})].$$

As $\Psi_{n+1}(v_{n+1}) = u_{n+1}$, then from (5.3), from the Mean Value Theorem, from $(6.5)_{n+1}$ and from (6.23) one obtains that for $v \in \tilde{B}^2(v_{n+1}, \tilde{\delta}'_{n+1})$

$$\begin{cases} \|\Psi_{n+1}(v)\| \leq \|\Psi_{n+1}(v_{n+1})\| + \|\Psi_{n+1}(v) - \Psi_{n+1}(v_{n+1})\| = \\ = \|(u_{n+1}, v_{n+1})\| + \|\Psi_{n+1}(v) - \Psi_{n+1}(v_{n+1})\| < 200(\kappa_{r,s})^{n+1}\|(u_0, v_0)\| + \\ + \delta'_{n+1} \cdot \eta^{n+1-n_1} < 300\delta^{\ell}_{r,s,\alpha,\gamma}(\kappa_{r,s})^{n+1} + (\xi')^{n+1} < \hat{\xi}^{n+1}. \end{cases}$$

Thus $a_{n+1} \leq \hat{\xi}^{n+1}$; i.e. $(6.4)_{n+1}$ is proved.

Finally the proof of $(6.2)_{n+1}$ proceeds along exactly the same line as the proof of $(6.2)_{n_1}$ described at the end of Step 1.

Proof of (6.1.2). First we prove $(6.8)_0$ and $(6.11)_0$. This allows us to deduce $(6.10)_0$. In the second step we prove that for every $n \geq 0$, $(6.8)_n$ and $(6.10)_n$ allows to define the mapping Ψ_{n+1} satisfying $(6.9)_n$, $(6.8)_{n+1}$, $(6.10)_{n+1}$ and $(6.11)_{n+1}$.

Step 1. The base of induction.

$(6.8)_0$ follows from (6.6), from the inequality $0 < \delta_0 \leq \frac{q_0}{4}$, from $(2.2.5)_I$, (6.1), $(6.7)_I$ and $(3.9)_I$. Indeed,

$$\|(\Psi_0(v), v)\| \leq \|\Psi_0(v)\| + \|v\| \leq \frac{q_0}{4} + \delta_0 \leq \frac{q_0}{2}.$$

Thus $\|(\Psi_0(v), v)\| \leq \frac{q_0}{2} < \delta^{\ell}_{r,s,\alpha,\gamma} \leq [Q(\ell,\alpha,\gamma)]^2$. $(6.10)_0$ and $(6.11)_0$ follow from (6.6) and (6.7) respectively.

Step 2. The inductive step in this proof is very similar to that in the proof of (6.1.1).

PROPOSITION 6.6. *Let* $n \geq 0$. *For every* $v^1, v^2 \in \tilde{B}^2(v_n, \delta'_n)$ *one has*

$$\|t_n(\Psi_n(v^1), v^1) - t_n(\Psi_n(v^2), v^2)\| \leq \Gamma q_0 e^{-10\varepsilon_{r,s} n}\|v^1 - v^2\|,$$

where (*cf. Sec.* 6.1) $\Gamma = \dfrac{2HA^{\ell}_{r,s}}{\gamma^{b_\nu}}$.

Proof. By the Mean Value Theorem one has

$$\begin{cases} \|h_n(\Psi_n(v^1),v^1) - h_n(\Psi_n(v^2),v^2)\| \le \\ \le \sup_{z\in I}\|dh_n(z)\|(\|\Psi_n(v^1) - \Psi_n(v^2)\| + \|v^1-v^2\|). \end{cases} \tag{6.32}$$

where I is the interval in $\mathbb{R}^m$ connecting $(\Psi_n(v^1),v^1)$ and $(\Psi_n(v^2),v^2)$. As $q(1,C) < \frac{\delta^{\ell}_{r,s,\alpha,\gamma}}{8} < \frac{Q^2(\ell,\alpha,\gamma)}{8}$ (see (6.1) and $(6.7)_I$), from $(6.10)_n$, $(6.5)_I$ and from the convexity of $B_n(w)$ it follows that $I \subset B_n(w)$.

$$\|dh_n(z)\| \le \|dt_n(z)\| = \|dt_n(z) - dt_n(0)\| < \frac{HA^{\ell}_{r,s}}{\gamma^b\nu(\alpha^b\nu)^n}\|z\|.$$

But $z \in I$ and thus by $(6.10)_n$ one obtains in the same way as in the proof of Lemma 5.1.

$$\begin{cases} \|z\| \le \max_{i=1,2}\{\|\Psi_n(v^i)\| + \|v^i\|\} \le (\frac{1}{4}+C)q_0\xi^n + \frac{q_0}{2\sqrt{2}}(\xi')^n \le \\ \le q_0(\xi')^n \le q_0[\alpha^{8b}\exp(-15\varepsilon_{r,s})]^n \end{cases}$$

because $\xi' < \alpha^{8b}\exp(-15\varepsilon_{r,s})$ and $0 < C \le 1$. Consequently one obtains

$$\|dh_n(z)\| \le \frac{HA^{\ell}_{r,s}}{\gamma^b\nu} q_0 e^{-10\varepsilon_{r,s}n} = \frac{1}{2}\Gamma q_0 e^{-10\varepsilon_{r,s}n}. \tag{6.33}$$

Now let $v^1,v^2 \in \tilde{B}^2(v_n,\delta'_n)$. From $(6.11)_n$ one obtains that

$$\|\Psi_n(v^1) - \Psi_n(v^2)\| \le C\eta^n\|v^1 - v^2\|$$

and consequently that

$$\begin{cases} \|\Psi_n(v^1) - \Psi_n(v^2)\| + \|v^1 - v^2\| \le (C\eta^n + 1)\|v^1 - v^2\| \le \\ \le (C+1)\|v^1 - v^2\|. \end{cases} \tag{6.34}$$

Finally from (6.32) - (6.34) and from $0 < C \le 1$ one gets

$$\|h_n(\Psi_n(v^1),v^1) - h_n(\Psi_n(v^2),v^2)\| \le \Gamma q_0 e^{-10n\varepsilon_{r,s}}\|v^1-v^2\|. \qquad \square$$

<u>Step 3</u> of the proof of (6.1.1) remains unchanged but now it applies to every $n \ge 0$. In the proof of counterpart of Proposition 6.4, we use the inequality

$$\hat{\xi} < \mu - \Gamma q_0 \tag{6.35}$$

which follows from (6.1).

Let us proceed now to the proof of $(6.11)_{n+1}$. This proof is exactly the same as the proof of $(6.5)_{n+1}$, except the final step which we are going to describe now. From (6.35), $\mu - \Gamma q_0 > 0$, and thus one obtains that

$$\left\{\begin{aligned} \frac{\|\Psi_{n+1}(v+\tau) - \Psi_{n+1}(v)\|}{\|\tau\|} &\le \frac{\lambda\dfrac{\|\Psi_n(\tilde{v}+\tilde{\tau})-\Psi_n(\tilde{v})\|}{\|\tilde{\tau}\|} + \Gamma q_0\exp(-10n\varepsilon_{r,s})}{\mu-\Gamma q_0\exp(-10n\varepsilon_{r,s})} \le \\ &\le \frac{\lambda C\eta^n+\Gamma q_0\exp(-10n\varepsilon_{r,s})}{\mu-\Gamma q_0\exp(-10n\varepsilon_{r,s})} \le \\ &\le C\eta^n \frac{\lambda+\Gamma q_0\exp(-3n\varepsilon_{r,s})}{\mu-\Gamma q_0\exp(-10n\varepsilon_{r,s})}\,. \end{aligned}\right.$$

To finish the proof of $(6.11)_{n+1}$ it is sufficient to see that $\frac{\lambda+\Gamma q_0}{\mu-\Gamma q_0} \le \eta$, or equivalently $q_0 \le \frac{\mu\eta-\lambda}{\Gamma(\frac{1}{C}+\eta)}$ which is satisfied by our assumption (cf. (6.1)). Thus $b_{n+1} \le C\eta^{n+1}$, i.e., $(6.11)_{n+1}$ is true.

The proof of $(6.10)_n$ goes similarly to the proof of $(6.4)_{n+1}$. Since $\|(u_0,v_0)\| \le \frac{q_0}{800\sqrt{2}}$ we obtain

$$\begin{cases} a_{n+1} \le \|\Psi_{n+1}(v_{n+1})\| + \sup\limits_{v\in\tilde{B}^2(v_{n+1},\delta'_{n+1})} \|\Psi_{n+1}(v) - \Psi_{n+1}(v_{n+1})\| \le \\ \le 200\ \kappa^{n+1}\|(u_0,v_0)\| + \delta'_{n+1} \sup\limits_{v\in\tilde{B}^2(v_{n+1},\delta'_{n+1})} \|d\Psi_{n+1}(v)\| \le \\ \le \kappa^{n+1}\frac{q_0}{4} + b_{n+1}\delta_0(\xi')^{n+1} \le \\ \le \frac{q_0}{4}\left(re^{4\varepsilon_{r,s}}\right)^{n+1} + C\eta^{n+1}\frac{q_0}{4}\left(re^{8\varepsilon_{r,s}}\right)^{n+1} \le \\ \le (\frac{1}{4}+C)q_0\ r\ \exp(4(n+1)\varepsilon_{r,s}) < (\frac{1}{4}+C)q_0\xi^n. \end{cases}$$

Finally, let us prove $(6.8)_{n+1}$. From $(6.11)_{n+1}$ one has that $\|\Psi_{n+1}(v) - \Psi_{n+1}(v_{n+1})\| < \|v-v_{n+1}\|$ and thus that

$$\{\Psi_{n+1}(v); v \in \tilde{B}^2(v_{n+1},\delta'_{n+1})\} \subset \tilde{B}^1(v_{n+1},\delta'_{n+1}).$$

Now, from Lemma 5.1.2 one obtains $(6.8)_{n+1}$. ∎

7. START OF THE PROOF - II

7.1. Let $y_1, y_2 \in \Lambda^1_{r,s,\alpha,\gamma}$ and let $\rho(y_1,y_2) \le \frac{\gamma^g}{4}$ (cf. Condition (Ba) from (Sec. 2.1_I). Let us define the smooth mapping I_{y_1,y_2} by the formula

$$I_{y_1,y_2} = \tau'_{y_1} \circ \exp^{-1}_{y_1} \circ \exp_{y_2} \circ (\tau'_{y_2})^{-1}. \tag{7.1}$$

This mapping is well defined on the set $\{h \in \mathbb{R}^m; \|h\| < \frac{\gamma^g}{2}\}$. Indeed, if $y \in \Lambda^\ell_{r,s,\alpha,\gamma}$, then the mapping $\exp_y$ is defined on the ball $\{u\in T_yN; \|u\|_y < \gamma^g\}$. Moreover, if $h \in \mathbb{R}^m$ and $\|h\| \le \frac{\gamma^g}{2}$, then by (2.9) $\|(\tau'_{y_2})^{-1}(h)\|_{y_2} < \frac{\gamma^g}{\sqrt{2}}$ and consequently

$$\begin{cases} \rho(y_1,\exp_{y_2} \circ (\tau'_{y_2})^{-1}(h)) \le \rho(y_1,y_2) + \rho(y_2,\exp_{y_2}\circ(\tau'_{y_2})^{-1}(h)) \le \\ \le \frac{\gamma^g}{4} + \frac{\gamma^g}{\sqrt{2}} < \gamma^g. \end{cases}$$

Thus the map $\exp_{y_1}^{-1} \circ \exp_{y_2} \circ (\tau'_{y_2})(h)$ is well defined.

<u>PROPOSITION 7.1</u>. <u>For every $\ell, r, s, \alpha, \gamma$ there exists a continuous nondecreasing function $\Omega = \Omega(q)$ defined for $0 \le q \le \frac{\gamma^g}{4}$ such that $\Omega(0) = 0$, $\Omega(q) > 0$ for $q > 0$ and such that for every $y_1, y_2 \in \Lambda^{\ell}_{r,s,\alpha,\gamma}$ satisfying $\rho(y_1, y_2) \le q$ and for every $h \in \mathbb{R}^m$, $\|h\| \le \frac{\gamma^g}{4}$, one has</u>

$$\|(dI_{y_1,y_2})(h) - \mathrm{Id}\| \le \Omega(q), \tag{7.2}$$

<u>where the differentiation of I_{y_1,y_2} is taken with respect to h.</u>

<u>Proof</u>. Let us note

$$\left.\begin{aligned} P = \{(y_1, y_2, h) \in \Lambda^{\ell}_{r,s,\alpha,\gamma} \times \Lambda^{\ell}_{r,s,\alpha,\gamma} \times \mathbb{R}^m ;\ \rho(y_1, y_2) \le \frac{\gamma^g}{4}, \\ \|h\| \le \frac{\gamma^g}{2}\}. \end{aligned}\right\}$$

As $\Lambda^{\ell}_{r,s,\alpha,\gamma}$ is compact, P is also compact. The map $dI_{y_1,y_2}(h)$ is continuous and consequently uniformly continuous on P. For every q, $0 \le q \le \operatorname{diam} \Lambda^{\ell}_{r,s,\alpha,\gamma}$ let us define the non-decreasing function $\Omega(q)$ by

$$\Omega(q) \overset{\text{def}}{=} \sup_{\substack{(y_1, y_2, h) \in P \\ (\bar{y}_1, \bar{y}_2, \bar{h}) \in P \\ \rho(y_1, \bar{y}_1) + \rho(y_2, \bar{y}_2) + \|h - \bar{h}\| \le q}} \|dI_{y_1,y_2}(h) - dI_{\bar{y}_1,\bar{y}_2}(\bar{h})\|.$$

It is clear that $0 \le \Omega(q) \le \Omega(\operatorname{diam} \Lambda^{\ell}_{r,s,\alpha,\gamma}) < +\infty$ and that Ω is continuous on $[0, \operatorname{diam} \Lambda^{\ell}_{r,s,\alpha,\gamma}]$.

Now our proposition follows if one takes $\bar{y}_1 = y_2 = \bar{y}_2$ and $h = \bar{h}$. ∎

7.2. As the next step towards the definition of $q^{\ell}_{r,s,\alpha,\gamma}(B)$ let us fix a number $q(2) = q^2(\ell,r,s,\alpha,\gamma)$ satisfying the following conditions (7.3) and (7.4).

$$0 < q(2) < \frac{\gamma^g}{4q} \quad \text{and} \quad 0 < \Omega(q(2)) < \frac{1}{5}. \tag{7.3}$$

(7.4). Consider in $\tilde{U}(x,q(2))$ an arbitrary smooth submanifold W such that $|W| \le \frac{1}{2}$ (see Sec. 4.1). Then, for every $w \in \tilde{U}(x,q(2)) \cap \cap \Lambda^{\ell}_{k,r,s,\alpha,\gamma}$, $(\tau'_w \circ \exp_w^{-1})(W) \cap \mathrm{Int}\, \tilde{B}(q(2))$ is the graph of the smooth mapping $\Psi_w\colon U_w \to \mathbb{R}^k$, where U_w is an open subset of $\mathbb{R}^p$. The possibility of such choice of $q(2)$ follows from the definition of the sets $\Lambda^{\ell}_{k,r,s,\alpha,\gamma}$ (cf. $(3.4)_I$).

Let

$$W = \exp_x(\tau'_x)^{-1}\{(\Psi(v),v); v \in \tilde{B}^2(q(2))\} \tag{7.4'}$$

be a submanifold transversal to the family $S^{\ell}_{k,r,s,\alpha,\gamma}(x)$ such that $|W| \le \frac{1}{2}$. We will write

$$\begin{cases} \|\Psi_w\| = \sup\limits_{v\in U_w} \|\Psi_w(v)\|, & \|\Psi\| = \sup\limits_{v\in \tilde{B}^2(q(2))} \|\Psi(v)\|, \\ \|d\Psi_w\| = \sup\limits_{v\in U_w} \|d\Psi_w(v)\|, & \|d\Psi_x\| = \sup\limits_{v\in \tilde{B}^2(q(2))} \|d\Psi(v)\|, \\ |W| = \|\Psi\| + \|d\Psi_x\|. & \end{cases} \tag{7.5}$$

7.3. In this section we show how to compare the representation of a transversal submanifold in local charts centered at x and at w respectively. We begin with the comparison of $\|d\Psi_x\|$ and $\|d\Psi_w\|$.

<u>PROPOSITION 7.2</u>. If $\|d\Psi_x\| \le 1$, then for every $w \in \tilde{U}(x,q(2)) \cap \cap \Lambda^{\ell}_{k,r,s,\alpha,\gamma}$

$$\|d\Psi_w\| \le \frac{1}{1-2\Omega(q(2))}\left[(1+\Omega(q(2)))\|d\Psi_x\| + \Omega(q(2))\right].$$

<u>Proof</u>. From (7.3), condition (Bb) of Sec. 2_I and (2.9) one has

$\rho(w,x) < \frac{\gamma^g}{2}$. Let us denote $I_{w,x}(u,v) = (H_1(u,v), H_2(u,v))$ (cf. 7.1). It is easy to see that

$$\Psi_w H_2(\Psi(v),v) = H_1(\Psi(v),v). \tag{7.6}$$

Let us differentiate (7.6). We obtain for $z = H_2(\Psi(v),v) \in U_w$

$$\begin{cases} d\Psi_w(z) \circ [D_1H_2(\Psi(v),v) \circ d\Psi(v) + D_2H_2(\Psi(v),v)] = \\ = D_1H_1(\Psi(v),v) \circ d\Psi(v) + D_2H_1(\Psi(v),v), \end{cases} \tag{7.7}$$

where D_1, D_2 denote the differentiation along $\mathbb{R}^k$ and $\mathbb{R}^p$ respectively. Furthermore we have from (7.2) and (2.9)

$$\begin{cases} \sup\limits_{\{h\in\mathbb{R}^m;\|h\|\le\frac{\gamma^g}{2}\}} \|dI_{w,x}(h) - Id\| = \\ = \sup\limits_{\{h\in\mathbb{R}^m;\|h\|\le\frac{\gamma^g}{2}\}} \left\| \begin{pmatrix} D_1H_1(h) - Id_k, & D_2H_1(h) \\ D_1H_2(h), & D_2H_2(h) - Id_p \end{pmatrix} \right\| \le \\ \le \Omega(q(2)), \end{cases} \tag{7.8}$$

$$\begin{cases} \|D_1H_2(\Psi(v),v) \circ d\Psi(v) + D_2H_2(\Psi(v),v) - Id_p\| \le \\ \le \|D_1H_2(\Psi(v),v)\|\cdot\|d\Psi(v)\| + \|D_2H_2(\Psi(v),v) - Id_p\| \le \\ \le \Omega(q(2))\|d\Psi(v)\| + \Omega(q(2)) \le \Omega(q(2))(\|d\Psi(v)\| + 1) \le \\ \le \Omega(q(2))(\|d\Psi_x\| + 1) \le 2\Omega(q(2)) < 1 \end{cases} \tag{7.9}$$

because $\|d\Psi_x\| \le 1$ and because (7.3) implies $2\Omega(q(2)) < 1$.

In particular, the linear mapping

$$D_1H_2(\Psi(v),v) \circ d\Psi(v) + D_2H_2(\Psi(v),v): \mathbb{R}^p \to \mathbb{R}^p$$

is invertible.

If for a linear operator A, $\|A - \mathrm{Id}\| < 1$ then

$$\|A^{-1}\| \le \frac{1}{1-\|A-\mathrm{Id}\|}\,.$$

Thus using (7.7) - (7.9) we obtain

$$\begin{cases} \|d\Psi_w(z)\| \le \|D_1H_1(\Psi(v),v) \circ d\Psi(v) + D_2H_1(\Psi(v),v)\| \cdot \\ \cdot \|(D_1H_2(\Psi(v),v) \circ d\Psi(v) + D_2H_2(\Psi(v),v))^{-1}\| \le \\ \le \frac{1}{1-2\Omega(q(2))}(\|D_1H_1(\Psi(v),v)\| \cdot \|d\Psi(v)\| + \|D_2H_1(\Psi(v),v)\|) \le \\ \le \frac{1}{1-2\Omega(q(2))}\,[(1+\Omega(q(2)))\|d\Psi_x\| + \Omega(q(2))]. \end{cases}$$ ∎

Proposition 7.2 and (7.3) imply the following statement.

<u>COROLLARY 7.1</u>. <u>For every</u> $w \in \Lambda^{\ell}_{k,r,s,\alpha,\gamma} \cap \tilde{U}(x,q(2))$ <u>one has</u>

$$\|d\Psi_w\| \le 2(|W|+\Omega(q(2))).$$

The next step is the comparison of $\|\cdot\|'_w$ and $\|\cdot\|'_x$.

Let $K = \sqrt{2}\, q^2 A^{\ell}_{r,s}$ where q comes from condition (Bb) of Sec. 1.1_I. As $q \ge 1$ and $A^{\ell}_{r,s} \ge \frac{1}{\sqrt{2}}$ then $K \ge 1$.

<u>PROPOSITION 7.3</u>. <u>For every</u> $w \in \Lambda^{\ell}_{k,r,s,\alpha,\gamma} \cap \tilde{U}(x,q(2))$ <u>and every</u> $z_1,z_2 \in \tilde{U}(x,q(2)) \cap \tilde{U}(w,q(2))$ <u>one has</u>

$$\begin{cases} \frac{1}{K}\|\exp_w^{-1}(z_1) - \exp_w^{-1}(z_2)\|'_w \le \|\exp_x^{-1}(z_1) - \exp_x^{-1}(z_2)\|'_x \le \\ \le K\|\exp_w^{-1}(z_1) - \exp_w^{-1}(z_2)\|'_w. \end{cases} \tag{7.10}$$

<u>Proof</u>. Since x and w appear symmetrically in (7.10) it is enough to prove the second inequality.

Let L be the interval in T_wN connecting the points $\exp_w^{-1}(z_1)$ and $\exp_w^{-1}(z_2)$. For every $u \in L$ we have

$$\rho(\exp_w u, w) \le \sqrt{2}\, q\|u\|'_w \le 2qq(2)$$

and thus

$$\rho(\exp_w u, x) \le \rho(\exp_w u, w) + \rho(w, x) \le 4qq(2),$$

By (7.3), $4qq(2) < \gamma^g$. Thus $\rho(\exp_w u, x) < \gamma^g$, so that from condition (Ba) of Sec. 1.1_I the mapping $\exp_x^{-1} \circ \exp_w$ is defined at u. By condition (Bb)

$$\|d[\exp_x^{-1} \circ \exp_w(u)]\| \le q^2. \tag{7.11}$$

Henceforth we can write using $(2.2.5)_I$, the Mean Value Theorem and (7.11)

$$\begin{cases} \|\exp_x^{-1}(z_1) - \exp_x^{-1}(z_2)\|'_x \le A^{\ell}_{r,s}\|\exp_x^{-1}(z_1) - \exp_x^{-1}(z_2)\|_x = \\ = A^{\ell}_{r,s}\|(\exp_x^{-1} \circ \exp_w) \circ \exp_w^{-1}(z_1) - (\exp_x^{-1} \circ \exp_w) \circ \exp_w^{-1}(z_2)\|_x \le \\ \le A^{\ell}_{r,s} \sup\limits_{u \in L}\|d(\exp_x^{-1} \circ \exp_w)(u)\| \cdot \|\exp_w^{-1}(z_1) - \exp_w^{-1}(z_2)\|_w \le \\ \le q^2 A^{\ell}_{r,s}\|\exp_w^{-1}(z_1) - \exp_w^{-1}(z_2)\|_w \le K\|\exp_w^{-1}(z_1) - \exp_w^{-1}(z_2)\|'_w. \end{cases}$$ ■

7.4. We will now specify several extra conditions on the number $q(2)$ introduced in Sec. 7.2. These supplementary conditions on $q(2)$ as well as new quantities $\varepsilon^{\ell}_{r,s,\alpha,\gamma}$ will depend on one extra parameter C.

Let $0 < C \le 1$. Let us define the numbers $\varepsilon(C) = \varepsilon^{\ell}_{r,s,\alpha,\gamma}(C)$ and $q(2,C) = q^2(\ell,r,s,\alpha,\gamma,C)$ such that the following three conditions (7.12)-(7.14) hold.

$$0 < \varepsilon(C) < \frac{1}{2}, \tag{7.12}$$

$$0 < q(2,C) \le \min\left(\frac{q(1,C)}{3600K}, q(2)\right) \tag{7.13}$$

$$\varepsilon(C) + \Omega(q(2,C)) \le \frac{C}{2} \tag{7.14}$$

where K comes from Proposition 7.2.

From now we will replace in (7.5) the parameter $q(2)$ by $q(2,C)$. Let us consider a manifold W of type (7.4) such that (cf. (7.5))

$$|W| \le \varepsilon(C). \tag{7.15}$$

Let $w \in \tilde{U}(x,q(2,C)) \cap \Lambda^{\ell}_{k,r,s,\alpha,\gamma}$ be such a point that

$$V(w) \cap W \cap \operatorname{Int} \tilde{U}(x,q(2,C)) \ne \emptyset.$$

(7.12) and the remark following (4.3) implies that this intersection consists of one point which we denote by y, and as in (5.1) $\tau'_w \circ \exp_w^{-1}(y) = (u_0,v_0)$.

Let

$$\begin{cases} q_C(w,W) \text{ be the supremum of all numbers } \delta_0 \text{ such that} \\ \delta_0 \le \dfrac{q(2,C)}{4} \text{ and } \tilde{B}^2(v_0,\delta_0) \subset U_w \cap \tilde{B}^2(q(2,C)) \text{ where} \\ U_w \text{ is defined in (7.3) and } [\exp_w \circ (\tau'_w)^{-1}](\tilde{B}((u_0,v_0),\delta_0)) \subset \\ \subset U(x,q(2,C)) \text{ (see (2.12)).} \end{cases} \tag{7.16}$$

As

$$\|(\tau'_x \circ \exp_x^{-1})(y) - (\tau'_x \circ \exp_x^{-1})(w)\| < \sqrt{2}\, q(2,C),$$

then from (7.13) and Proposition 7.2 it follows that

$$\|(u_0,v_0)\| \le \frac{q(1,C)}{800\sqrt{2}}. \tag{7.17}$$

Moreover, as the graph of the mapping Ψ lies in $\tilde{B}(q(2,C))$, exactly the same arguments imply that

$$\sup_{v \in \tilde{B}^2(v_0,\delta_0)} \|\Psi_w(v)\| \le \frac{q(1,C)}{800\sqrt{2}}. \tag{7.18}$$

Finally from Corollary 7.1 and from (7.13)-(7.15) one obtains that

$$\sup_{v \in \tilde{B}^2(v_0,\delta_0)} \| d\Psi_w(v) \| \le C. \qquad (7.19)$$

The above remarks (7.17) - (7.19) prove that for every $0 < C \le 1$, for $q_0 = q(1,C)$ defined above, for every δ_0 such that $0 < \delta_0 \le q_C(w,W)$ and for $\Psi_0 = \Psi_w|_{\tilde{B}^2(v_0,\delta_0)}$ all assumptions of Lemma 6.1.2 are fulfilled.

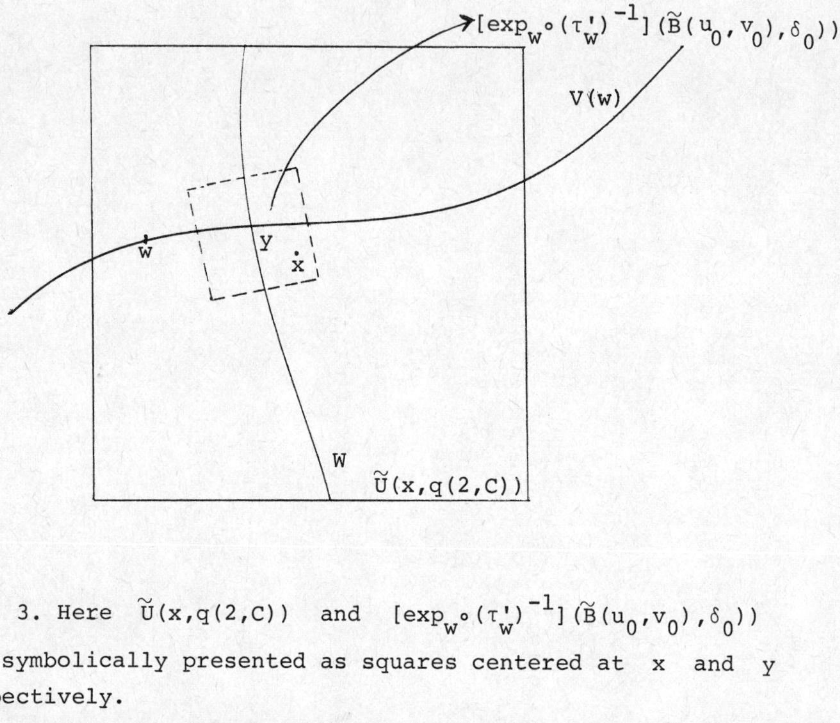

Fig. 3. Here $\tilde{U}(x,q(2,C))$ and $[\exp_w \circ (\tau_w')^{-1}](\tilde{B}(u_0,v_0),\delta_0))$ are symbolically presented as squares centered at x and y respectively.

Thus from (6.1.2) one obtains for every $n \ge 0$, the existence of mapping $\Psi_n: \tilde{B}^2(v_n,\delta_n') \to \mathbb{R}^k$ satisfying conditions $(6.8)_n$ - $(6.11)_n$.

7.5. We pass now to the description of the main construction of the proof, which leads to the submanifolds $\{W_{in}^1\}_{1\le i\le p(n)}$ mentioned in Sec. 4.2.

As in Sec. 7.4, let $y = V(w) \cap W \cap \text{Int}\, \tilde{U}(x,q(2,C))$, $y = [\exp_w \circ (\tau_w')^{-1}](u_0,v_0)$.

For any $n \ge 0$ and any r, $0 < r < q_C(w,W)(\xi')^n$, let us define the following p dimensional submanifold of N.

$$\widetilde{W}_n(w,y,r) = \{\exp_{\Phi^n(w)} \circ (\tau'_{\Phi^n(w)})^{-1}(\Psi_n(v),v); v \in \widetilde{B}^2(v_n,r)\}. \tag{7.20}$$

In particular, for δ_0, $0 < \delta_0 < q_C(w,W)$, one can consider the submanifolds $\widetilde{W}_n(w,y,\delta_n')$ where as before $\delta_n' = \delta_0(\xi')^n$.

This definition immediately implies that for every $n \geq 1$ one has

$$\widetilde{W}_n(w,y,\delta_n') \subset \Phi(\widetilde{W}_{n-1}(w,y,\delta_{n-1}')). \tag{7.21}$$

This last property is very important for the future. Let us emphasize that if one uses the Riemannian metric instead of Lyapunov metric in (7.20), the analog of (7.21) is not true anymore.

Let us recall that for every $w \in \Lambda^{\ell}_{k,r,s,\alpha,\gamma}$ and $n \geq 0$

$$V_n(w) = \{y \in M; \rho(y,\Phi^n(w)) < (Q(\ell,\alpha,\gamma)a^{2bn})^2\} \qquad (3.9)_I$$

where $Q(\ell,\alpha,\gamma)$ is defined by $(3.11)_I$.

Moreover, for every $n \geq 0$ one has

$$\widetilde{W}_n(w,y,\delta_n') \subset \widetilde{U}(\Phi^n(w),q_n) \subset V_n(w). \tag{7.22}$$

The first inclusion follows directly from (5.3) and $(6.10)_n$. The second one follows directly from (2.9), (2.13) and $(3.9)_I$.

<u>PROPOSITION 7.4</u>. $\operatorname{diam}(\Phi^{-n}(\widetilde{W}_n(w,y,\delta_n'))) \leq \sqrt{2}(1+C)\,\dfrac{\delta_0}{\exp(4n\varepsilon_{r,s})}$.

<u>Proof</u>. This is an almost immediate corollary of Proposition 6.4, which holds in our framework for every $n \geq 0$ (see step 3 of the proof of (6.1.2)).

By $(2.2.5)_I$ it is enough to prove that

$$\begin{cases} \lim\limits_{n\to\infty} \operatorname{diam}((\Phi'_{0w})^{-1}\circ\cdots\circ(\Phi'_{(n-1)w})^{-1}(M_n(w,y,\delta_n')) \leq \\ \leq (1+C)\,\dfrac{\delta_0}{\exp(4n\varepsilon_{r,s})}, \end{cases} \tag{7.23}$$

where $M_n(w,y,\delta_n') = \{(\Psi_n(v),v);\ v \in \widetilde{B}^2(v_n,\delta_n')\}$.

Let us denote by π' the projection of $\mathbb{R}^m = \mathbb{R}^k \oplus \mathbb{R}^p$ onto $\mathbb{R}^p$. From Proposition 6.4 and from the definition of $M_n(w,y,\delta_n')$ one has

$$\pi'[(\Phi'_{(n-1)w})^{-1}M_n(w,y,\delta_n')] \subset \tilde{B}^2(v_{n-1}, \frac{\delta_n'}{\hat{\xi}}) \subset \tilde{B}^2(v_{n-1}, \delta_{n-1}'),$$

$$\begin{cases} \pi'[(\Phi'_{(n-2)w})^{-1} \circ (\Phi'_{(n-1)w})^{-1}(M_n(w,y,\delta_n'))] \subset \tilde{B}^2(v_{n-2}, \frac{\delta_n'}{(\hat{\xi})^2}) \subset \\ \subset \tilde{B}^2(v_{n-2}, \delta_{n-2}'), \text{ etc.} \end{cases}$$

Finally one obtains that

$$\begin{cases} \pi'[(\Phi'_{\circ w})^{-1} \circ \ldots \circ (\Phi'_{(n-1)w})^{-1}(M_n(w,y,\delta_n'))] \subset \tilde{B}^2(v_0, \frac{\delta_n'}{(\hat{\xi})^n}) \subset \\ \subset \tilde{B}^2(v_0, \frac{\delta_0}{\exp(4n\varepsilon_{r,s})}). \end{cases}$$

Now, (7.23) follows from the construction of manifolds $\tilde{W}_n(w,y,\delta_n')$ and from $(6.10)_n$. ■

8. PROJECTION AND COVERING LEMMAS

8.1. For a fixed δ_0, $0 < \delta_0 < q_C(w,W)$, and $n \geq 0$, let us introduce the following notations.

$$\hat{W}_n(w,y,\delta_n') = \exp^{-1}_{\Phi^n(w)}(\tilde{W}_n(w,y,\delta_n')),$$

and if $z \in \tilde{W}_n(w,y,\delta_n')$ then $\hat{z} = \exp^{-1}_{\Phi^n(w)}(z)$.

We denote $\exp_{\Phi^n(w)} \circ (\tau'_{\Phi^n(w)})^{-1}$ by H_n. This mapping is well defined on $B_n(w)$ because $w \in \Lambda^{\ell}_{k,r,s,\alpha,\gamma}$. As in Sec. 7.5, let us denote $M_n(w,y,\delta_n') = H_n^{-1}(\tilde{W}_n(w,y,\delta_n'))$.

Let us recall that $H_n^{-1}(y) = (u_n, v_n) \in \mathbb{R}^k \oplus \mathbb{R}^p$ and that (cf. (7.23))

$$M_n(w,y,\delta_n') = \{(\Psi_n(v),v);\ v \in \tilde{B}^2(v_n,\delta_n')\}.$$

For any $z \in \widetilde{W}_n(w,y,\delta'_n)$ let $\overline{z} = H_n^{-1}(z)$.

Let $\widetilde{\rho}$ (respectively $\hat{\rho}$ or $\overline{\rho}$) be the metric induced on $\widetilde{W}_n(w,y,\delta'_n)$ (respectively on $\hat{W}_n(w,y,\delta'_n)$ or on $M_n(w,y,\delta'_n)$) by the metric ρ on N (respectively by the scalar product $(\cdot,\cdot)_{\Phi^n(w)}$ provided by the Riemannian metric ρ on N or by Euclidean metric on $\mathbb{R}^m$).

When $z \in \Phi^n(W)$ (cf. (7.5)) and when $q > 0$ by $Q(z,q)$ we will denote the closed ball in $\Phi^n(W)$ of radius q and centered at z with respect to the induced Riemannian metric on $\Phi^n(W)$. If $z \in \widetilde{W}^1_n(w,y,\delta'_n)$ then we will denote $\overline{z} = (\Psi_n(v_z),v_z)$ where $v_z \in \widetilde{B}^2(v_n,\delta'_n)$.

For fixed δ_0, ℓ, r, s let us define

$$\begin{cases} d_0 = \dfrac{\delta_0 R}{12q(A^{\ell}_{r,s}+1)}. \\ \\ d_n = d_0\xi^n, \quad n \geq 1, \end{cases} \tag{8.1}$$

where R comes from condition (Ba) from Sec. 1.1_I, $0 < R < 1$.

<u>PROPOSITION 8.1</u>. <u>For any</u> $n \geq 0$,

(8.1.1) <u>If</u> $z \in \widetilde{W}_n(w,y,\frac{1}{2}\delta'_n)$ <u>then</u> $\widetilde{\rho}(z,\partial\widetilde{W}_n(w,y,\frac{3}{4}\delta'_n)) > 3d_n$

(8.1.2) <u>If</u> $z \in \widetilde{W}_n(w,y,\frac{3}{4}\delta'_n)$ <u>then</u> $\widetilde{\rho}(z,\partial\widetilde{W}_n(w,y,\delta'_n)) > 3d_n$.

<u>Proof</u>. First of all we will prove that for every $n \geq 0$ and every $z \in \widetilde{W}_n(w,y,\delta'_n)$ one has

$$K(z,3d_n) \subset V_n(w) \tag{8.2}$$

where $K(z,q) = \{y \in M;\ \rho(y,z) \leq q\}$ and where $V_n(w)$ is defined by $(3.9)_I$.

Let $y \in K(z,3d_n)$. Thus

$$\rho(y,\Phi^n(w)) \leq 3d_n + \rho(z,\Phi^n(w)).$$

By (2.9) and $(6.10)_n$ one obtains that

$$\begin{cases} \rho(z,\Phi^n(w)) \le \sqrt{2}\ \|\bar{z}\| \le \sqrt{2}\ (\|\Psi_n(v_z)\| + \|v_z\|) \le \\ \le \sqrt{2}\ (\sup\limits_{v\in\tilde{B}^2(v_n,\delta_n')} \|\Psi_n(v)\| + \delta_n') \le \sqrt{2}\ (q_0\xi^n + \delta_0(\xi')^n). \end{cases}$$

Consequently,

$$\begin{cases} \rho(z,\Phi^n(w)) \le 3d_n + \sqrt{2}(q_0\xi^n + \delta_0(\xi')^n) < \\ < 3(d_0\xi^n + q_0\xi^n + \delta_0(\xi')^n) \le 6q_0(\xi')^n, \end{cases}$$

because $\xi < \xi'$, $d_0 < \frac{\delta_0}{12}$ (cf. (8.1)) and $\delta_0 < \frac{q_0}{4}$.

Now, as $\xi' < \alpha^{2b}$ (cf. $(6.5)_I$) and as $6q_0 \le 6q(1,C) < [Q(\ell,\alpha,\gamma)]^2$ (cf. (6.1) and $(6.7)_I$) one has $6q_0(\xi')^n < (Q(\ell,\alpha,\gamma)\alpha^{2bn})^2$ which proves (8.2).

Let us prove now (8.1.1). The proof of (8.1.2) is completely analogous.

In virtue of (8.2), for every $u \in Q(z,6d_n) \subset K(z,6d_n)$ one can consider the point $\bar{u} = H_n^{-1}(u) \in \mathbb{R}^m$.

From Condition (Bb) of Sec. 2_I and from (2.9), it follows that for every $n \ge 0$, the mapping H_n satisfies the Lipschitz condition with constant $q\sqrt{2}$ and the mapping H_n^{-1} satisfies this condition with constant $qA^{\ell}_{r,s}e^{n\varepsilon_{r,s}}$. Thus

$$\begin{cases} \|\bar{u} - \bar{z}\| \le qA^{\ell}_{r,s}e^{n\varepsilon_{r,s}}\|u - z\|_{\Phi^n(w)} \le 3qA^{\ell}_{r,s}e^{n\varepsilon_{r,s}}d_n < \\ < 3qd_0rA^{\ell}_{r,s}e^{7n\varepsilon_{r,s}} < \frac{1}{4}\ \delta_0 re^{7n\varepsilon_{r,s}} < \frac{1}{4}\ \delta_n', \end{cases} \tag{8.3}$$

because from (8.1) it follows that $3qd_0A^{\ell}_{r,s} < \frac{1}{4}\ \delta_0$. Consequently $\|\pi'(\bar{u}) - \pi'(z)\| \le \|\bar{u} - \bar{z}\| < \frac{1}{4}\ \delta_n'$, where π' denotes the canonical projection of $\mathbb{R}^m = \mathbb{R}^k \oplus \mathbb{R}^p$ onto $\mathbb{R}^p$. Now, as $z \in \tilde{W}_n(w,y,\frac{1}{2}\delta_n')$, then $\|\pi'(\bar{z})\| \le \frac{1}{2}\ \delta_n'$ and finally one obtains that $\|\pi'(\bar{u})\| < \frac{3}{4}\ \delta_n'$.

In other words: $\pi'(H_n^{-1}(Q(z,6d_n))) \subset \text{Int } \tilde{B}^2(v_n,\frac{3}{4}\delta_n')$. The last inclusion implies immediately (8.1.1). ∎

8.2. Let $F \subset T_{\Phi^n(w)}N$ be a k-dimensional subspace of $T_{\Phi^n(w)}N$ transversal to the subspace $E_{2\Phi^n(w)}$ and such that for n big enough (i.e. $n \geq n_{20}(\ell,r,s,F)$)

$$\gamma(F,E_{2\Phi^n(w)}) \geq \frac{1}{\ell \exp(n\varepsilon_{r,s})} \tag{8.4}$$

where the angle $\gamma(\cdot,\cdot)$ is defined by (2.15).

Two principal examples of such subspaces are the Riemannian orthogonal complement to $E_{2\Phi^n(w)}$ and $E_{1\Phi^n(w)}$ (cf. (iv) from Section 2.2_I together with (2.16) and $(2.1.3)_I$ respectively).

Let us denote by t_F^n the projection of $T_{\Phi^n(w)}N$ onto $E_{2\Phi^n(w)}$ parallel to the subspace F. Let us define $\hat{z}_F = t_F^n(\hat{z})$. For $z \in \tilde{W}_n(w,y,\frac{3}{4}\delta_n')$ and q, $0 < q \leq 3d_n$, let us denote $\hat{Q}(z,q) = \exp_{\Phi^n(w)}^{-1}(Q(z,q))$ and $H(\hat{z},q) = \{x \in \hat{W}(w,y,\delta_n');\ \hat{\rho}(x,z) \leq q\}$.

8.3. Let us note by $\hat{B}_n^2(v,q)$ the ball of radius q, centered at v in the space $E_{2\Phi^n(w)}$ provided with Riemannian norm in $T_{\Phi^n(w)}$, noted $\|\cdot\|_n$.

The following Lemma about projection is one of the cornerstones of the Absolute Continuity Theorem proof; it is like a second main lemma.

<u>LEMMA 8.1. (III)</u>. <u>For every</u> ε, $0 < \varepsilon < 1$, <u>there exists</u> $n_2 = n_2(\varepsilon) = n_2(\ell,r,s,\alpha,\gamma,\varepsilon)$ <u>such that for any</u> $n \geq n_2$, <u>any</u> $z \in \tilde{W}_n(w,y,\frac{3}{4}\delta_n')$, <u>any</u> q, $0 < q \leq 3d_n$, <u>and any subspace</u> $F \subset T_{\Phi^n(w)}N$ <u>satisfying</u> (8.4), <u>one has</u>

$$\hat{B}_n^2(\hat{z}_F,(1-\varepsilon)q) \subset t_F^n(\hat{Q}(z,q)) \subset \hat{B}_n^2(\hat{z}_F,(1+\varepsilon)q).$$

<u>Remark</u>. Without loss of generality we will assume that n_2 is a decreasing function of ε.

Proof. Step 1. Let us remark that for every $\varepsilon > 0$ there exists $n_{21} = n_{21}(\alpha,\gamma,\varepsilon)$ such that for every $n \geq n_{21}$, $z \in \tilde{W}_n(w,y,\frac{3}{4}\delta_n')$, $0 < q \leq 3d_n$, and $F \subset T_{\Phi^n(w)}$ satisfying (8.4), one has

$$H(\hat{z},(1-\tfrac{\varepsilon}{2})q) \subset \hat{Q}(z,q) \subset H(\hat{z},(1+\tfrac{\varepsilon}{2})q). \tag{8.5}$$

Indeed, from Proposition 8.1 and (7.22) one obtains that for all $n \geq 0$ and q, $0 < q \leq 3d_n$, one has

$$Q(z,q) \subset \tilde{W}_n(w,y,\delta_n') \subset V_n(x).$$

Let $n_{21} = n_{21}(\alpha,\gamma,\varepsilon)$ be defined as a number such that for all $n \geq n_{21}$

$$\left(Q(\ell,\alpha,\gamma)\alpha^{2bn}\right)^2 \leq \min\left(r_{\frac{\varepsilon}{8}},\gamma\right) \leq R_{\frac{\varepsilon}{8}}(w)$$

where $r_{\frac{\varepsilon}{8}}$ and $R_{\frac{\varepsilon}{8}}$ come from Condition (Bc) from Sec. 2.2.

Let $n \geq n_{21}$. Let $y \in \partial Q(z,q)$ and let Γ_ε be a smooth curve belonging to $Q(z,q)$, connecting z and y, whose length is less than $(1+\frac{\varepsilon}{8})q$. Thus the curve $\exp^{-1}_{\Phi^n(w)}(\Gamma_\varepsilon)$ belongs to $\hat{Q}(z,q)$, its ends are $\hat{z}$ and $\hat{y}$. In virtue of condition (Bc) from Sec. 1.1_I, its $\hat{\rho}$-length is less than $\left(1+\frac{\varepsilon}{8}\right)^2 q < \left(1+\frac{\varepsilon}{2}\right)q$, because $0 < \varepsilon < 1$. This proves the second inclusion of (8.5). The proof of the first one follows exactly the same lines, if only one takes into account that for any smooth curve $\Gamma \subset Q(z,q)$, Γ connecting z and y, one has $\ell(\Gamma) \geq q$ and that by Condition (Bc) one has

$$\|d(\exp^{-1}_{\Phi^n(w)})(y)\|_n \geq \frac{1}{\|d\exp_{\Phi^n(w)}(\hat{y})\|_n} \geq \frac{1}{1+\frac{\varepsilon}{8}} > 1-\frac{\varepsilon}{8}.$$

Step 2. Now, to finish the proof it suffices to show that for every ε, $0 < \varepsilon < 1$, there exists $n_{22} = n_{22}(\ell,r,s,\varepsilon)$ such that for every $n \geq n_{22}$, for every $z \in \tilde{W}_n(w,y,\frac{3}{4}\delta_n')$, for every q, $0 < q \leq 3d_n$ and for every subspace $F \subset T_{\Phi^n(w)}N$ satisfying (8.4) one has

$$t_F^n(H(\hat{z},(1+\tfrac{\varepsilon}{2})q) \subset \hat{B}_n^2(\hat{z}_F,(1+\varepsilon)q) \tag{8.6}$$

and

$$\hat{B}_n^2(\hat{z}_F,(1-\varepsilon)q) \subset t_F^n(H(\hat{z},(1-\tfrac{\varepsilon}{2})q). \tag{8.7}$$

We will prove here only (8.6). The proof of (8.7) is exactly along the same lines.

As $\hat{z}-\hat{z}_F \in F$, then the set

$$G(\hat{z},(1+\tfrac{\varepsilon}{2})q) = \{x \in T_{\Phi^n(w)}N;\ x+(\hat{z}-\hat{z}_F) \in H(\hat{z},(1+\tfrac{\varepsilon}{2})q)\}$$

is such that

$$t_F^n(G(\hat{z},(1+\tfrac{\varepsilon}{2})q)) = t_F^n(H(\hat{z},(1+\tfrac{\varepsilon}{2})q)).$$

Thus to prove (8.6) it suffices to show that for $n \geq n_{22} = n_{22}(\ell,r,s,\varepsilon)$ one has

$$t_F^n(G(\hat{z},(1+\tfrac{\varepsilon}{2})q)) \subset \hat{B}_n^2(\hat{z}_F,(1+\varepsilon)q) \tag{8.8}$$

Let us remark that $\hat{z}_F \in G(\hat{z},(1+\frac{\varepsilon}{2})q)$.

Let $y \in G(\hat{z},(1+\frac{\varepsilon}{2})q)$. Two possibilities occur

(a) $y \in E_{2\Phi^n(w)}$ and (b) $y \notin E_{2\Phi^n(w)}$.

Let us examine separately these two cases.

(a) In this case $\|y-\hat{z}_F\|_n \leq \hat{\rho}_G(y,\hat{z}_F) \leq (1+\frac{\varepsilon}{2})q$, where $\hat{\rho}_G$ denotes the distance induced by the Riemannian scalar product in $T_{\Phi^n(w)}N$ on $G(\hat{z},(1+\frac{\varepsilon}{2})q)$. Consequently $y = t_F^n(y) \in \tilde{B}_n^2(\hat{z}_F,(1+\varepsilon)q)$.

(b) The proof in this case is a bit longer but it uses only the elementary trigonometry, (8.4) and the inequality $(6.11)_n$, i.e.

$$\max_{v\in\tilde{B}^2(v_n,\delta_n')} \|d\Psi_n(v)\| \leq C\eta^n \leq \eta^n.$$

Let $0 < \delta \le \frac{\pi}{2}$. Let us denote by $C_n(\delta)$ (respectively $C_n'(\delta)$) the union of all straight lines $K \subset T_{\Phi^n(w)}N$, passing through $\hat{z}_F$ such that for some straight line $L \subset E_{2\Phi^n(w)}$ also passing through $\hat{z}_F$ one has $\gamma_n(K,L) \le \delta$ (respectively $\gamma_n'(K,L) \le \delta$), where $\gamma_n(K,L)$ (respectively $\gamma_n'(K,L)$) denotes the angle between the straight lines K and L measured with respect to the Riemannian norm $\|\cdot\|_n$ (respectively measured with respect to the Lyapunov norm $\|\cdot\|_n'$) in $T_{\Phi^n(w)}N$.

The inequalities (2.9) and (2.16) imply that for every $n \ge 0$ and every δ, $0 < \delta \le \frac{\pi}{2}$ one has

$$C_n\left(\frac{\delta}{2\sqrt{2}A^{\ell}_{r,s}\exp(n\varepsilon_{r,s})}\right) \subset C_n'(\delta) \subset C_n(2\sqrt{2}\,A^{\ell}_{r,s}\exp(n\varepsilon_{r,s})). \tag{8.9}$$

Now, (2.16) and $(6.11)_n$ imply that for every $n \ge 1$ one has $G(\hat{z},(1+\frac{\varepsilon}{2})q) \subset C_n'(\eta^n)$. Consequently, as $\eta = \exp(-7\varepsilon_{r,s})$, by (8.9) one has that for every $n \ge 1$

$$G(\hat{z},(1+\frac{\varepsilon}{2})q) \subset C_n(2\sqrt{2}\,A^{\ell}_{r,s}\exp(-6n\varepsilon_{r,s})). \tag{8.10}$$

Beginning from this moment up to the end of the proof, all angles and lengths will be measured exclusively with respect to the Riemannian norm in $T_{\Phi^n(w)}N$.

If $y \in G(z,(1+\frac{\varepsilon}{2})q)$ then $\|y-\hat{z}_F\|_n \le \hat{\rho}(y,z_F) \le (1+\frac{\varepsilon}{2})q$. Thus, in virtue of (8.10), to prove (8.8) it suffices to prove that for all $n \ge n_{22}(\ell,r,s,\varepsilon)$ one has

$$\begin{cases} \text{if } y \in C_n(2\sqrt{2}\,A^{\ell}_{r,s}\exp(-6n\varepsilon_{r,s}),\ y \notin E_{2\Phi^n(w)} \\ \text{and } \|y-\hat{z}_F\|_n \le (1+\frac{\varepsilon}{2})q \text{ then } \|t^n_w(y)-\hat{z}_F\|_n \le (1+\varepsilon)q. \end{cases} \tag{8.11}$$

Let $n_{23} = n_{23}(\ell,r,s)$ be a number such that for every $n \ge n_{23}$ one has

$$2\sqrt{2}\, A^{\ell}_{r,s} \exp(-6n\varepsilon_{r,s}) < \frac{\pi}{4} \quad \text{and} \quad \frac{1}{\ell \exp(n\varepsilon_{r,s})} < \frac{\pi}{4}. \tag{8.12}$$

and that (8.4) is true for the Riemannian orthogonal complement to $E_{2\Phi^n(w)}$.

Let us consider a point y as in (8.11). Let us note by y' the orthogonal projection of y to the supspace $E_{2\Phi^n(w)}$.

As follows from (8.12), for $n \geq n_{23}$, the straight line passing through y and $\hat{z}_F$ is never orthogonal to $E_{2\Phi^n(w)}$. Moreover, as $y \notin E_{2\Phi^n(w)}$, the points y, $t^n_F(y)$ and $\hat{z}_F$ are not colinear (see Fig. 4) and when $F \neq E^{\perp}_{2\Phi^n(w)}$, the points $\hat{z}_F$, y, $t^n_F(y)$ and y' are never in the same plane except the case where the points $\hat{z}_F$, $t^n_F(y)$ and y' are colinear. When $F = E^{\perp}_{2\Phi^n(w)}$ then $y' = t^n_F(y)$.

Let us consider the tetrahedron which may be sometimes reduced to a triangle with the vertices $\hat{z}_F$, y, $t^n_F(y)$ and y'. An example of such a tetrahedron is presented in Fig. 4.

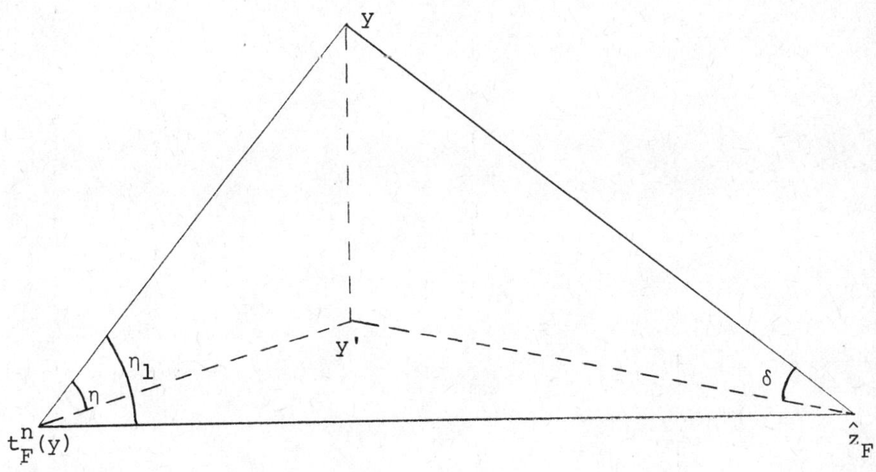

Fig. 4 $\frac{1}{\ell \exp(n\varepsilon_{r,s})} \leq \eta$, $\eta_1 \leq \pi - \frac{1}{\ell \exp(n\varepsilon_{r,s})}$ (cf. (8.4)),

$\delta \leq 2\sqrt{2}\, A^{\ell}_{r,s} \exp(-6n\varepsilon_{r,s})$ and $\|y - \hat{z}_F\|_n \leq (1 + \frac{\varepsilon}{2})q$.

Now, it is easy to see that $\|t_F^n(y) - \tilde{z}_F\|_n$ is never bigger than the length of the side AB of the triangle ABC presented in Fig. 5.

Let us denote the length of the interval $[a,b]$, by $|ab|$.

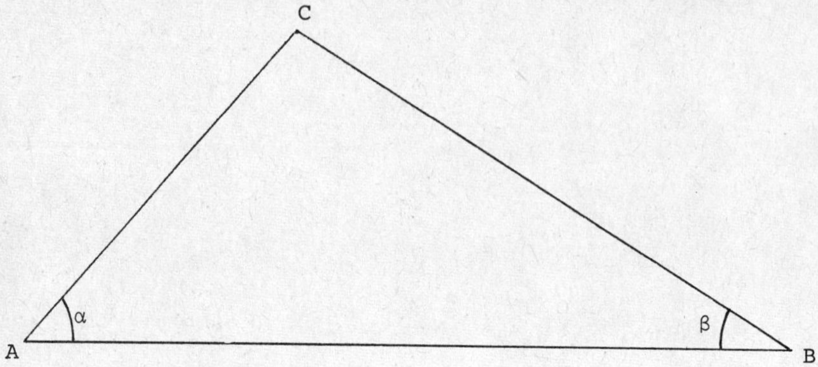

Fig. 5. $\alpha = \frac{1}{\ell \exp(n\varepsilon_{r,s})} < \frac{\pi}{4}$, $\beta = 2\sqrt{2}\, A_{r,s}^{\ell} \exp(-6\varepsilon_{r,s} n) < \frac{\pi}{2}$,

$|CB| = (1+\frac{\varepsilon}{2})q$.

The straightforward computation shows that

$$\begin{cases} |AB| = (1+\frac{\varepsilon}{2})q(\cos\beta + \frac{\sin\beta}{tg\alpha}) \le (1+\frac{\varepsilon}{2})q(1+2\,\frac{\beta}{\alpha}) \le \\ \le (1+\frac{\varepsilon}{2})q(1 + 4\sqrt{2}\, A_{r,s}^{\ell}\, \ell \exp(-n\varepsilon_{r,s})). \end{cases}$$

Thus for $n \ge n_{23}$ one has

$$\|t_F^n(y) - \hat{z}_F\|_n \le (1+\frac{\varepsilon}{2})(1 + 4\sqrt{2}\, A_{r,s}^{\ell}\, \ell \exp(-n\varepsilon_{r,s}))q. \tag{8.13}$$

But for $n \ge n_{24} = n_{24}(\ell,r,s,\varepsilon)$ one has that

$$(1+\frac{\varepsilon}{2})(1+4\sqrt{2}\, A_{r,s}^{\ell}\, \ell \exp(-n\varepsilon_{r,s})) < 1 + \varepsilon.$$

Consequently, (8.13) implies that for all $n \ge n_{22} \overset{\text{def}}{=\!=} \max(n_{23}, n_{24})$ one has (8.11) and thus (8.6).

Finally, let us note that in the proof of (8.7), one uses substantially the fact that $\frac{1}{\ell \exp(n\varepsilon_{r,s})} \le \eta_1 \le \pi - \frac{1}{\ell \exp(n\varepsilon_{r,s})}$. ∎

Let us denote by π the orthogonal projection of $T_{\Phi^n(w)}N$ onto $E_{2\Phi^n(w)}$, with respect to the Riemannian scalar product in $T_{\Phi^n(w)}N$.

PROPOSITION 8.2. <u>There exists a number</u> $n_3 = n_3(\ell,r,s,\alpha,\gamma)$ <u>such that for every</u> $n \ge n_3$ <u>and</u> $z \in \widetilde{W}_n(w,y,\frac{3}{4}\delta'_n)$ <u>one has</u>

(8.2.1) <u>for every</u> q, $0 < q \le \frac{7}{3} d_n$, <u>if</u> $y \in \partial H(\hat{z},q)$, <u>then</u> $\hat{\rho}(z,y) = q$,

(8.2.2) <u>there exists</u> C^1 <u>mapping</u> $\Psi_{\pi n} : \widetilde{B}^2_n(\pi(\hat{z}),\frac{7}{3} d_n) \to E_{2\Phi^n(w)}$ <u>such that</u> $H(\hat{z},\frac{7}{3} d_n) \subset \text{graph } \Psi_{\pi n} \subset \hat{Q}(z,3d_n)$,

(8.2.3) $\|d\Psi_{\pi n}(y)\| \le 4\sqrt{2}\, A^{\ell}_{r,s}\exp(-6n\varepsilon_{r,s})$ <u>for any</u> $y \in \hat{B}^2_n(\hat{z},\frac{7}{3} d_n)$.

<u>Proof</u>. It easily follows from (8.5), Lemma 8.1 applied to $F = E^{\perp}_{2\Phi^n(w)}$, (8.9) and from the inequality $\operatorname{tg}\alpha \le 2\alpha$ valid for $0 < \alpha \le \frac{\pi}{4}$. ∎

Let us remark that in Proposition 8.1, the multiple $\frac{7}{3}$ can be replaced by any other number t, $0 < t < 3$. In this case the number n_3 will also depend on t.

8.4. Let us denote by ν_n and μ_n the p dimensional Riemannian volume on $\widetilde{W}_n(w,y,\delta'_n)$ and $\hat{W}_n(w,y,\delta'_n)$ respectively.

For $z \in \widetilde{W}_n(w,y,\frac{3}{4}\delta'_n)$ and θ, $0 < \theta < 1$, let us denote

$$A(z,\theta) = \{y \in \Phi^n(W);\ 2d_n(1-\theta) \le \widetilde{\rho}(y,z) \le 2d_n\}. \tag{8.14}$$

It follows from (8.1.2) that $A(z,\theta) \subset \widetilde{W}_n(w,y,\delta'_n)$.

All numbers n_4 etc., which will appear, can depend among others on P,β, h, ℓ, r, s,α,γ, the dependence on these parameters will not be marked in the future.

The aim of the present section is to deduce from Lemma 8.1 and Proposition 8.2 the following.

Lemma 8.2. There exists an absolute constant C_2 such that for every θ, $0 < \theta < \frac{1}{6}$, there exists $n_4 = n_4(\theta)$ such that for every $n \geq n_4$ and every $z \in \tilde{W}_n(w,y,\frac{3}{4}\delta_n')$ one has

$$\frac{\nu_n(A(z,\theta))}{\nu_n(Q(z,d_n))} \leq C_2\theta. \tag{8.15}$$

Proof. Step 1. Taking the inverse to the exponential map and applying condition (Bb) of Sec. 1.1_I, one easily sees that (8.15) is equivalent to

$$\frac{\mu_n(\hat{A}(z,\theta))}{\mu_n(\hat{Q}(z,d_n))} \leq C_{21}\theta \tag{8.16}$$

where $\hat{A}(z,\theta) = \exp^{-1}_{\Phi^n(w)}(A(z,\theta))$ and C_{21} a constant. (8.16) will be easily deduced from

$$\hat{B}_n^2(\pi(\hat{z}),\ d_n(1-\theta)^2) \subset \pi(\hat{Q}(z,d_n)) \tag{8.17}$$

and

$$\begin{cases} \pi(\hat{A}(z,\theta)) \subset \{x \in E_{2\Phi^n(w)};\ 2d_n(1-2\theta)(1-\theta) \leq \|x - \pi(\hat{z})\| \leq \\ \qquad \leq 2d_n(1+\theta) \overset{\text{def}}{=\!=} R(z,\theta)\}. \end{cases} \tag{8.18}$$

Let us note that $2d_n(1+\theta) < \frac{7}{3}d_n$, because $\theta < \frac{1}{6}$.

Lemma 8.1 applied to $F = E^{\perp}_{2\Phi^n(w)}$ immediately implies that (8.17) is true for $n \geq n_{41}(\theta)$. Now let us prove that (8.18) is true for $n \geq n_{42}(\theta)$.

First, let us remark that from condition (Bc) from Sec. 1.1_I, from (8.14), in the same way as in the proof of (8.5), one deduces that there exists $n_{43} = n_{43}(\theta)$ such that for every $n \geq n_{43}$ one has

$$\begin{cases} \hat{A}(z,\theta) \subset \{y \in \hat{W}_n(w,y,\delta_n');\ 2d_n(1-\theta)(1-\theta)^2 \leq \hat{\rho}(y,z) \leq \\ \qquad \leq 2d_n(1+\theta)^2\}. \end{cases} \tag{8.19}$$

Now, the inclusion $\pi(\hat{A}(z,\theta)) \subset \hat{B}_n^2(\pi(\hat{z}),2d_n(1+\theta))$ is evident.

Thus it is enough to show that there exists a number $n_{42} = n_{42}(\theta) \geq n_{43}$ such that for $n \geq n_{42}$

$$\hat{B}_n^2(\pi(\hat{z}),2d_n(1-2\theta)(1-\theta)) \subset \pi(\hat{Q}(z,2d_n(1-\theta)(1-\theta^2))). \tag{8.20}$$

Let $n_{42} = n_{42}(\theta) \geq \max(n_2,n_3,n_{43})$ (where n_2 comes from Lemma 8.1 and n_3 which comes from Proposition 8.2) be such a number that for $n \geq n_{42}$

$$\|d\Psi_{\pi n}(y)\| \leq \theta \tag{8.21}$$

for all $y \in \hat{B}_n^2(\pi(\hat{z}),\frac{7}{3}d_n)$ (cf. (8.23)).

Let us fix $n \geq n_{42}$ and suppose that (8.20) is not true. Thus there exists a point $z_n \in \partial\hat{Q}(z,2d_n(1-\theta)(1-\theta^2))$ such that

$$\|\pi(z) - \pi(z_n)\|_n \leq 2d_n(1-2\theta)(1-\theta). \tag{8.22}$$

Let I be the interval in $E_{2\Phi^n(w)}$ connecting $\pi(\hat{z})$ and $\pi(z_n)$. As $n_{42} \geq n_1$, then by Lemma 8.1 applied to $F = E^{\perp}_{2\Phi^n(w)}$ and to $\varepsilon = \theta$ one obtains by (8.22) that $I \subset \pi(\hat{Q}(z,2d_n(1-2\theta)))$.

Let us consider the curve

$$\Gamma = \{(\Psi_{\pi n}(v),v);\ v \in I\} \subset \hat{Q}(z,2d_n(1-2\theta))$$

connecting z and z_n. Now, from (8.21) it is easy to see that its length does not exceed $2d_n(1-2\theta)(1+\theta)$.

On the other hand, it follows from Lemma 8.1 that for $n \geq n_{42}$ one has

$$\hat{B}_n^2(\pi(\hat{z}),2d_n(1-\theta)^2(1-\theta^2)) \subset \pi(\hat{Q}(z,2d_n(1-\theta)(1-\theta^2))).$$

Thus, for $n \geq n_{42}$ one obtains that $\ell(\Gamma) \geq 2d_n(1-\theta)^2(1-\theta^2)$.

Consequently $(1-\theta)^2(1-\theta^2) < (1-2\theta)(1+\theta)$, i.e. $(1-\theta^2)(1-\theta) < 1-2\theta$. For $\theta > 0$, this last inequality is equivalent to $1+\theta^2 < \theta$ and we obtain the contradiction which proves (8.20) and consequently (8.18).

<u>Step 2</u>. As $0 < \theta < \frac{1}{6}$, (8.21) and (3.2) imply now that there exists

a constant $C_{22} > 0$ such that for every measurable subset $V \subset \pi(\hat{Q}(z,\frac{7}{3}d_n))$, $z \in \tilde{W}_n(w,y,\frac{3}{4}\delta_n')$, one has

$$|V| \leq \nu_n(\hat{Q}(z,\tfrac{7}{3}d_n) \cap \pi^{-1}(V)) \leq C_{22}|V| \tag{8.23}$$

where $|V|$ denotes the Lebesgue measure of V induced by the Riemannian scalar product in $T_{\phi^n(w)}N$.

Now, (8.17), (8.18), (8.23) and the standard factorization of $a^p - b^p$ implies that for $n \geq n_3$ one has

$$\begin{cases} \dfrac{\mu_n(\hat{A}(z,\theta))}{\mu_n(\hat{Q}(z,d_n))} \leq \dfrac{\mu_n[\hat{Q}(z,2d_n(1+\theta)) \cap \pi^{-1}(R(z,\theta))]}{\mu_n(\pi^{-1}(\hat{B}_n^2(\pi(\hat{z}),d_n(1-\theta)^2)))} \\ \leq C_{22}\dfrac{|R(z,\theta)|}{|\hat{B}_n^2(\pi(\hat{z}),\frac{25}{36}d_n)|} \leq C_{23}\dfrac{[2d_n(1+\theta)]^p-[2d_n(1-2\theta)(1-\theta)]^p}{d_n^p} \\ = C_{24}\{(1+\theta)^p - [(1-2\theta)(1-\theta)]^p\} \leq \\ \leq C_{25}[(1+\theta) - (1-2\theta)(1-\theta)] \leq C_{21}\theta, \end{cases}$$

where $C_{22} - C_{25}$ are some constants. Thus (8.16) and consequently (8.15) is proved with $n_4 = \max(n_2,n_3,n_{41},n_{42})$. ■

8.5. Let us return to the submanifold W (cf. (7.5) and (7.16)). If $z \subset \text{Int}\,W$, we will denote by $Q(z,h)$ the closed ball in W, with respect to the induced Riemannian metric on W, centered at z and of radius h. When $h > 0$ is small enough, i.e. $0 < h < h_z$, $Q(z,h)$ does not touch the boundary of W.

Let us note for the sake of brevity

$$\begin{cases} \Lambda(x,C) = \Lambda^{\ell}_{k,r,s,\alpha,\gamma} \cap \tilde{U}(x,q(2,C)), \\ \hat{\Lambda}(x,C) = \hat{\Lambda}^{\ell}_{k,r,s,\alpha,\gamma}(x,q(2,C)), \end{cases} \tag{8.24}$$

where the last symbol is defined in (4.1).

Let $z \in \operatorname{Int} W$ and let $0 < h < h_z$. Let us denote $D(z,h) = \hat{\Lambda}(x,C) \cap Q(z,h)$. As W is compact, then $Q(z,h)$ and consequently $D(z,h)$ are also compact subsets of N.

Let us remark that although in general Φ^n is not defined on W for $n \geq 2$, nevertheless Φ^n is defined on $D(z,h)$ for all $n \geq 1$, because Φ^n is well defined on $\hat{\Lambda}(x,C)$ for all $n \geq 1$.

Nevertheless, for $Y \subset W$ we will use the notation $\Phi^n(Y)$ for the set $\Phi^n\{y \in Y;\ \Phi^n(y)$ is defined$\}$.

LEMMA 8.3.(IV). For every $P \in \operatorname{Int} W$ and every $\beta > 0$ there exists $\delta_{P,\beta} > 0$ such that for every δ_0, $0 < \delta_0 < \delta_{P,\beta}$, every h, $0 < h < h_P$ and every $n \geq 1$ there exists $n_5 = n_5(n,P,\beta,\delta_0,h)$ and the points $w_i \in \Lambda^{\ell}_{k,r,s,\alpha,\gamma} \cap \tilde{U}(x,q(2,C))$, $1 \leq i \leq n_5$, such that for every i one has

(8.3.1) $\quad V(w_i) \cap W \neq \emptyset.$

Let us note $y_i = V(w_i) \cap W$.

(8.3.2) The submanifold $\tilde{W}_n(w_i,y_i,\delta'_n)$ is well defined and

$$\Phi^n(D(P,h)) \subset \overline{W}_n \overset{\text{def}}{=\!=} \bigcup_{i=1}^{n_5} \tilde{W}_n(w_i,y_i,\tfrac{1}{2}\delta'_n) \subset \overline{\overline{W}}_n$$

$$\overset{\text{def}}{=\!=} \bigcup_{i=1}^{n_5} \tilde{W}_n(w_i,y_i,\delta'_n) \subset \Phi^n(Q(P,h+\beta)).$$

Proof. Step 1. As $|W| \leq \varepsilon^{\ell}_{r,s,\alpha,\gamma}(C)$, then (2.9) applied to $n = 0$ implies there exists a number $Y^{\ell}_{r,s,\alpha,\gamma}(C) > 0$ and a strictly positive and a strictly increasing function $t = t(Y)$ depending only on ℓ, r, s, α, γ, C and defined for $0 < Y < Y^{\ell}_{r,s,\alpha,\gamma}(C)$, satisfying the inequality $t(Y) \leq Y$ and such that the following condition holds.

For every $w \in \Lambda(x,C)$ (cf. (8.24)) such that $V(w) \cap W \neq \emptyset$ and $y = V(w) \cap W \in \operatorname{Int} W$ one has that for any Y, $0 < Y < \min(q_0(w,W), Y^{\ell}_{r,s,\alpha,\gamma}(C))$ (cf. (7.16))

$$\tilde{W}_0(w,y,t(Y)) \subset Q(y,Y). \tag{8.25}$$

Let us fix $P \in \text{Int } W$ and define

$$A(P) = \inf\{q_C(w,W);\ w \in \Lambda(x,C) \text{ and } V(w) \cap W \in Q(P,h)\}.$$

As follows from the definition of $q_C(w,W)$, $A(P) > 0$.

Let us define

$$\delta_{P,\beta} = \min\left(t\left[\min\left(\frac{\beta}{4}, Y^{\ell}_{r,s,\alpha,\gamma}\right)\right], A(P)\right). \tag{8.26}$$

and fix numbers $n \geq 1$, $\beta > 0$, and δ_0 such that $0 < \delta_0 < \delta_{P,\beta}$.

Step 2. As $0 < h < h_P$, then one can consider the covering of the set $\Phi^n(D(P,h))$ by the following open subsets of $\Phi^n(W)$,

$$\{\text{Int } \widetilde{W}_n(w,y,\tfrac{1}{2}\delta'_n)\}_{w\in\Lambda(x,C)}.$$

From the compactness of $\Phi^n(D(P,h))$, there exists a finite covering of that type containing the minimal possible number, say sets: $\widetilde{W}_n(w_i,y_i,\frac{1}{2}\delta'_n)$, $1 \leq i \leq n_5 = n_5(n,P,\delta_0,\beta,h)$.

Now it remains only to prove that

$$\overline{\overline{W}}_n = \bigcup_{i=1}^{n_5} \widetilde{W}_n(w_i,y_i,\delta'_n) \subset \Phi^n(Q(P,h+\beta)). \tag{8.27}$$

From (7.21) it follows that if $y \in \widetilde{W}_n(w_i,y_i,\delta'_n)$, then $\Phi^{-n}(y)$ is well defined and $\Phi^{-n}(y) \in \widetilde{W}_0(w_i,y_i,\delta_0) \subset W$. Thus (8.27) is equivalent to

$$\Phi^{-n}[\widetilde{W}_n(w_i,y_i,\delta'_n)] \subset Q(P,h+\beta) \tag{8.28}$$

for all i, $1 \leq i \leq n_5$.

If (8.28) is not true, then there exists a point q and some i, $1 \leq i \leq n_5$, such that $q \in \Phi^{-n}[\widetilde{W}_n(w_i,y_i,\delta'_n)]$ and $q \notin Q(P,h+\beta)$. But as $\Phi^{-n}[\widetilde{W}_n(w_i,y_i,\delta'_n)] \cap D(P,h) \neq \emptyset$, then $\Phi^{-n}[\widetilde{W}_n(w_i,y_i,\delta'_n)] \cap \cap Q(P,h) \neq \emptyset$. Now, as the set $\Phi^{-n}[\widetilde{W}_n(w_i,y_i,\delta'_n)]$ is connected, then if a point q exists, there exists necessarily a point $z \in \partial Q(P,h+\beta) \cap \Phi^{-n}[\widetilde{W}_n(w_i,y_i,\delta'_n)]$. Thus from (7.21), (8.25) and (8.26) one has

$$\begin{cases} z \in \partial Q(P,h+\beta) \cap \Phi^{-n}[\widetilde{W}_n(w_i,y_i,\delta_n')] \subset \\ \\ \subset \partial Q(P,h+\beta) \cap \widetilde{W}_0(w_i,y_i,\delta_0) \subset \partial Q(P,h+\beta) \cap Q(y_i,\frac{\beta}{4}). \end{cases} \tag{8.29}$$

On the other hand, as $D(P,h) \cap \Phi^{-n}[\widetilde{W}_n(w_i,y_i,\delta_n')] \neq \emptyset$, by (7.21), (8.25) and (8.26) one has

$$\begin{cases} \emptyset \neq D(P,h) \cap \Phi^{-n}[\widetilde{W}_n(w_i,y_i,\delta_n')] \subset D(P,h) \cap \widetilde{W}_0(w_i,y_i,\delta_0) \subset \\ \\ \subset D(P,h) \cap Q(y_i, \frac{\beta}{4}). \end{cases} \tag{8.30}$$

Finally, (8.29) and (8.30) give (see Fig. 6) that

$$\begin{cases} \partial Q(P,h+\beta) \cap Q(y_i, \frac{\beta}{4}) \neq \emptyset \\ \\ D(P,h) \cap Q(y_i, \frac{\beta}{4}) \neq \emptyset. \end{cases} \tag{8.31}$$

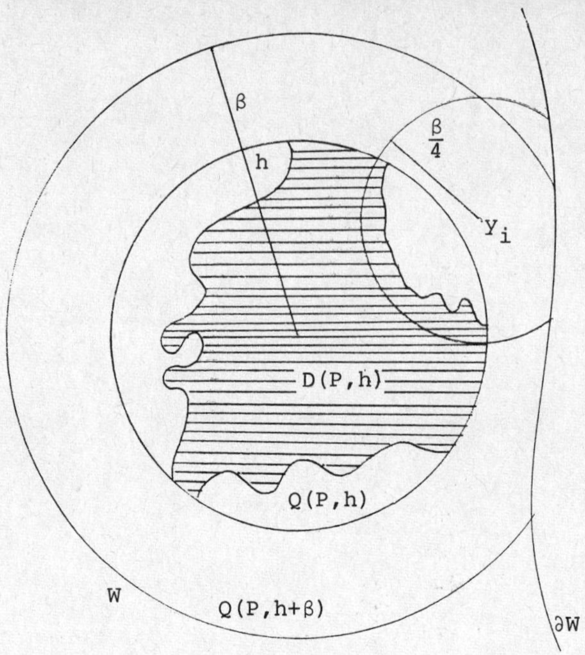

Fig. 6. In general D(P,h) is a very complicated set of Cantor type.

But $\rho(D(P,h), \partial Q(P,h+\beta)) \geq \beta$, because $D(P,h) \subset Q(P,h)$. On the other hand diam $Q(y_i, \frac{\beta}{4}) \leq \frac{\beta}{2}$ and we obtain the contradiction with (8.31). ■

In the future we will simply write $n_5 = n_5(n)$.

8.6. Let us consider $n \geq 1$ and i, $1 \leq i \leq n_5(n)$. We will use the following notations. If $z \in \widetilde{W}_n(w_i, y_i, \delta_n')$ then $\hat{z} = \exp^{-1}_{\Phi^n(w_i)}(z)$, $t^n_{w_i} = t^n_{E_{1\Phi^n(w_i)}}$ and $\tilde{z} = t^n_{w_i}(\hat{z})$. For $Q(z,r) \subset \widetilde{W}_n(w_i, y_i, \delta_n')$ let us note also $\hat{Q}(z,r) = \exp^{-1}_{\Phi^n(w_i)} Q(z,r)$, where $0 < r \leq 3d_n$. Although z may belong simultaneously to $\widetilde{W}_n(w_i, y_i, \delta_n')$ and $\widetilde{W}_n(w_j, y_j, \delta_n')$, where $i \neq j$, nevertheless these notations will never lead to ambiguities.

Let $\{A_i\}_{i \in I}$ be a family of subsets of set X and let $Y \subset X$,

$Y \subset \bigcup_{i\in I} A_i$. We will say that the multiplicity of the covering $\{A_i\}_{i\in I}$ of Y is not bigger than L if for any $y \in Y$, $\#\{i\in I;\ y \in A_i\} \le L$.

Now we can prove the following refinement of Lemma 8.2.

LEMMA 8.4(V). *For every* δ_0, $0 < \delta_0 < \delta_{P,\beta}$, *there exist* d_0, $0 < d_0 < \delta_0$, $L > 0$, $n_6 = n_6(P,\beta,\delta_0,h)$ *such that for every* $n \ge n_6$ *there exist* $n_7 = n_7(n,P,\beta,\delta_0,h)$ *and the points* $\{z_j\}_{1\le j\le n_7} \subset \Phi^n(W)$ *such that:*

(8.4.1) *for every* j, $1 \le j \le n_7$, *there exists* i, $1 \le i \le n_5$, *such that* $Q(z_j, 2d_n) \subset \widetilde{W}_n(w_i, y_i, \delta_n')$

$$(8.4.2)\quad \left.\begin{aligned} \widetilde{\widetilde{W}}_n &= \bigcup_{i=1}^{n_5} \widetilde{W}_n(w_i, y_i\ \tfrac{1}{2}\ \delta_n') \subset \bigcup_{j=1}^{n_7} Q(z_j, d_n) \subset \\ &\subset \bigcup_{j=1}^{n_7} Q(z_j, 2d_n) \subset \overline{\overline{W}}_n = \bigcup_{i=1}^{n_5} \widetilde{W}_n(w_i, y_i, \delta_n'). \end{aligned}\right\}$$

(8.4.3) *The multiplicity of the covering of* $\widetilde{\widetilde{W}}_n$ *by the balls* $Q(z_j, 2d_n)$, $1 \le j \le n_7$ *is not bigger than* L.

Proof. Let $n \ge 1$. As $\widetilde{\widetilde{W}}_n$ is compact, then we can find a finite set of points $\{z_j\}_{1\le j\le n_7} \subset \widetilde{\widetilde{W}}_n$ such that $\widetilde{\rho}(z_i, z_j) \ge d_n$ for all $i \ne j$, $1 \le i$, $j \le n_7$ and that for any point $z \in \widetilde{\widetilde{W}}_n$ there exists some j, $1 \le j \le n_7$ such that $\widetilde{\rho}(z, z_j) < d_n$. Let us note that such set is not unique and its cardinality may depend on the choice of points.

We will prove that the set $\{z_j\}_{1\le j\le n_7}$ satisfies conditions (8.4.1) - (8.4.3). First of all let us remark that (8.4.1) is satisfied in virtue of Proposition 8.1. Now, the first two inclusions of (8.4.2) are evidently satisfied and the last inclusion of (8.4.2) follows from (8.4.1). Thus, it remains only to prove (8.4.3).

Let us remark that by the definition of points $\{z_j\}_{1\le j\le n_7}$, for every $i \neq j$, $1 \leq i$, $j \leq n_7$ one has

$$Q\left(z_i, \frac{d_n}{3}\right) \cap Q\left(z_j, \frac{d_n}{3}\right) = \emptyset. \tag{8.32}$$

For the sake of brevity we will denote $Q_j = Q(z_j, d_n)$, $1 \leq j \leq n_7$.

We want to prove that there exist a number K and $n_6 = n_6(P, \beta, \delta_0, h)$ such that for every $n \geq n_6$ and every j, $1 \leq j \leq n_7$, one has

$$\#\{1 \leq a \leq n_7;\ Q_a \cap Q_j \neq \emptyset\} \overset{\text{def}}{=\!=} K(n,j) \leq K. \tag{8.33}$$

(8.33) implies immediately (8.4.3) with $L = K + 1$.

For some j, $1 \leq j \leq n_7$, let us consider $Q(z_j, d_n)$, $z_j \in \widetilde{W}_n(w_i, y_i, \frac{1}{2}\delta_n')$ for some $i = i(j)$, $1 \leq i \leq n_5$. Let us denote by $1 \leq a_1 < a_2 < \ldots\ldots < a_s \leq n_7$ all the indices such that $Q_{a_k} \cap Q_j \neq \emptyset$, $1 \leq k \leq s$. As $\operatorname{diam} Q_j \leq 2d_n$, then $Q_{a_k} \subset Q(z_j, 3d_n)$, $1 \leq k \leq s$. Now, to prove (8.33) it is sufficient to prove that there exist a number $K' > 0$ and a number n_6 such that for every $n \geq n_6$ and every j, $1 \leq j \leq n_7$, one has

$$\#\{1 \leq f \leq n_7;\ Q_f \cap Q(z_j, 3d_n)\} \overset{\text{def}}{=\!=} \gamma_j \leq K'. \tag{8.34}$$

In virtue of (8.32), to prove (8.34) it is sufficient to prove that there exists a number $K'' > 0$ that for n and j as above, the maximal number of mutually disjoint open balls of radius $\frac{d_n}{3}$ contained in $Q(z_j, 3d_n)$, noted by $K'(n,j)$, does not exceed K''.

Let us consider for some z, $Q(z, \frac{d_n}{3}) \subset Q(z_j, 3d_n)$. In virtue of (8.4.1) and of Proposition 8.1 one obtains $Q(z, \frac{d_n}{3}) \subset \widetilde{W}_n(w_i, y_i, \delta_n')$. Thus $Q(z, \frac{d_n}{3})$ is well defined and $\hat{Q}(z, \frac{d_n}{3}) \subset \hat{Q}(z_j, 3d_n) \subset \exp^{-1}_{\Phi^n(w_i)}[\widetilde{W}_n(w_i, y_i, \delta_n)]$.

As $Q(z, \frac{d_n}{3}) \subset Q(z_j, 3d_n)$ and as $z_j \in \widetilde{W}_n(w_i, y_i, \frac{1}{2}\delta_n')$, then from Proposition 8.1.1. one obtains that $z \in \widetilde{W}_n(w_i, y_i, \frac{3}{4}\delta_n')$. Now, from Lemma 8.1 applied to $\varepsilon = \frac{1}{2}$ and to $Q(z,\frac{d_n}{3})$ and to $Q(z_j,3d_n)$ respectively, one sees that there exists a number $n_6 = n_6(P,\beta,\delta_0,h)$ such that for every $n \geq n_6$ one has

$$\hat{B}_n^2(\tilde{z}, \frac{d_n}{6}) \subset t_{w_i}^n (\hat{Q}(z, \frac{d_n}{3})) \subset t_{w_i}^n (\hat{Q}(z_j,3d_n)) \subset \hat{B}_n^2(\tilde{z}_j, \frac{9d_n}{2}), \qquad (8.35)$$

where $\hat{B}_n^2(v,r)$ is defined at the beginning of Sec. 8.3.

(8.35) immediately implies that

$$K'(n,j) \leq \frac{\operatorname{Vol}\left(B^p\left(\frac{9}{2}\, d_n\right)\right)}{\operatorname{Vol}\left(B^p\left(\frac{d_n}{6}\right)\right)} = 27^p \stackrel{\text{def}}{=\!=} K''.$$

where $B^p(r)$ denotes the p-dimensional Euclidean ball of radius r and Vol its volume. Thus (8.33) and consequently (8.4.3) is proved.

■

9. COMPARISON OF THE VOLUMES

9.1. Let us consider a point $w \in \Lambda^{\ell}_{k,r,s,\alpha,\gamma} \cap \tilde{U}(x,q(2,C))$ such that the TLSM $V(w) \cap \tilde{U}(x,q(2,C))$ is not an isolated element of the family of TLSM $\hat{\Lambda}(x,C)$ (cf. (8.24)), i.e. that the set $\hat{\Lambda}(x,C)\backslash V(w)$ is not closed. Naturally the absolute continuity question becomes trivial for TLSM $V(w)$ which does not satisfy the above condition.

Let us consider two submanifolds W^1 and W^2 transversal to the family $S^{\ell}_{k,r,s,\alpha,\gamma}(x)$ in $\tilde{U}(x,q(2,C))$ such that $|W^1| \leq \frac{1}{2}$, $|W^2| \leq \frac{1}{2}$ and satisfying all the conditions from Sec. 7.4. In order to avoid the problems related to the domain of the definition of Poincaré map, as in Sec. 4.1. we will consider the open submanifolds $\tilde{W}^1$ and $\tilde{W}^2$. Let us suppose that the submanifolds $\tilde{W}^1$ and $\tilde{W}^2$ intersect $V(w)$ at the points y^1 and y^2 respectively, i.e. $y^1 = V(w) \cap \tilde{W}^1$, $y^2 = V(w) \cap \tilde{W}^2$ where $y^1,y^2 \in \operatorname{Int} \tilde{U}(x,q(2,C))$.

Let us fix two numbers δ_{i0}, $i = 1,2$, such that

$$0 < \delta_{i0} < \frac{1}{2} \min(q_C(w,W^1),\, q_C(w,W^2)) \stackrel{\text{def}}{=\!=} q_C(w,W^1,W^2), \qquad (9.1)$$

where $q_C(w,W)$ is defined by (7.16).

Now we can apply to the manifolds $\widetilde{W}^1$ and $\widetilde{W}^2$ the construction described in Sec. 7.5 and obtain for $i = 1,2,$ and $n \geq 0$ the manifolds

$$\widetilde{W}_n^i = \widetilde{W}_n(w,y^i,\delta'_{in}) = \Big\{\exp_{\Phi^n(w)}(\tau'_{\Phi^n(w)})^{-1}(\Psi_n^i(v),v);\quad v \in \widetilde{B}^2(v_n,\delta'_n)\Big\} \tag{9.2}$$

where $\delta'_{in} = \delta_{i0}(\xi')^n$, $i = 1,2$. Let us note that for the functions $\{\Psi_j^i\}_{j\geq 0}$ defined by (9.2) the estimations (6.6) - (6.11) hold.

Let $z^1 = z^1(n) \neq y^1$ be a point such that

$$z^1 \in \Phi^{-n}(\widetilde{W}_n^1) \cap \hat{\Lambda}(x,C) \subset \widetilde{W}_0^1 \cap \hat{\Lambda}(x,C).$$

Consequently $z^1 \in V(\overline{w})$ for some $\overline{w} \in \Lambda^{\ell}_{k,r,s,\alpha,\gamma} \cap \widetilde{U}(x,q(2,C))$. We can assume that z^1 is so close to y^1 that $V(\overline{w})$ meets $\widetilde{W}_0^2$ at some point which we denote by z^2 (cf. Fig. 7).

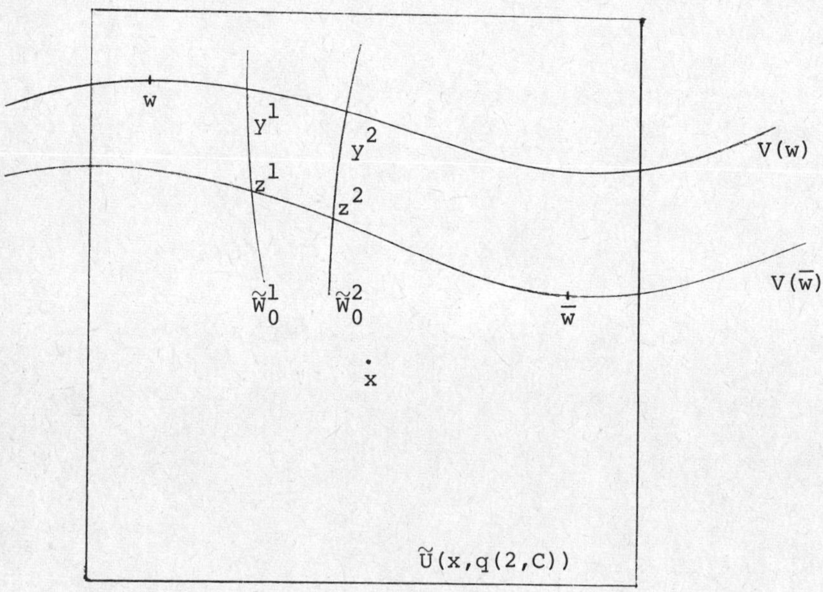

Fig. 7

PROPOSITION 9.1. *For any* $n \geq 0$ *and* k, $0 \leq k \leq n$ *one has*

$$\rho(\Phi^k(z^2),\ \Phi^k(w)) < (Q(\varepsilon,\alpha,\gamma)\alpha^{2bk})^2$$

i.e. $\Phi(z^2) \subset V_k(w)$ (*cf.* $(3.9)_I$).

Proof. Using $(7.1.1)_I$, $(6.11)_n$, condition (Bb) from Sec. 1.1_I, (2.9), (7.14), (6.1) and inequality $0 < C \leq 1$, we have

$$\begin{cases} \rho(\Phi^k(z^2),\Phi^k(w)) \leq \rho(\Phi^k(z^2),\Phi^k(\bar{w})) + \rho(\Phi^k(\bar{w}),\Phi^k(z^1)) + \\ \quad + \rho(\Phi^k(z^1),\Phi^k(y^1)) + \rho(\Phi^k(y^1),\Phi^k(w)) \leq \\ \quad \leq 3 \cdot 200\sqrt{2}\, A^{\ell}_{r,s}(\kappa_{r,s})^k \operatorname{diam} \widetilde{U}(x,q(2,C)) + \\ \quad + \operatorname{diam} \widetilde{W}^1_k(w,y^1,\delta'_{1k}) \leq 2400\sqrt{2}\, q\, A^{\ell}_{r,s}(\kappa_{r,s})^k q(2,C) + \\ \quad + 2\sqrt{2}\, q\delta_0(\xi')^k(1+C) \leq 2400\sqrt{2}\, q\, A^{\ell}_{r,s}(\kappa_{r,s})^k q(2,C) + \\ \quad + qq(1,C)(\xi')^k < (2A^{\ell}_{r,s}+1)qq(1,C)(\xi')^k \leq \\ \quad \leq \delta^{\ell}_{r,s,\alpha,\gamma}(\xi')^k \leq (Q(\ell,\alpha,\gamma)\alpha^{2bk})^2. \end{cases}$$

The last inequality follows from (6.1), $(6.7)_I$ and $(6.5)_I$ because $\xi' < \alpha^{4b}$.

9.2. According to (5.1) and (5.2) we will denote for $i = 1,2$ and $1 \leq k \leq n$

$$\begin{cases} P^i_0 = (\tau'_w \circ \exp^{-1}_w)(y^i) = (u^i_0,v^i_0) \\ P^i_k = (\tau'_{\Phi^k(w)} \circ \exp^{-1}_{\Phi^k(w)})(\Phi^k(y^i)) = \left(\prod_{j=0}^{k-1} \Phi'_j\right)(P^i_0) = \\ \quad = (u^i_k,v^i_k), \end{cases} \tag{9.3}$$

and by Proposition 9.1. we can also define

$$\begin{cases} \overline{P}^i_0 = (\tau'_w \circ \exp_w^{-1})(z^i) = (\overline{u}^i_0, \overline{v}^i_0) \\ \overline{P}^i_k = (\tau'_{\Phi^k(w)} \circ \exp^{-1}_{\Phi^k(w)})(\Phi^k(z^i)) = \left(\prod_{j=0}^{k-1} \Phi'_j\right)(\overline{P}^i_0) \\ \quad = (\overline{u}^i_k, \overline{v}^i_k). \end{cases} \tag{9.4}$$

Let us notice that these points actually depend on n.

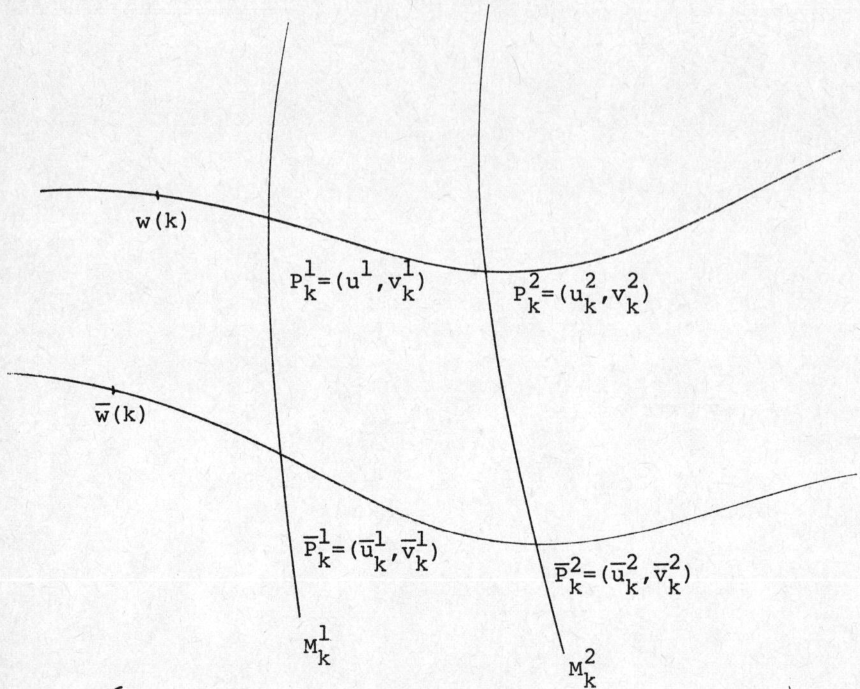

Fig. 8. $\begin{cases} w(k) = (\tau'_{\Phi^k(w)} \circ \exp^{-1}_{\Phi^k(w)})(\Phi^k(w)), \\ \overline{w}(k) = (\tau'_{\Phi^k(w)} \circ \exp^{-1}_{\Phi^k(w)})(\Phi^k(\overline{w})) \quad , \\ M^i_k = (\tau'_{\Phi^k(w)} \circ \exp^{-1}_{\Phi^k(w)})(\widetilde{W}^i_k),\ i = 1,2,\ 0 \le k \le n. \end{cases}$

<u>LEMMA 9.1</u>. <u>For every</u> $\widetilde{C} > 0$ <u>and</u> $t > 0$ <u>there exists</u> $n_8(t) = n_8(\ell,r,s,\alpha,\gamma,t) > 0$ <u>such that for every</u> $n \ge n_8(t)$ <u>and</u> $z^1 \in \Phi^{-n}(\widetilde{W}^1_n) \cap \hat{\Lambda}(x,C)$ <u>if</u> $\|\overline{v}^1_n - v^1_n\|_n \le \widetilde{C}\xi^n$ <u>then</u> $\overline{v}^2_n \in B^2_n(v^2,(\widetilde{C}+t)\xi^n)$ (<u>cf</u>. (2.2) <u>and</u> (2.11)).

Proof. Using (5.4) as well as (2.11) we have

$$\begin{cases} \|\bar{v}_n^2 - v_n^2\|_n \le \|\bar{v}_n^2 - \bar{v}_n^1\| + \|\bar{v}_n^1 - v_n^1\| + \|v_n^1 - v_n^2\| \le \\ \le \tilde{C}\xi^n + \sqrt{2}\|(\bar{u}_n^2,\bar{v}_n^2)\| + \|(\bar{u}_n^1,\bar{v}_n^1)\| + \|(u_n^1,v_n^1)\| + \|(u_n^2,v_n^2)\| \le \\ \le \tilde{C}\xi^n + \sqrt{2}\ 200(\|(\bar{u}_0^2,\bar{v}_0^2)\| + \|(\bar{u}_0^1,\bar{v}_0^1)\| + \|(u_0^1,v_0^1)\| + \\ + \|(u_0^2,v_0^2)\|)(\kappa_{r,s})^n \le \tilde{C}\xi^n + \mathrm{Const}(\kappa_{r,s})^n . \end{cases}$$

Since $\kappa_{r,s} < \xi$ the Lemma is proved. ∎

9.3. We proceed now to the lemmas concerning the comparison of the volumes which will play a crucial role in the proof of Theorem 4.1.

At this stage it is convenient to work in the tangent spaces $T_{\Phi^n(w)}N$. This requires a few extra notations. Namely, let

$$\hat{W}_n^i = \exp^{-1}_{\Phi^n(w)}(\tilde{W}_n^i),\ i = 1,2$$

and for $y \in \Phi^{-n}(\tilde{W}_n^i)$ and $j = 0,1,\cdots\cdots,n$ let

$$F_j^i(y) = T_{\exp^{-1}_{\Phi^j(w)}(\Phi^j(y))}(\hat{W}_j^i) \quad \text{and} \quad y_j = \exp^{-1}_{\Phi^j(w)}(\Phi^j(y)).$$

We will also use the notation from Sec. 2.6.

Beginning from this moment we will assume for convenience that $\delta_{10},\delta_{20} \le C$.

<u>LEMMA 9.2.(VI)</u>. <u>There exists</u> $C_3 = C_3(\ell,r,s,\alpha,\gamma)$ <u>such that for</u> $n \ge 1$ <u>and every</u> $a \in \Phi^{-n}(\tilde{W}_n^1)$, $b \in \Phi^{-n}(\tilde{W}_n^2)$

$$\left| \frac{\left|d\Phi^{-n}_{(n-1)w}(a_n)\Big|_{F_n^1(a)}\right|}{\left|d\Phi^{-n}_{(n-1)w}(b_n)\Big|_{F_n^2(b)}\right|} - 1 \right| \le C_3 C .$$

Proof. <u>Step 1</u>. Clearly, it is enough to prove the Lemma for two particular cases:

(i) two transversal manifolds $\widetilde{W}^1$ and $\widetilde{W}^2$ coincide,

(ii) $a,b \in V(w)$, i.e. in notations of Sec. 7.5. $a = y^1$, $b = y^2$.

For, since $\dfrac{\left|d\Phi^{-n}_{(n-1)w}(a_n)\right| F^1_n(a) \Big|}{\left|d\Phi^{-n}_{(n-1)w}(b_n)\right| F^2_n(b) \Big|} = \dfrac{\left|d\Phi^{-n}_{(n-1)w}(a_n)\right| F^1_n(a) \Big|}{\left|d\Phi^{-n}_{(n-1)w}(y^1_n)\right| F^1_n(y^1) \Big|} \cdot$

$\cdot \dfrac{\left|d\Phi^{-n}_{(n-1)w}(y^1_n) \; F^1_n(y^1)\right|}{\left|d\Phi^{-n}_{(n-1)w}(y^2_n) \; F^2_n(y^2)\right|} \cdot \dfrac{\left|d\Phi^{-n}_{(n-1)w}(y^2_n) \; F^1_n(y^2)\right|}{\left|d\Phi^{-n}_{(n-1)w}(b_n) \; F^2_n(b)\right|}$ those two cases imply

the general one via the following general inequality: for $A,B,C > 0$

$$|ABC - 1| \le |A-1|BC + |B-1|C + |C-1|. \tag{9.5}$$

Step 2. We proceed to the proof in case (i). We have

$$\begin{cases} L_n(a,b) = \dfrac{\left|d\Phi^{-n}_{(n-1)w}(a_n)\right| F^1_n(a) \Big|}{\left|d\Phi^{-n}_{(n-1)w}(b_n)\right| F^1_n(b) \Big|} = \displaystyle\prod_{j=1}^{n} \dfrac{\left|d\Phi^{-1}_{(j-1)w}(a_j)\right| F^1_j(a) \Big|}{\left|d\Phi^{-1}_{(j-1)w}(b_j)\right| F^1_j(b) \Big|} \le \\ \le \displaystyle\prod_{j=1}^{n} \left(1 + \dfrac{\Big| \left|d\Phi^{-1}_{(j-1)w}(a_j)\right| F^1_j(a) \Big| - \left|d\Phi^{-1}_{(j-1)w}(b_j)\right| F^1_j(b) \Big| \Big|}{\left|d\Phi^{-1}_{(j-1)w}(b_j)\right| F^1_j(b) \Big|} \right). \end{cases} \tag{9.6}$$

We will estimate the numerator and the denominator of the last expression. We obtain using Lemma 3.2.

$$\begin{cases} \Big| \left|d\Phi^{-1}_{(j-1)w}(a_j)\right| F^1_j(a) \Big| - \left|d\Phi^{-1}_{(j-1)w}(b_j)\right| F^1_j(b) \Big| \Big| \le \\ \le C_1 \Big(\sup\limits_{y \in \Lambda_j(w)} \| d\Phi^{-1}_{(j-1)w}(y) \| \Big)^p \cdot \\ \cdot \; [\| d\Phi^{-1}_{(j-1)w}(a_j) - d\Phi^{-1}_{(j-1)w}(b_j) \| + \Gamma(F^1_j(a), F^1_j(b))]. \end{cases} \tag{9.7}$$

From Mean Value Theorem and $(3.3.6)_I$ one has

$$\begin{cases} \|d\Phi^{-1}_{(j-1)w}(a_j) - d\Phi^{-1}_{(j-1)w}(b_j)\| \le \\ \le \|a_j - b_j\|_{\Phi^j(w)} \cdot (\sup_{u\in\Lambda_j(w)} \|d^2\Phi^{-1}_{(j-1)w}(u)\|) \le \\ \le \operatorname{diam}(\hat{W}^1_j) \cdot \dfrac{c_8\ell^3}{\gamma^b\alpha^{4b(j-1)}} . \end{cases} \tag{9.8}$$

Step 3. We will now estimate separately $\operatorname{diam}(\hat{W}^1_j)$ and the aperture $\Gamma(F^1_j(a), F^1_j(b))$.

From condition (Bb) from Sec. 2.1_I, from (2.9) and from $(6.11)_n$ we have

$$\begin{cases} \operatorname{diam}(\hat{W}^1_j) \le q\sqrt{2} \cdot \operatorname{diam}'(\hat{W}^1_j) \le \\ \le \sqrt{2}\, q[2\delta'_j + (\sup_{v\in\tilde{B}^2(v_j,\delta'_j)} \|d\Psi^1_j(v)\|) 2\delta'_j] = \\ = \sqrt{2}\, q\, 2\delta_{10}(\xi')^j(1 + C\eta^n) \le 4\sqrt{2}\, q\delta_{10}(\xi')^j. \end{cases} \tag{9.9}$$

Substituting (9.9) into (9.8) one obtains

$$\|d\Phi^{-1}_{(j-1)w}(a_j) - d\Phi^{-1}_{(j-1)w}(b_j)\| \le 4\sqrt{2}\, q\, c_8\ell^3 \frac{\delta_{10}\alpha^{4b}}{\gamma^b}(\frac{\xi'}{\alpha^{4b}})^j. \tag{9.10}$$

In order to estimate the aperture we first estimate it with respect to the Lyapunov norm using Proposition 3.4 and (6.11). We use (3.4) to pass from aperture Γ to aperture Γ' computed with respect to the Lyapunov norm.

$$\begin{cases} \Gamma(F^1_j(a), F^1_j(b)) \le \sqrt{2}\, A^\ell_{r,s}\exp(j\varepsilon_{r,s})\Gamma'(F^1_j(a), F^1_j(b)) \le \\ \le 4\sqrt{2}\, A^\ell_{r,s}\exp(j\varepsilon_{r,s}) \cdot \sup_{v\in\tilde{B}^2(v_j,\delta'_j)} \|d\Psi^1_j(v)\| \le \\ \le 4\sqrt{2}\, CA^\ell_{r,s}\eta^j\exp(j\varepsilon_{r,s}) = \\ = 4\sqrt{2}\, CA^\ell_{r,s}\exp(-6j\varepsilon_{r,s}), \end{cases} \tag{9.11}$$

where Ψ_j^1 is defined in (9.2).

Finally, substituting (9.10) and (9.11) into (9.7), and applying $(3.3.5)_I$ one obtains

$$\begin{cases} \left| \left| d\Phi^{-1}_{(j-1)w}(a_j) \right| F_j^1(a) \right| - \left| d\Phi^{-1}_{(j-1)w}(b_j) \right| F_j^1(b) \right| \Big| \le \\ \le C_1 \frac{c_7 \ell \alpha^b}{\alpha^{bj}} (4\sqrt{2}\ q\ c_8 \ell^3 \frac{\delta_{10} \alpha^{4b}}{\gamma^b} (\frac{\xi'}{\alpha^{4b}})^j + \\ + 4\sqrt{2}\ CA^{\ell}_{r,s} \exp(-6j\varepsilon_{r,s})) \le C_{31}(\delta_{10} + C)\exp(-6j\varepsilon_{r,s}), \end{cases} \tag{9.12}$$

because from $(6.5)_I$, $\xi' < \alpha^{4b}\exp(-6j\varepsilon_{r,s})$.

<u>Step 4</u>. It remains now only to estimate the denominator in (9.6). From estimate $(3.3.2)_I$ one obtain that

$$\begin{cases} \left| d\Phi^{-1}_{(j-1)w}(b_j) \left| F_j^1(b) \right| \right|^{-1} = \left| d\Phi_{(j-1)w}(b_{j-1}) \left| F_{j-1}^1(b) \right| \right| \le \\ \le \| d\Phi_{(j-1)w}(b_{j-1}) \|^p \le \\ \le \left(\frac{c_5 \ell}{\alpha^{b(j-1)}} \right)^p . \end{cases} \tag{9.13}$$

Thus finally, substituting (9.12) and (9.13) into (9.6) we have for $\theta = \dfrac{1}{\alpha^b \exp(6\varepsilon_{r,s})} < 1$ (cf. (2.8))

$$L_n(a,b) \le \prod_{j=1}^{n} (1 + C_{32}(\delta_{10} + C)\theta^{j-1}) \le 1 + C_{33}(\delta_{10} + C), \tag{9.14}$$

where $C_{33} = C_{33}(\ell,r,s,\alpha,\gamma)$. Indeed this follows from the following easy fact. For any θ, $0 < \theta < 1$, there exists a number $K = K(\theta)$ such that for every a, $0 < a < 2C_{32}$, one has $\prod_{i=0}^{\infty}(1+a\theta^i) \le 1 + Ka$.

Since a and b appear symmetrically in all our considerations we also have

$$L_n(b,a) = \frac{1}{L_n(a,b)} \le 1 + C_{33}(\delta_{10} + C).$$ Finally one obtains

and thus $L_n(a,b) \ge 1 - C_{33}(\delta_{10} + C)$.

$|L_n(a,b) - 1| \le C_{33}(\delta_{10} + C) \le 2C_{33}C$, because $\delta_{10} \le C$.

Step 5. Proof of (ii) follows exactly the same line, namely we write the analogue of (9.6) and proceed in the same way as above. The only difference appears in producing the estimate of $\|y_j^1 - y_j^2\|_{\Phi^j(w)}$, i.e. the counterpart of (9.9).

For that purpose we note that $y^1, y^2 \in V(w) \cap \tilde{U}(x, q(2,C))$. This allows us to use (5.3) and Proposition 7.2 along with our usual tool - inequality (2.9). Thus for every $j \ge 1$ we have

$$\left\{\begin{aligned} &\|y_j^1 - y_j^2\|_{\Phi^j(w)} \le \sqrt{2}\|y_j^1 - y_j^2\|'_{\Phi^j(w)} \le \\ &\le \sqrt{2}(\|y_j^1\|'_{\Phi^j(w)} + \|y_j^2\|'_{\Phi^j(w)}) \le 200\sqrt{2}(\kappa_{r,s})^j(\|y_0^1\|'_w + \|y_0^2\|'_w) = \\ &= 200\sqrt{2}(\kappa_{r,s})^j(\|\exp_w^{-1}(y^1) - \exp_w^{-1}(w)\|_w + \\ &+ \|\exp_w^{-1}(y^2) - \exp_w^{-1}(w)\|'_w \le \\ &\le 200\sqrt{2}\, K(\kappa_{r,s})^j(\|\exp_x^{-1}(y^1) - \exp_x^{-1}(w)\|'_x + \\ &+ \|\exp_x^{-1}(y^2) - \exp_x^{-1}(w)\|'_x) \le \\ &\le 800\sqrt{2}\, K(\chi_{r,s})^j q(2,C) \le C_{34}(\ell,r,s,\alpha,\gamma)(\chi_{r,s})^j C. \end{aligned}\right. \tag{9.15}$$

Here we use the inequality

$$q(2,C) \le \frac{\mu\eta - \lambda}{1600\Gamma K}\, C$$

which follows from (6.1) and (7.13). Indeed, (6.1) implies that

$q(1,C) \le \frac{\mu\eta-\lambda}{\Gamma(\frac{1}{C}+\eta)}$. Inequality (9.15) allows us to produce the analogy of (9.14), because $\kappa_{r,s} < \xi'$. ∎

9.4. Let us denote by ν_n^i the measure (p-dimensional volume) induced on $\tilde{W}_n^i$ by the Riemannian structure.

The following statement may be considered as the third main lemma in the proof of Theorem 4.1.

<u>LEMMA 9.3(VII)</u>. <u>There exists</u> $C_4 = C_4(\ell,r,s,\alpha,\gamma)$ <u>such that for any</u> β, $0 < \beta < 1$, <u>there exists</u> $n_9 = n_9(\ell,r,s,\alpha,\gamma,C,\beta)$ <u>such that if</u> $n \ge n_9$, $A^1 \subset \tilde{W}_n^1$, $A^2 \subset \tilde{W}_n^2$, $\nu_n^2(A^2) > 0$ <u>and</u>

$$\left|\frac{\nu_n^1(A^1)}{\nu_n^2(A^2)} - 1\right| < \beta \tag{9.16}$$

<u>then</u>

$$\left|\frac{\nu_0^1(\Phi^{-n}(A^1))}{\nu_0^2(\Phi^{-n}(A^2))} - 1\right| \le C_4(\beta + C).$$

<u>Proof</u>. Let μ_n^i be the measure on $\hat{W}_n^i$ induced by the Riemannian structure in $T_{\Phi^n(w)}N$. Obviously for $Y^i = \exp^{-1}_{\Phi^n(w)}(A^i)$, $i = 1,2$, we can write

$$\frac{\mu_n^1(Y^1)}{\mu_n^2(Y^2)} = \frac{\mu_n^1(Y^1)}{\nu_n^1(A^1)} \cdot \frac{\nu_n^1(A^1)}{\nu_n^2(A^2)} \cdot \frac{\nu_n^2(A^2)}{\mu_n^2(Y^2)}. \tag{9.17}$$

As follows from condition (Bc) of Sec. 2.2. the first and the last multiplier in (9.17) go to 1 as $n \to \infty$, uniformly on $A^1 \subset \tilde{W}_n^1$, $A^2 \subset \tilde{W}_n^2$, with a speed depending on C but independent of W^1 and W^2. Thus if n is big enough, i.e. $n \ge n_{91}(\ell,r,s,\alpha,\gamma,C,\beta)$ by (9.5)

$$\left|\frac{\mu_n^1(Y^1)}{\mu_n^2(Y^2)} - 1\right| \le 2\beta. \tag{9.18}$$

From Proposition 3.1 and Mean Value Theorem we have

$$\begin{cases} \mu_0^i(\Phi_{(n-1)w}^{-n}(Y^i)) = \int_{Y_i} \left| d\Phi_{(n-1)w}^{-n} \Big| T_m \hat{W}_n^i \right| d\mu_n^i(m) = \\ \qquad = \left| d\Phi_{(n-1)w}^{-n} \Big| T_{m_i} \hat{W}_n^i \right| \cdot \mu_n^i(Y^i), \ i = 1,2 \end{cases} \tag{9.19}$$

for some points $m_1 \in Y^1$, $m_2 \in Y^2$.

Therefore, (9.18), (9.19) and Lemma 9.2. imply that

$$\left| \frac{\mu_0^1(\Phi_{(n-1)w}^{-n}(Y^1))}{\mu_0^2(\Phi_{(n-1)w}^{-n}(Y^2))} - 1 \right| \le C_3 C(1+2\beta) + 2\beta. \tag{9.20}$$

Similarly to (9.17) we can write

$$\frac{\nu_0^1(\Phi^{-n}(A^1))}{\nu_0^2(\Phi^{-n}(A^2))} = \frac{\nu_0^1(\Phi^{-n}(A^1))}{\mu_0^1(\Phi_{(n-1)w}^{-n}(Y^1))} \cdot \frac{\mu_0^1(\Phi_{(n-1)w}^{-n}(Y^1))}{\mu_0^2(\Phi_{(n-1)w}^{-n}(Y^2))} \cdot \frac{\mu_0^2(\Phi_{(n-1)w}^{-n}(Y^2))}{\nu_0^2(\Phi^{-n}(A^2))} .$$

As above, taking into account condition (Bc) of Sec. 2.2. and also Proposition 7.4, we see that the first and last multiplier go to 1 as $n \to \infty$, uniformly on $A^1 \subset \tilde{W}_n^1$, $A^2 \subset \tilde{W}_n^2$, with a speed depending on C but independent of W^1 and W^2, so that the Lemma follows from (9.20) and (9.5). ∎

10. THE PROOF OF THE ABSOLUTE CONTINUITY THEOREM

10.1. We will apply now the results of Sec. 8 to the transversal submanifold W^1. Thus we will consider the submanifolds $\tilde{W}_n^1(w_i, y_i^1, \delta_n')$ (cf. Sec. 8.1), the sets $\bar{\bar{W}}_n^1$ (cf. (8.3.1.)), $\approx{W}_n^1$ (cf. (8.4.2)), $Q(z_j, d_n) \subset \tilde{W}_n^1(w_i, y_i^1, \delta_n')$ (cf. Sec. (8.5)) etc., without any additional explanations.

Let $n \ge n_6$ (cf. Lemma 8.4). Let us consider j, $1 \le j \le n_7$. Let i, $1 \le i \le n_5$, be such that $Q(z_j, d_n) \cap \tilde{W}_n^1(w_i, y_i^1, \frac{1}{2}\delta_n') \ne \emptyset$ and $Q(z_j, d_n) \subset \tilde{W}_n^1(w_i, y_i^1, \delta_n')$.

Let us consider the covering of the ball $\hat{B}_n^2(\tilde{z}_j, (2-2\theta)d_n) \subset$

$\subset E_{2\Phi^n(w_i)}$ by the closed p-dimensional cubes $\hat{D}_{mj} \subset E_{2\Phi^n(w_i)}$, $\subset$ $1 \le m \le N_j$, of diameter θd_n (with respect to the norm $\|\cdot\|_{\Phi^n(w)}$) with disjoint interiors.

Let ℓ be the length of an edge of the cube $\hat{D}_{mj}$. Let us denote by $(\hat{D}_{mj})_h$ the concentric cube with edge $\ell + h$.

By Vol we will now denote the p-dimensional Riemannian volume (measure) in $E_{2\Phi^n(w)}$.

Let us remark that if $0 < \omega < \dfrac{\theta d_0}{\sqrt{p}}$, then for every $n \ge 0$ and for $\omega_n = \omega\xi^n$ one has

$$\left| \frac{\mathrm{Vol}((\hat{D}_{mj})_{\omega_n})}{\mathrm{Vol}(\hat{D}_{mj})} - 1 \right| < 2^p \frac{\omega}{\theta d_0} . \tag{10.1}$$

10.2. Let us define

$$\begin{cases} D^1_{mj} = \exp_{\Phi^n(w_i)} \circ (\tau'_{\Phi^n(w_i)})^{-1}\{(\Psi^1_n(v),v);\ v \in \tau'_{\Phi^n(w_i)}(\hat{D}_{mj})\}, \\ \bar{D}^2_{mj} = \exp_{\Phi^n(w_i)} \circ (\tau'_{\Phi^n(w_i)})^{-1}\{(\Psi^2_n(v),v);\ v \in \tau'_{\Phi^n(w_i)}(\hat{D}_{mj})_{\omega_n}\}, \\ D^1_{mj}(e) = \exp^{-1}_{\Phi^n(w_i)}(D^1_{mj}), \\ \bar{D}^2_{mj}(e) = \exp^{-1}_{\Phi^n(w_i)}(D^2_{mj}), \end{cases} \tag{10.2}$$

where Ψ^r_{in} is the mapping which defines the submanifold $\widetilde{W}^r_n(w_i,y^r_i,\delta'_n)$ (cf. (7.20) and Sec. 9.1), $r = 1,2$. To be sure that Ψ^r_{in} is well defined on $v \in \tau'_{\Phi^n(w_i)}((\hat{D}_{mj})_{\omega_n})$, we will suppose beginning from now that $0 < \omega < \frac{1}{2} q_C(w,W^1,W^2)$ (cf. (9.1)). This implies that $\bar{D}^2_{mj} \subset \widetilde{W}^2_n(w_i,y^2_i, \frac{1}{2} q_C(w,W^1,W^2)(\xi')^n)$. Let us note also that the intersection of the interior of any two sets $D^1_{m_1j}$ and $D^1_{m_2j}$ is empty, when $m_1 \ne m_2$.

<u>PROPOSITION 10.1</u>. <u>Let</u> $0 < \theta < \frac{1}{6}$. <u>For all</u> $n \geq n_4(\theta)$ (<u>cf</u>. <u>Lemma</u> 8.2) <u>and all</u> j, $1 \leq j \leq n_7$, <u>one has</u>

$$Q(z_j,d_n) \subset \bigcup_{m=1}^{N_j} D^1_{mj} \subset Q(z_j,2d_n). \tag{10.3}$$

<u>Proof</u>. By application of $\exp^{-1}_{\Phi^n(w_i)}$ to (10.3) one immediately sees that (10.3) is equivalent to

$$t^n_{w_i}(\hat{Q}(z_j,d_n)) \subset \bigcup_{m=1}^{N_j} \hat{D}^1_{mj} \subset t^n_{w_i}(\hat{Q}(z_j,2d_n)), \tag{10.4}$$

where, let us recall, $\hat{Q}(z_j,r) = \exp^{-1}_{\Phi^n(w_i)} Q(z_j,r)$.

From Lemma 8.1 applied to $F = E_{1\Phi^n(w)}$ we have that $t^n_{w_i}(\hat{Q}(z_j,d_n)) \subset \hat{B}^2_n(\hat{z}_j,(1+\theta)d_n)$ for all $n \geq n_4 = n_4(\theta) \geq n_2(\theta)$. Thus to prove the first inclusion of (10.4) it suffices to prove that for all $n \geq n_4$ one has $\hat{B}^2_n(\hat{z}_j,(1+\theta)d_n) \subset \bigcup_{m=1}^{N_j} \hat{D}^1_{mj}$.

By the definition of $\{\hat{D}^1_{pj}\}$ one has $\hat{B}^2_n(z_j,(2-2\theta)d_n) \subset \bigcup_{m=1}^{N_j} D^1_{mj}$. But as $0 < \theta < \frac{1}{6}$, $\hat{B}^2(\hat{z}_j,(1+\theta)d_n) \subset \hat{B}^2(\hat{z}_j,(2-2\theta)d_n)$ and thus the first inclusion of (10.4) is proved.

Now let us proceed to the proof of the second inclusion of (10.4). The diameter of $\hat{D}^1_{mj}$ is equal to θd_n and as $\hat{D}^1_{mj} \cap \hat{B}^2_n(\hat{z}_j,(2-2\theta)d_n) \neq \emptyset$ one deduces that $\bigcup_{m=1}^{N_j} \hat{D}^1_{mj} \subset \hat{B}^2_n(\hat{z}_j,(2-\theta)d_n)$. Thus, the second inclusion follows from the first inclusion in Lemma 8.1 applied to $F = E_{1\Phi^n(w)}$, $r = 2d_n$ and $\varepsilon = \theta$. ∎

By Proposition 10.1 and Lemma 8.4 we obtain that

$$\tilde{\tilde{W}}^1_n \subset \bigcup_{j=1}^{n_7} \bigcup_{m=1}^{N_j} D^1_{mj} \subset \bar{\bar{W}}^1_n. \tag{10.5}$$

Moreover, from (8.4.3) it follows that there exists some number $L > 0$ such that for every $n \geq 1$, the covering of $\tilde{\tilde{W}}^1_n$ by the sets

$\{D^1_{mj}\}_{\substack{1\le j\le n_7\\ 1\le m\le N_j}}$ is of multiplicity at most L. We will denote this covering by $\mathfrak{A}$.

10.2. We will describe now how one can choose a subcover of $\mathfrak{A}$ which has multiplicity one, except a set of a very small measure. To obtain it we proceed consecutively from the ball $Q(z_j,2d_n)$ to the ball $Q(z_{j+1},2d_n)$, $j = 1,2,\cdots\cdots,n_7 - 1$ and in $(j+1)$th step we eliminate all sets D^1_{mj+1} such that $D^1_{mj+1} \subset \bigcup_{k=1}^{j} \bigcup_{m=1}^{N_k} D_{mk}$. Let $\{D^1_i\}_{1\le i\le N}$ be the covering of $\widetilde{\widetilde{W}}{}^1_n$ formed by all remaining elements of $\mathfrak{A}$. Clearly, the multiplicity of covering $\{D^1_i\}_{1\le i\le N}$ is not bigger than L and

$$\bigcup_{i=1}^{N} D^1_i = \bigcup_{j=1}^{n_7} \bigcup_{m=1}^{N_j} D^1_{mj}, \tag{10.6}$$

$$\widetilde{\widetilde{W}}{}^1_n \subset \bigcup_{i=1}^{N} D^1_i \subset \overline{\overline{W}}{}^1_n. \tag{10.7}$$

For $D^1_i = D_{mj}$ we will note $\hat{D}_i = \hat{D}_{mj}$, $\overline{D}^2_i = \overline{D}^2_{mj}$ etc. (see (10.2)).

The following Lemma is the last of our eight basic lemmas. Unlike the previous lemmas it is essentially global.

<u>LEMMA 10.1 (VIII)</u>. <u>There exists</u> $C_5 = C_5(\ell,k,r,s,\alpha,\gamma)$ <u>such that for every</u> θ, $0 < \theta < \min(\frac{1}{6}, \frac{1}{C_2})$ (<u>cf</u>. <u>Lemma</u> 8.2), <u>and for every</u> $n \ge n_4(\theta)$

$$\left| \frac{\sum_{i=1}^{N} \nu^1_0(\Phi^{-n}(D^1_i))}{\nu^1_0(\Phi^{-n}(\bigcup_{i=1}^{N} D^1_i))} - 1 \right| \le C_5(\theta + C), \tag{10.8}$$

<u>where</u> C <u>comes from Sec</u>. 7.4.

<u>Proof</u>. Let us divide the set of $\{1,\ldots,N\}$ into two disjoint subsets G and B (G=good, B=bad)

$$\begin{cases} i \in G \quad \text{iff} \quad D_i^1 \in Q(z_j, 2(1-\theta)d_n) \quad \text{for some} \quad j \\ i \in B \quad \text{otherwise.} \end{cases}$$

It is clear that if $i \in G$, then $\mathrm{Int}(D_i^1 \cap D_j^1) = \emptyset$ for every $i \neq j$.

Moreover, it follows from Proposition 8.2 that for $n \geq n_4(\theta) \geq n_3$ (where $n_4(\theta)$ arise from Lemma 8.2) and for $1 \leq i \leq N$ one has $\mathrm{diam}\, D_i^1 \leq 2\theta d_n$. Thus, from Proposition 10.1 one deduces that for such n

$$\bigcup_{i \in B} D_i^1 \subset \bigcup_{j=1}^{n_7} \{z \in Q(z_j, 2d_n); \tilde{\rho}(z, \partial Q(z_j, 2d_n)) \leq 4\theta d_n\}, \tag{10.9}$$

where we note by $\tilde{\rho}$ the metric induced on $\Phi^n(\tilde{W}^1)$ by the Riemannian metric ρ on N.

We will denote $A(z_j, 2\theta)$ defined by (8.14) by $A_j(\theta)$, so that

$$A_j(\theta) = \{z \in Q(z_j, 2d_n); \tilde{\rho}(z, \partial Q(z_j, 2d_n) \leq 4\theta d_n\}, \quad 1 \leq j \leq n_7.$$

Since the multiplicity of covering $\{D_i\}_{1 \leq i \leq N}$ does not exceed L, one has

$$\begin{cases} \sum_{i=1}^{N} \nu_0^1(\Phi^{-n}(D_i^1)) = \sum_{i \in G} \nu_0^1(\Phi^{-n}(D_i^1)) + \sum_{i \in B} \nu_0^1(\Phi^{-n}(D_i^1)) \leq \\ \qquad \leq \sum_{i \in G} \nu_0^1(\Phi^{-n}(D_i^1)) + L\nu_0^1(\bigcup_{i \in B} \Phi^{-n}(D_i^1)) = \\ \qquad = \nu_0^1(\Phi^{-n}(\bigcup_{i \in G} D_i^1)) + L\nu_0^1(\Phi^{-n}(\bigcup_{i \in B} D_i^1)) \leq \\ \qquad \leq \nu_0^1(\Phi^{-n}(\bigcup_{i=1}^{N} D_i^1)) + L\nu_0^1(\Phi^{-n}(\bigcup_{i \in B} D_i^1)). \end{cases}$$

Thus consequently

$$\frac{\sum_{i=1}^{N} \nu_0^1(\Phi^{-n}(D_i^1))}{\nu_0^1(\Phi^{-n}(\bigcup_{i=1}^{N} D_i^1))} \le 1 + L \frac{\nu_0^1(\Phi^{-n}(\bigcup_{i\in B} D_i^1))}{\nu_0^1(\Phi^{-n}(\bigcup_{i=1}^{N} D_i^1))} .$$

Clearly, (10.8) is equivalent to

$$\frac{\nu_0^1(\Phi^{-n}(\bigcup_{i\in B} D_i^1))}{\nu_0^1(\Phi^{-n}(\bigcup_{i=1}^{N} D_i^1))} \le C_5(\theta+C). \tag{10.10}$$

To prove (10.10) we will use the following evident remark.

$$\begin{cases} \text{If } a_1,\dots,a_N > 0 \text{ and } b_1,\dots,b_N > 0 \text{ and if } \dfrac{a_i}{b_i} \le h \\ \text{for all } i,\ 1 \le i \le N, \text{ then } \dfrac{a_1+\dots+a_N}{b_1+\dots+b_N} \le h. \end{cases} \tag{10.11}$$

Let us pass now to the proof of (10.10). In virtue of (10.3), (10.9) and of the fact that the multiplicity of the covering $\{D_i\}_{1\le i\le N}$ is not bigger than L, one has

$$\begin{cases} \dfrac{\nu_0^1(\Phi^{-n}(\bigcup_{i\in B} D_i^1))}{\nu_0^1(\Phi^{-n}(\bigcup_{i=1}^{N} D_i^1))} \le \dfrac{\nu_0^1(\Phi^{-n}(\bigcup_{j=1}^{n_7} A_j(\theta)))}{\nu_0^1(\Phi^{-n}(\bigcup_{j=1}^{n_7} Q(z_j,d_n)))} \le L \dfrac{\nu_0^1(\Phi^{-n}(\bigcup_{j=1}^{n_7} A_j(\theta)))}{\sum_{j=1}^{n_7} \nu_0^1(\Phi^{-n}(Q(z_j,d_n)))} \le \\ \le L \dfrac{\sum_{i=1}^{n_7} \nu_0^1(\Phi^{-n}(A_j(\theta)))}{\sum_{j=1}^{n_7} \nu_0^1(\Phi^{-n}(Q(z_j,d_n)))} . \end{cases}$$

Thus in virtue of (10.11), to prove (10.10) it is sufficient to prove that for some number $C_5 = C_5(\ell,r,s,\alpha,\gamma)$ and for all j, $1 \le j \le n_7$ one has for all $n \ge n_9$

$$\frac{\nu_0^1(\Phi^{-n}(A_j(\theta)))}{\nu_0^1(\Phi^{-n}(Q(z_j,d_n)))} \le C_5(\theta + C). \tag{10.12}$$

Let us fix j, $1 \le j \le n_7$, and denote $A^1 = A_j(\theta) \cup Q(z_j,d_n)$, $A^2 = Q(z_j,d_n)$. From Lemma 8.2 one obtains that for all $n \ge n_4(\theta)$ one has

$$1 \le \frac{\nu_n^1(A^1)}{\nu_n^1(A^2)} = 1 + \frac{\nu_n^1(A_j(\theta))}{\nu_n^1(Q(z_j,d_n))} \le 1 + C_2\theta.$$

In other words, for all $n \ge n_4(\theta)$

$$\left| \frac{\nu_n^1(A^1)}{\nu_n^1(A^2)} - 1 \right| \le C_2\theta.$$

Thus, by application of Lemma 9.3 to $\beta = C_2\theta < 1$ one deduces that for all $n \ge n_4(\theta) \ge \max(n_2(\theta), n_3, n_4(\theta), n_9(\theta))$, one has

$$\frac{\nu_0^1(\Phi^{-n}(A_j(\theta))}{\nu_0^1(\Phi^{-n}(Q(z_j,d_n))} \le C_4(C_2\theta + C).$$

This proves (10.12) and consequently (10.8). ∎

10.3. Now, it follows from Lemma 9.1 applied to $t = \omega$ (cf. Sec. 10.1) that for all i, $1 \le i \le N$, (p denote the Poincaré map)

$$p(D_i^1 \cap \hat{\Lambda}(x,C)) \subset \bar{D}_i^2 \tag{10.13}$$

if only $n \ge n_{10} = n_{10}(\omega,\theta) \overset{\text{def}}{=\!=} \max(n_4(\theta), n_8(\omega))$. Consequently, by Lemmas 8.3, 8.4 and (10.7) one obtains that for $n \ge n_{10}$ and for $D = D(P,h) \subset \tilde{W}^1$ (cf. Sec. 8.5) one has

$$\Phi^n(p(D)) = p(\Phi^n(D)) \subset p\left(\bigcup_{i=1}^N D_i^1\right) = \bigcup_{i=1}^N p(D_i^1) \subset \bigcup_{i=1}^N \bar{D}_i^2. \tag{10.14}$$

Beginning from now, C, θ, δ_{10} and δ_{20} will be assumed so small that all our preceding assumptions about them, in particular those enumerated in consecutive lemmas, are satisfied. In addition we will suppose that

$$0 < \omega < \frac{\theta d_0}{\sqrt{p}}, \tag{10.15}$$

PROPOSITION 10.2. *There exists a constant* $C_6 = C_6(p)$ *and* $n_{11} = n_{11}(\theta)$ *such that for any* $\theta, \omega, n \geq n_{11}$ *and* i, $1 \leq i \leq N$, *one has*

$$\left| \frac{\nu_n^2(\bar{D}_i^2)}{\nu_n^1(D_i^1)} - 1 \right| \leq C_6\left(\theta + \frac{\omega}{\theta d_0}\right)\left(1 + \frac{\omega}{\theta d_0}\right). \tag{10.16}$$

Proof. Using the notations introduced in (10.2) one can write

$$\left\{ \begin{aligned} \frac{\nu_n^2(\bar{D}_i^2)}{\nu_n^1(D_i^1)} &= \frac{\nu_n^2(\bar{D}_i^2)}{\mu_n^2(\bar{D}_i^2(e))} \cdot \frac{\mu_n^2(\bar{D}_i^2(e))}{\mathrm{Vol}((\hat{D}_i)_{\omega_n})} \cdot \frac{\mathrm{Vol}((\hat{D}_i)_{\omega_n})}{\mathrm{Vol}(\hat{D}_i)} \cdot \\ &\quad \cdot \frac{\mathrm{Vol}(\hat{D}_i)}{\mu_n^1(D_i^1(e))} \cdot \frac{\mu_n^1(D_i^1(e))}{\nu_n^1(D_i^1)} . \end{aligned} \right. \tag{10.17}$$

We will now estimate separately each of the multiples in this product. From condition (Bc) from Sec. 2.2 one obtains that there exists $n_{111} = n_{111}(\theta)$ such that for $n \geq n_{111}$ one has

$$1 - \theta \leq \frac{\nu_n^2(\bar{D}_i^2)}{\mu_n^2(\bar{D}_i^2(e))} \leq 1 + \theta. \tag{10.18}$$

Let us estimate the second multiple. From Proposition 8.2 and from (3.2) it follows immediately that there exists $n_{112} = n_{112}(\theta)$ such that for $n \geq n_{112}$ one has

$$1 \leq \frac{\mu_n^2(\bar{D}_i^2(e))}{\mathrm{Vol}((\hat{D}_i)_{\omega_n})} \leq 1 + 2^p\theta. \tag{10.19}$$

By (10.15), the third multiple is estimated in (10.1). The fourth and the fifth multiples are estimated in the same way as the second and first one respectively. Indeed,

$$1 - 2^{p}\theta \le \frac{\mathrm{Vol}(D_i)}{\mu_n^1(D_i^1(e))} \le 1, \tag{10.20}$$

$$1 - \theta \le \frac{\mu_n^1(D_i^1(e))}{\nu_n^1(D_i^1)} \le 1 + \theta. \tag{10.21}$$

Finally, (10.18) - (10.21) together with the counterpart of (9.5) for five multiple imply that there exists $n_{11}(\theta)$ such that for $n \ge n_{11}(\theta)$, (10.16) is true, with a constant C_6 depending only on p. ∎

From now on we will suppose that C, θ, d_0 and ω satisfy also the following condition

$$C_6(\theta + \frac{\omega}{\theta d_0})(1 + \frac{\omega}{\theta d_0}) < 1 \tag{10.22}$$

so that the right hand side of (10.16) is smaller than 1. Thus, one can apply to $A^1 = \bar{D}_i^2$ and $A^2 = D_i^1$ the Lemma 9.3 and one obtains that for $n \ge n_{12} = (\max n_9, n_{11})$

$$\left| \frac{\nu_0^2(\Phi^{-n}(\bar{D}_i^2))}{\nu_0^1(\Phi^{-n}(D_i^1))} - 1 \right| \le C_7 \ [C + (\theta + \frac{\omega}{\theta d_0})(1 + \frac{\omega}{\theta d_0})], \tag{10.24}$$

where $C_7 = C_7(\ell,k,r,s,\alpha,\gamma)$.

10.4. Now, we can pass to the proof of (4.1.2). In the spirit of the notations of Sec. 4.1 we will note now $\nu_i = \nu_{W^i}$, $i = 1,2$.

Let us remark that as the Poincaré map

$$p : \tilde{W} \cap \hat{\Lambda}^{\ell}_{k,r,s,\alpha,}(x,C) \to \tilde{W}^2 \cap \hat{\Lambda}^{\ell}_{k,r,s,\alpha,\gamma}(x,C)$$

is a homeomorphism, then $p(D)$, where $D = D(p,h)$ (cf. Sec. 8.5) is a measurable subset of $\tilde{W}^2$. Thus, in virtue of (10.14) one obtains that for all $n \geq n_{10}$

$$\nu_2(p(D)) \leq \nu_2\left(\Phi^{-n}\left(\bigcup_{i=1}^{N} \bar{D}_i^2\right)\right) \leq \sum_{i=1}^{N} \nu_2(\Phi^{-n}(\bar{D}_i^2)).$$

Consequently, from (10.24) and Lemma 10.1 it follows that for $n \geq n_{13} = \max(n_4, n_{10}, n_{12})$ one has

$$\left\{\begin{aligned} &\nu_2(p(D)) \leq \\ &\leq \{1 + C_7[C + (\theta + \frac{\omega}{\theta d_0})(1 + \frac{\omega}{\theta d_0})]\} \sum_{i=1}^{N} \nu_1(\Phi^{-n}(D_i^1)) \leq \\ &\leq \{1 + C_7[C + (\theta + \frac{\omega}{\theta d_0})(1 + \frac{\omega}{\theta d_0})]\} \cdot \\ &\cdot \{1 + C_5(\theta + C)\} \cdot \nu_1(\Phi^{-n}(\bigcup_{i=1}^{N} D_i^1)). \end{aligned}\right. \tag{10.25}$$

Let us take $\omega = (\theta d_0)^2$. When n increases indefinitely, $\varepsilon > 0$, $d_0 > 0$ and θ may be chosen as small as we like. Since $0 < C \leq \frac{1}{2}$, one deduces from (10.25) that

$$\left\{\begin{aligned} &\nu_2(p(D)) \leq (1 + C_7C)(1 + C_5C)\nu_1\left(\Phi^{-n}\left(\bigcup_{i=1}^{N} D_i^1\right)\right) \leq \\ &\qquad\qquad \leq (1 + C_8C)\nu_1\left(\Phi^{-n}\left(\bigcup_{i=1}^{N} D_i^1\right)\right), \end{aligned}\right. \tag{10.26}$$

where $C_8 = C_8(\ell, r, s, \alpha, \gamma) = C_5 + C_7 + C_5C_7$.

From Lemmas 8.3 and 8.4 one has that for every $n \geq 1$

$$\bigcup_{i=1}^{N} D_i \subset \bigcup_{i=1}^{n_5} \tilde{W}_n^1(w_i, y_j^1, \delta_n') \subset \Phi^n(Q(P, h + \omega)).$$

Consequently, from (10.26) one obtains that

$$\nu_2(p(D)) \leq (1 + C_8C)\nu_1(Q(P, h + \omega)).$$

As $\omega > 0$ may be chosen as small as we like, this implies that

$$\nu_2(p(D)) \le (1 + C_8 C)\nu_1(Q(P,h)) \tag{10.27}$$

i.e. that

$$\nu_2(p(\hat{\Lambda}(x,C) \cap Q(P,h)) \le (1 + C_8 C)\nu_1(Q(P,h)).$$

But as $Q(P,h)$ is an arbitrary sufficiently small ball in $\tilde{W}^1$, the absolute continuity of the mapping p is proved. Thus, (4.1.2) is completely proved, if one sets

$$q^{\ell}_{r,s,\alpha,\gamma} = q(2,C) \quad \text{and} \quad \varepsilon^{\ell}_{r,s,\alpha,\gamma} = \varepsilon(C),$$

where $q(2,C)$ and $\varepsilon(C)$ are arbitrary pairs of numbers satisfying all our preceding conditions, i.e. (7.12) - (7.14). ∎

Assertion (4.1.1) follows now from (6.1.1), from the definition of the sets $V^{\ell}_{r,s,\alpha,\gamma}$ and from the proof of (4.1.2). Indeed, let $w \in R^{\ell}_{k,r,s,\alpha,\gamma}$ and let $V(w) \in S^{\ell}_{k,r,s,\alpha,\gamma}(x)$. Let $w^i = W^i \cap V(w)$, $i = 1,2$. From (6.1.1) and from the definition of the sets $V^{\ell}_{k,r,s,\alpha,\gamma}$ one deduces that there exists $n^1 > 0$ such that

$\Phi^{n^1}(w) \in \Lambda^{\ell}_{k,r,s,\alpha,\gamma}$ and that the sufficiently small pieces $\bar{W}^i$ of $\Phi^{n^1}(W^i)$ containing the points $\Phi^{n^1}(w_i)$, $i = 1,2,$ satisfies all the assumptions of (4.1.2) with respect to the point $x^1 = \Phi^{n^1}(w)$. Now it remains only to prove the absolute continuity of the mapping $p_{\bar{W}^1,\bar{W}^2}$, where this Poincaré map is defined with respect to the family $\Phi^{n^1}(R^{\ell}_{k,r,s,\alpha,\gamma})$ of TLSM. With few unessential modifications, the proof is exactly the same as the proof of (4.1.2). ∎

Let us underline that the above proof requires the use of the iterations Φ^n with an arbitrary large n.

10.5. Up to now in the proof of absolute continuity we never needed the assumption that $\nu(\Lambda^{\ell}_{k,r,s,\alpha,\gamma}) > 0$. This assumption becomes essential in the proof of (4.1.3). Incidentally the invariance of μ never directly appear in the proof of Theorem 4.1.

10.6. We pass now to the proof of (4.1.3) i.e. to the estimation of $|J(p)(y) - 1|$.

To this end we will consider only C so small that in addition to all our preceding conditions

$$0 < C < \frac{1}{2C_8} . \tag{10.28}$$

Let $x \in \Lambda^{\ell}_{k,r,s,\alpha,\gamma}$ be a density point of the set $\Lambda^{\ell}_{k,r,s,\alpha,\gamma}$.

For any δ, $0 < \delta < q(2,C)$ let us define the submanifold W_δ, transversal to the family $S^{\ell}_{k,r,s,\alpha,\gamma}(x)$, by the formula

$$W_\delta = (\exp_x \circ (\tau'_x)^{-1})\{(\delta,v);\ v \in \widetilde{B}^2(q(2,0))\}.$$

As x is a density point of the set $\Lambda^{\ell}_{k,r,s,\alpha,\gamma}$, then by the Fubini Theorem we can find δ, $0 < \delta < \varepsilon^{\ell}_{r,s,\alpha,\gamma}$ such that the p-dimensional Riemannian volume $\nu_{W_\delta}(W_\delta \cap \Lambda(x,C))$ is positive and consequently $\nu_{W_\delta}(W_\delta \cap \hat{\Lambda}(x,C)) > 0$ (cf. (8.24)).

Let us consider the mappings $p_{\widetilde{W}^1,\widetilde{W}_\delta}$ and $p_{\widetilde{W}_\delta,\widetilde{W}^2}$. Clearly

$p = p_{\widetilde{W}^1,\widetilde{W}^2} = p_{\widetilde{W}_\delta,\widetilde{W}_2} \circ p_{\widetilde{W}^1,\widetilde{W}_\delta}$.

As $\nu_{W_\delta}(W_\delta \cap \hat{\Lambda}(x,C)) > 0$ and as mappings $p_{\widetilde{W}_\delta,\widetilde{W}^i} = (p_{\widetilde{W}^i,\widetilde{W}_\delta})^{-1}$ and $p_{\widetilde{W}^i,\widetilde{W}_\delta}$ are absolutely continuous in virtue of (4.1.2), then one obtains that $\nu_i(\widetilde{W}^i \cap \hat{\Lambda}(x,C)) > 0$, $i = 1,2$.

Let us note by T the set of all points $w \in \widetilde{W}^1 \cap \hat{\Lambda}(x,C)$ such that w is a point of density of the set $\widetilde{W}^1 \cap \hat{\Lambda}(x,C)$ with respect to the measure ν_1 and that the point $p(w)$ is a density point of the set $\widetilde{W}^2 \cap \hat{\Lambda}(x,C)$ with respect to the measure ν_2. As ν_1-almost all points of $\widetilde{W}^1 \cap \hat{\Lambda}(x,C)$ are of density and as p^{-1} is absolutely continuous, then ν_2- almost all points of $\widetilde{W}^2 \cap \hat{\Lambda}(x,C)$ belong to $p(T)$.

Let us consider $y \in T$. It follows from the definition of the points of density that for every $\varepsilon > 0$ there exists $h(\varepsilon) > 0$ such that for every h, $0 < h < h(\varepsilon)$, one has

$$\nu_1(Q(y,h)) \leq (1+\varepsilon)\nu_1(T \cap Q(y,h)). \qquad (10.29)$$

Now, (10.27) and (10.29) imply that if $0 < h < h(\varepsilon)$, one has

$$\nu_2(p(T \cap Q(y,h)) \leq (1+\varepsilon)(1 + C_8 C)\nu_1(T \cap Q(y,h))$$

i.e. that

$$\frac{\nu_2(p(T \cap Q(y,h))}{\nu_1(T \cap Q(y,h))} \leq (1+\varepsilon)(1 + C_8 C). \qquad (10.30)$$

As $\varepsilon > 0$ can be chosen as small as we like, (10.30) implies by Lebesgue - Vitali Theorem (cf. Sec. 3.1) that

$$J(p)(y) \leq 1 + C_8 C. \qquad (10.31)$$

As $y \in T$, then $p(y)$ is a density point of the set $\tilde{W}^2 \cap \hat{\Lambda}(x,C)$. Now, in our considerations p and p^{-1} play completely symmetrical roles and hence

$$J(p^{-1})(p(y)) \leq 1 + C_8 C. \qquad (10.32)$$

But $J(p)(y) = \dfrac{1}{J(p^{-1})(p(y))}$ (cf. Sec. 3.1) and $C_8 C < \frac{1}{2}$ (cf. (10.28)), then from (10.31) and (10.32) one deduces that

$$|J(p)(y) - 1| \leq 2C_8 C.$$

Let us fix C_0: $0 < C_0 \leq \frac{1}{2}$ which is so small that one can find $q(2,C_0)$ and $\varepsilon(C_0)$ such that all our proceding conditions, i.e. (7.12) - (7.14) and (10.28), are satisfied. Then it is sufficient to take

$$q^{\ell}_{r,s,\alpha,\gamma}(B) = \begin{cases} q(2, \frac{B}{2C_8}) & \text{for } 0 < B \leq 2C_8 C_0 \\ q(2,C_0) & \text{for } B > 2C_8 C_0 \end{cases}$$

$$\varepsilon^{\ell}_{r,s,\alpha,\gamma}(B) = \begin{cases} \varepsilon(\frac{B}{2C_8}) & \text{for } 0 < B \le 2C_8C_0 \\ \varepsilon(C_0) & \text{for } B > 2C_8C_0. \end{cases}$$

This finishes the proof of Theorem 4.1. ∎

11. ABSOLUTE CONTINUITY OF CONDITIONAL MEASURES

11.1. The aim of the present section is to prove an important theorem (Theorem 11.1) which is due in fact to Ja.G. Sinai (cf. Lecture 5 of $[\text{Ano}]_2$) and which is an easy consequence of the Absolute Continuity Theorem and of Fubini Theorem. Roughly speaking, this theorem asserts that the conditional measures induced on local stable manifolds by a smooth measure on N, are absolutely continuous on them.

11.2. Let $(X,\mathfrak{M},\sigma)$ be a Lebesgue space, where we note as usually by $\mathfrak{M}$ the sigma-field of subsets of X on which the probability measure σ is defined.

A partition β of X is called measurable if the factor-space X/β is a Lebesgue space. Then the measure σ can be decomposed into a family of conditional measures σ_c, $c \in \beta$, so that

$$\sigma = \int_{X/\beta} \sigma_c.$$

These measures are unique up to a set of σ- measure zero. For detailed discussion on Lebesgue spaces, measurable partition and conditional measures see $[\text{Roc}]_{1-3}$.

For $x \in X$ let $C_\beta(x)$ denote the element of β containing x.

The following proposition is a straightforward corollary of the definition of conditional measures.

PROPOSITION 11.1. *Let $(X,\mathfrak{M},\sigma)$ be a Lebesgue space and let β be a measurable partition of X. Let $\bar{\sigma}$ be another probability measure on $\mathfrak{M}$, absolutely continuous with respect to σ so that $\frac{d\bar{\sigma}}{d\sigma} = f$.*

Then for $\bar{\sigma}$ almost all $x \in X$, the measure $\bar{\sigma}_{C_\beta(x)}$ is absolutely continuous with respect to $\sigma_{C_\beta(x)}$ and

$$\frac{d\overline{\sigma}_{C_\beta(x)}}{d\sigma_{C_\beta(x)}} = \frac{f\Big|_{C_\beta(x)}}{\int_{C_\beta(x)} f d\sigma_{C_\beta(x)}} .$$

11.3. Beginning from now we will suppose once and for all that the assumptions of (4.1.3) are fulfilled. As before, ν denotes the measure induced on N by the Riemannian metric ρ. For the sake of brevity we will note $q^{\ell}_{r,s,\alpha,\gamma}(B) = \hat{q}(B)$ and $\varepsilon^{\ell}_{r,s,\alpha,\gamma}(B) = \varepsilon(B)$.

If $X \subset N$, $0 < \nu(X) < +\infty$, then by ν^X we will denote the normalized measure $\frac{\nu}{\nu(X)}$ on X.

If β is a measurable partition of X formed by the smooth, compact submanifolds, then by ν^{β}_{x}, $x \in X$, we will denote the normalized measure on $C_\beta(x)$ induced by the Riemannian metric ρ.

Let us consider the set $\tilde{B}(\hat{q}(B)) = \tilde{B}^1(\hat{q}(B)) \times \tilde{B}^2(\hat{q}(B)) \subset \mathbb{R}^m$ and its measurable partition $\tilde{\eta} = \{y \times \tilde{B}^2(\hat{q}(B))\}_{y \in \tilde{B}^1(\hat{q}(B))}$.

Let us note by μ the normalized Lebesgue measure on $\tilde{B}(\hat{q}(B))$. Fubini Theorem implies that the conditional measure $\mu_{C_{\tilde{\eta}}(z)}$, $z \in \tilde{B}(\hat{q}(B))$, coincides with the normalized p-dimensional Lebesgue measure on $C_{\tilde{\eta}}(z)$.

Now, let us consider the measurable partition $\eta = (\exp_x \circ (\tau'_x)^{-1})\tilde{\eta}$ of $\tilde{U}(x,\hat{q}(B)) \overset{\text{def}}{=} \tilde{U}$. From Proposition 11.1 it follows directly that the conditional measure $\nu^{\tilde{U}}_{C_\eta(y)}$ is equivalent to the measure ν^{η}_{y} for ν almost all $y \in \tilde{U}$. Moreover, from condition (Bc) from Sec 1.1_I one deduces that there exists a positive function ϕ such that

$$\begin{cases} \phi \text{ depends only on } \ell, r, s, \alpha, \gamma, \\ \phi \text{ is defined for } 0 < t \le 1, \ \lim_{t \to 0} \phi(t) = 0 \end{cases} \tag{11.1}$$

and that

$$\left| \frac{d\nu^{\tilde{U}}_{C_\eta(y)}}{d\nu^{\eta}_{y}} - 1 \right| < \phi(B). \tag{11.2}$$

11.3. Let us recall that $\hat{\Lambda}(x,C)$ is defined by (8.24).

The following Proposition is a direct consequence of Corollary 7.1_I.

PROPOSITION 11.2. If $y \in \hat{\Lambda}(x,\frac{\hat{q}(B)}{2})$, then

$$(\tau'_x \circ \exp_x^{-1})(V(y) \cap \tilde{U}(x,\hat{q}(B)) = \{(u,\phi_y(u));\ u \in \tilde{B}^1(\hat{q}(B))\},$$

where $\phi_y\colon \tilde{B}^1(\hat{q}(B)) \to \tilde{B}^2(\hat{q}(B))$ is a C^1 mapping.

11.4. Let

$$A \subset C_\eta(x) \cap \hat{\Lambda}(x\,,\frac{\hat{q}(B)}{2}) \tag{11.3}$$

be a measurable subset of $C_\eta(x)$ of positive $\nu_{C_\eta(x)}$- measure or equivalently of positive ν_x^η- measure.

Let us note $\bar{\bar{A}} = \{z \in \Lambda(x,\hat{q}(B));\ V(z) \cap A \neq \emptyset\}$

and

$$[A] = \bigcup_{z\in\bar{\bar{A}}} (V(z) \cap \tilde{U}(x,\hat{q}(B)).$$

Let us consider the partition ξ of [A] into the sets $\{V(z) \cap \tilde{U}(x,\hat{q}(B))\}_{z\in\bar{\bar{A}}}$. From Absolute Continuity Theorem, Fubini Theorem and $(7.1.3)_I$ one easily sees that [A] is measurable, that $\nu([A]) > 0$ and that ξ is a measurable partition of [A].

Beginning from now, we will suppose that in (4.1.3), $\hat{q}(B) = \varepsilon(B)$; this last assumption does not present any restriction of generality. Under this assumption we will now prove

THEOREM 11.1. *For* ν *-almost every* $a \in [A]$, *the measure* $\nu^{[A]}_{C_\xi(a)}$ *is equivalent to the measure* ν^ξ_a. *Moreover, there exists a positive function* Ψ *satisfying* (11.1) *such that* ν^ξ_a *-almost everywhere on* $C_\xi(a)$ *one has*

$$\left|\frac{d\nu^{[A]}_{C_\xi(a)}}{d\nu^\xi_a} - 1\right| \le \Psi(B).$$

Proof. Step 1. For $a \in \Lambda^{\ell}_{k,r,s,\alpha,\gamma} \cap \tilde{U}(x,\hat{q}(B))$ and $K \subset V(a)$, we denote

$$K(\eta) = \bigcup_{z \in K} C_\eta(z) \quad \text{and} \quad \mu_a(K) = \nu^{[A]}(K(\eta)). \tag{11.4}$$

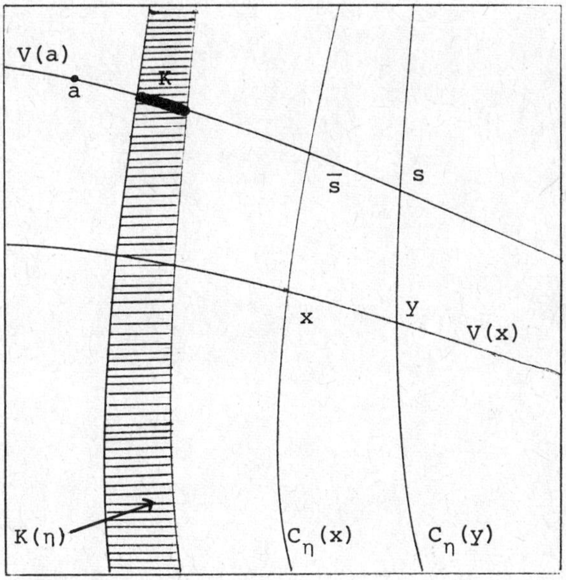

Fig. 9

Let $s \in [A]$. Then $s = C_\eta(y) \cap C_\xi(\bar{s})$, where $y \in V(x)$ and $\bar{s} \in A \subset C_\eta(x)$; y and $\bar{s}$ are uniquely determined by s. Thus we will use y and $\bar{s}$ as coordinates of $s \in [A]$ and we will sometimes write $(y,\bar{s})$ instead of s.

In virtue of Proposition 11.2 and (11.3) it follows that for every $y \in C_\xi(x)$ the Poincaré map $p_{xy} \overset{\text{def}}{=\!=} p_{C_\eta(x)C_\eta(y)}$

$$p_{xy}: A = C_\eta(x) \cap [A] \to C_\eta(y) \cap [A]$$

is well defined.

Let us denote by ηA the measurable partition of [A] into the sets $\{C_\eta(y) \cap [A]\}_{y \in [A]}$. Let us denote also $\nu_y^{\eta A} = \frac{1}{\nu_y^\eta(C_\eta(y) \cap [A])} \nu_y^\eta$.

Let $y \in C_\xi(x)$ and $s \in C_\eta(y) \cap [A]$. We define

$$\frac{d\nu_{C_\eta(y)}^{[A]}}{d\nu_y^{\eta A}}(s) = t_y(s), \tag{11.5}$$

$$\frac{d\mu_x}{d\nu_x^\xi}(y) = h(y) \tag{11.6}$$

(cf. (11.4). The functions t and h are measurable functions of y and s and of y respectively.

Let us note $p_{x\bar{s}}^\eta : C_\xi(x) \to C_\xi(\bar{s})$ the Poincaré map induced by the foliation η, i.e. $p_{x\bar{s}}^\eta(y) = s$. As η is a smooth foliation, $p_{x\bar{s}}^\eta$ is also a smooth mapping. Thus one can define

$$\frac{d\nu_x^\xi}{d(\nu_{\bar{s}}^\xi \circ p_{x\bar{s}}^\eta))}(y) = T_{\bar{s}}(y). \tag{11.7}$$

The function T is a measurable function of $\bar{s}$ and y.

Moreover, there exists a positive function g satisfying (11.1) (with $\phi = g$) such that

$$|t_y(s) - 1| \le g(B) \tag{11.8}$$

for $y \in C_\xi(x)$ and $s \in C_\eta(y) \cap [A]$, that

$$|h(y) - 1| \le g(B) \tag{11.9}$$

for $y \in C_\xi(x)$ and that

$$|T_{\bar{s}}(y) - 1| \le g(B) \tag{11.10}$$

for $y \in C_\xi(x)$ and $\bar{s} \in A$.

Indeed, (11.8) follows immediately from (11.2) and from Proposition 11.1. (11.9) follows from the fact that η is a smooth foliation and from the Condition (Bc) of Sec. 1.1_I. (11.10) follows from the same as above, together with the Theorem $7.1.3_I$.

As one admits that $\hat{q}(B) = \varepsilon(B)$, then in virtue of Absolute Continuity Theorem, for any $y \in C_\xi(x)$ and $\bar{s} \in A$ one can define

$$\frac{d(\nu_y^{\eta A} \circ p_{xy})}{d\nu_x^{\eta A}}(\bar{s}) = H_y(\bar{s}). \tag{11.11}$$

The function H is a measurable function of y and $\bar{s}$ and from (4.5) one deduces that

$$|H_y(\bar{s}) - 1| \le L(B), \tag{11.12}$$

where L is a function satisfying (11.1) (with $\phi = L$).

Finally let us define the probability measure α on A by the formula: if D is a measurable subset of A, then

$$\alpha(D) = \nu^{[A]}([D]),$$

where $[D]$ is defined like $[A]$. Absolute Continuity Theorem easily implies that the measure α is equivalent to the measure $\nu_x^{\eta A}$. Let us define

$$\frac{d\nu_x^{\eta A}}{d\alpha}(\bar{s}) = \beta(\bar{s}). \tag{11.13}$$

From (4.5) and condition (Bc) of Sec. 1.1_I one deduces that for $\bar{s} \in A$, the function β satisfies the inequality

$$|\beta(\bar{s}) - 1| \le f(B),$$

where f is a function satisfying (11.1) (with $\phi = f$).

Step 2. Let $Q \subset [A]$ be an arbitrary measurable subset. It follows from the uniqueness of the set of conditional measures that to prove the Theorem 11.1 it is sufficient to see that

$$\nu^{[A]}(Q) = \int_{[A]} \Big[\int_{C_\xi(a)} 1_{Q \cap C_\xi(a)}(r) G_a(r) d\nu_a^\xi(r)\Big] d\nu^{[A]}(a), \tag{11.14}$$

where 1_Z denote the characteristic function of the set Z and where the function G_a is such that

$$|G_a(r) - 1| \leq \Psi(B), \tag{11.15}$$

for ν_a^ξ almost all $r \in [A] \cap V(a)$.

From the definition of conditional measures $\{\nu^{[A]}_{C_\eta(y)}\}$, from (11.4)-(11.7), (11.11), (11.13) and from the Fubini Theorem, one obtains that

$$
\left\{
\begin{aligned}
\nu^{[A]}(Q) &= \int_{[A]}\left[\int_{C_\eta(a)} 1_{Q\,\cap\, C_\eta(a)}(s)\,d\nu^{[A]}_{C_\eta(a)}(s)\right] d\nu^{[A]}(a) = \\
&= \int_{[A]}\left[\int_{C_\eta(y)} 1_{Q\,\cap\, C_\eta(y)}(s)\,d\nu^{[A]}_{C_\eta(y)}(s)\right] d\mu_x(y) = \\
&= \int_{[A]}\left[\int_{C_\eta(y)} 1_{Q\,\cap\, C_\eta(y)}(s)\,d\nu^{[A]}_{C_\eta(y)}(s)\right] h(y)\,d\nu_x^\xi(y) = \\
&= \int_{[A]}\left[\int_{C_\eta(y)} 1_{Q\,\cap\, C_\eta(y)}(s)\,t_y(s)\,d\nu_y^{\eta A}(s)\right] h(y)\,d\nu_x^\xi(y) = \\
&= \int_{[A]} 1_Q(y,\bar{s})\,t_y(s)\,h(y)\,d\nu_y^{\eta A}(s)\,d\nu_x^\xi(y) = \\
&= \int_{[A]} 1_Q(y,\bar{s})\,t_y(p_{xy}(\bar{s}))\,H_y(\bar{s})\,h(y)\,d\nu_x^{\eta A}(\bar{s})\,d\nu_x^\xi(y) = \\
&= \int_{[A]}\left[\int_{C_\xi(\bar{s})} 1_{Q\cap C_\xi(\bar{s})}(y,\bar{s})\,t_y(p_{xy}(\bar{s}))\,H_y(\bar{s})\,h(y)\,d\nu_x^\xi(y)\right] d\nu_x^{\eta A}(\bar{s}) = \\
&= \int_{[A]}\left[\int_{C_\xi(\bar{s})} 1_{Q\,\cap\, C_\xi(\bar{s})}(y,\bar{s})\,t_y(p_{xy}(\bar{s}))\,H_y(\bar{s})\,h(y)\,\cdot \right. \\
&\qquad \left. \cdot\, T_{\bar{s}}(y)\,d\nu_{\bar{s}}^\xi(y)\right] d\nu_x^{\eta A}(\bar{s}) = \\
&= \int_{[A]}\left[\int_{C_\xi(\bar{s})} 1_{Q\,\cap\, C_\xi(\bar{s})}(y,\bar{s})\,L(y,\bar{s})\,d\nu_{\bar{s}}^\xi(y)\right] d\nu_x^{\eta A}(\bar{s}) = \\
&= \int_{[A]}\left[\int_{C_\xi(\bar{s})} 1_{Q\cap C_\xi(\bar{s})}(y,\bar{s})\,L(y,\bar{s})\,\beta(\bar{s})\,d\nu_{\bar{s}}^\xi(y)\right] d\nu^{[A]}(y,\bar{s}),
\end{aligned}
\right. \tag{11.16}
$$

with $L(t,\bar{s}) = t_y(p_{xy}(\bar{s}))H_y(\bar{s})h(y)T_{\bar{s}}(y)$.

The fact that the last integral is equal to $\nu^{[A]}(Q)$ is nothing else but (11.14) written in a slightly different manner.

(11.15) follows now from (11.8)-(11.10), (11.12), (11.13) and from the counterpart of inequality (9.5) for five multiples. ∎

11.4. From (11.16) one immediately deduces the following

COROLLARY 11.1. *For* ν_x^{η} *almost all points* $y \in A$ *and every point* $z \in \Lambda^{\ell}_{k,r,s,\alpha,\gamma} \cap \tilde{U}(x,\hat{q}(B))$ *such that* $V(z) \cap A = y$ *one has*

$\nu_{C_\xi(z)}(C_\xi(z)) = \nu_{C_\xi(z)}(V(z) \cap \tilde{U}(x,\hat{q}(B)) = 1.$

11.5. If $z \in \Lambda^{\ell}_{k,r,s,\alpha,\gamma}$ then by ν_z we will denote the measure induced by the Riemannian metric ρ on local stable manifold $V(z)$.

Now, we will suppose that the Φ-invariant probability measure μ (cf. Secs. 1_I and 4) is equivalent to the measure ν. Let us recall that

$$\Lambda_{k,r,s,\alpha} = \bigcup_{\substack{\ell>2\\ \gamma>0}} \Lambda^{\ell}_{k,r,s,\alpha,\gamma}.$$

Theorem 11.1 implies the following

COROLLARY 11.2. *If* $\nu(\Lambda_{k,r,s,\alpha}) > 0$ *then for* ν *almost every point* $z \in \Lambda^{\ell}_{k,r,s,\alpha,\gamma}$ *where* ℓ *is big enough and* $\gamma > 0$ *is small enough, one has*

$$\nu_z(V(z) \cap \Lambda_{k,r,s,\alpha} \cap \tilde{U}(z,\hat{q}(B)) = \nu_z(V(z) \cap \tilde{U}(z,\hat{q}(B)). \qquad (11.17)$$

Proof. Let H denote the set of regular points (cf. Sec. 2.2_I) and $G = M\backslash H$. As the Φ invariant measure μ is equivalent to the measure ν, then $\mu(G) = \nu(G) = 0$.

For $z \in \Lambda^{\ell}_{k,r,s,\alpha,\gamma}$ let us define $A_z = \{y \in V(z);\ y \in G\}$. Let us define also $G_1 = \bigcup_{\substack{\ell>2\\ \gamma>0}} \{z \in \Lambda^{\ell}_{k,r,s,\alpha,\gamma};\ \nu_z(A_z) > 0\}$. G_1 is measurable. First we will prove that $\nu(G_1) = 0$.

Let us suppose that $\nu(G_1) > 0$. Thus one can find ℓ and γ such that $\nu(G_1 \cap \Lambda^{\ell}_{k,r,s,\alpha,\gamma}) > 0$. Let x be a point of density (with respect to ν) of the set $G_1 \cap \Lambda^{\ell}_{k,r,s,\alpha,\gamma}$.

As we suppose that $\nu(G_1) > 0$, then in virtue of Theorem 11.1 one sees that on the set of $z \in \tilde{U}(x,\hat{q}(B))$ of positive ν measure one has $\nu_z(V(z) \cap G_1 \cap \tilde{U}(x,\hat{q}(B))) > 0$. This implies that for such z, $\nu_z(V(z) \cap G \cap \tilde{U}(z,\hat{q}(B))) > 0$ and consequently from Theorem 11.1

one obtains that $\nu(G) > 0$. This contradiction proves that $\nu(G_1) = 0$.

Now, let us note $G_2 = G \cup G_1$; $\nu(G_2) = 0$. From $(7.2.1)_I$, and from $(7.2.2)_I$ it follows that (11.17) is true for all $z \in \Lambda_{k,r,s,\alpha} \backslash G_2$, because $\Lambda_{k,r,s,\alpha} \backslash G_2 \subset H$. ∎

12. INFINITE DIMENSIONAL CASE

12.1. We will now quickly indicate how, using some results from Secs. 1-3 of $[\text{Rue}]_3$, one can modify our previous arguments to obtain the infinite dimensional counterpart of Theorems 6.1_I (cf. $[\text{Mañ}]_2$ and $[\text{Rue}]_3$) and 4.1. The last result has been proved independently by M. Brin and Z. Nitecki (see $[\text{Bri}]_2$). We thank sincerely M. Brin, F. Ledrappier and P. Collet for the discussions we had about the subject. We will be very sketchy and we leave the details to the reader.

12.2. Before recalling Ruelle's framework (cf. $[\text{Rue}]_3$) which allows the existence of LSM in infinite dimensional Hilbert space, let us emphasize that even in finite dimensional case considered up to now, the approach of $[\text{Rue}]_3$ is more advantageous compared to the Pesin approach (cf. Sec. 2_I and (T.1) from Appendix 2). Indeed, the condition (cf. $(1.2)_I$)

$$\int_M \log^+ \|d\Phi_x^{-1}\| d\mu < +\infty$$

is in fact unnecessary. Thus for the purposes of Parts I-III we can consider also the mappings with LCE equal to $-\infty$. Moreover, for the purpose of Part I, the assumption of invertibility of $\Phi: N \to \Phi(N)$ can be dropped. Indeed, in Part I, the invertibility of Φ is used only for the construction of Lyapunov metric and, in what follows, we will construct a kind of Lyapunov metric whose existence is independent of the invertibility of Φ.

12.3. Let E be a separable, infinite dimensional Hilbert space and $U \subset E$ be an open subset. Let $\Phi: U \to E$ be a C^1 mapping.

Let us suppose that $K \subset U$ is a compact subset such that $\Phi(K) \subset K$, and that for some $n_0 > 0$ and every $x \in K$, the operator $d\Phi^{n_0}(x)$ is compact. Let μ be a Φ-invariant Borel probability measure on K. The well known Krylov-Bogoliubov theorem guarantees that at least one such measure exists.

Let us suppose that

$$\int_K \log^+\|d\Phi_x\| d\mu(x) < +\infty. \tag{12.1}$$

For every $x \in K$ and $u \in E$ let us define the LCE $\chi(x,u)$ by

$$\chi(x,u) = \limsup_{n\to+\infty} \frac{1}{n} \log\|d\Phi_x^n(u)\|. \tag{12.2}$$

The following theorem, summarizes all the necessary facts proved in Secs. 1-3 of $[\mathrm{Rue}]_3$ that we need. It generalizes to the infinite dimensional case the Oseledec Multiplicative Ergodic Theorem formulated in Appendix 2. Let us note that in $[\mathrm{Rue}]_3$ these facts are formulated in a somewhat different manner unlike the rest of the book. In this section, the LCE will be enumerated in decreasing way; i.e. $\chi_i(x) > \chi_{i+1}(x)$.

THEOREM 12.1. *There exists a Borel subset* Λ *of* K *such that* $\Phi(\Lambda) \subset \Lambda$, $\mu(\Lambda) = 1$ *and such that for every* $x \in \Lambda$:

(12.1.1) *The exact limit in* (12.2), *perhaps equal to* $-\infty$, *exists for every* $u \in E$.

When u *varies in* $E\setminus\{0\}$, $\chi(x,u)$ *takes at most a countable number of distinct values* $\{\chi_i(x)\}$ *including* $-\infty$,

$$\chi_1(x) > \chi_2(x) > \cdots\cdots > \chi_{s(x)}(x) = -\infty,$$

where $s(x) = +\infty$ *is not excluded. Moreover, there exists a uniquely determined sequence of closed linear subspaces* $\{L_i(x)\}_{1\le i\le s(x)}$ *of* E, *of finite codimension if* $i < +\infty$,

$$E = L_1(x) \supset L_2(x) \supset \cdots\cdots \supset L_{s(x)}(x) \supset L_{s(x)+1}(x) \overset{\text{def}}{=} \{0\}$$

such that for every i *and for every* $u \in L_i(x)\setminus L_{i+1}(x)$ *one has* $\chi(x,u) = \chi_i(x)$.

(12.1.2) *Let us fix* $T < 0$. *For every* $n \ge 0$ *let us define the closed linear subspace* $V_n = V_n(x)$ *of* E *by*

$$V_n = \{u \in E;\ \lim_{m\to+\infty} \frac{1}{m} \log\|d\Phi^m_{\Phi^n(x)} u\| = \chi(\Phi^n(x),u) \le T\} \tag{12.3}$$

As $\Phi(\Lambda) \subset \Lambda$, this exact limit exists in virtue of (12.1.1).

Then for every linear subspace $L \subset E$ *satisfying* $L \oplus V_0(x) = E$, *one has* $d\Phi_x^n(L) \oplus V_n(x) = E$.

(12.1.3) *For every* $\varepsilon > 0$ *there exists a positive measurable function* ω_ε *defined on* Λ *such that for every* $n,m \geq 0$ *and every* $v \in V_n(x)$

$$\|d\Phi^m_{\Phi^n(x)} v\| \leq \|v\| \, \omega_\varepsilon(x) \exp(m\,T) \exp((n+m)\varepsilon).$$

(12.1.4) *If* L *is a linear subspace of* E *such that* $L \cap V_0(x) = \{0\}$, *then*

$$\lim_{n\to\infty} \frac{1}{n} \log \gamma(d\Phi_x^n(L), V_n(x)) = 0,$$

where the angle $\gamma(\cdot,\cdot)$ *was defined by* (2.15).

Let us note that (12.1.1) and (12.1.2) imply that the subspaces $V_n(x)$, $n \geq 0$, are all of the same finite codimension in E. Moreover for every $n \geq 0$, $d\Phi_x^n : L \to d\Phi_x^n(L)$ is an isomorphism.

Let us note also that (12.1.1) follows from Corollary 2.2 from $[\text{Rue}]_3$; (12.1.2)-(12.1.4) are directly related to the conditions (S.2)-(S.4) from Sec. 3.1 of $[\text{Rue}]_3$ respectively.

12.4. We will now describe in our framework the counterpart of Sec. 2.3_I

Let us fix $T < 0$. Let us define the set

$$\Lambda_T = \{x \in \Lambda;\ \chi_{\ell(x)}(x) > T \geq \chi_{\ell(x)+1}(x) \quad \text{for some} \quad \ell(x) \geq 1\}.$$

Obviously Λ_T is measurable and $\Phi(\Lambda_T) \subset \Lambda_T$. Let us note that in the introduced notation $V_n(x) = L_{\ell(x)+1}(\Phi^n(x))$ for $x \in \Lambda_T$.

For $x \in \Lambda_T$ let us define

$$\begin{cases} \lambda(x) = \exp \chi_{\ell(x)+1}(x), \\ \mu(x) = \exp \chi_{\ell(x)}(x). \end{cases}$$

Clearly $\lambda(x) < 1$.

Let r and s be positive numbers, $r < 1$ and $r < s$. Let us define $\varepsilon_{r,s}$ exactly as in Sec. 2.3_I by $(2.5)_I$. For every ε, $0 < \varepsilon \le \varepsilon_{r,s}$ let us define the set

$$\Lambda_{r,s,\varepsilon} = \{x \in \Lambda_T;\ r\exp(-\varepsilon) \le \lambda(x) \le r < s \le \mu(x)\} \tag{12.5}$$

when $\lambda(x) > 0$. For $\lambda(x) = 0$, let us define

$$\Lambda_{r,s,\varepsilon} = \Lambda_{r,s} = \{x \in \Lambda_T;\ \lambda(x) \le r < s \le \mu(x)\}.$$

This last case will be not considered here and we leave it to the reader as an easy exercise.

Obviously $\Lambda_{r,s,\varepsilon}$ is measurable and $\Phi(\Lambda_{r,s,\varepsilon}) \subset \Lambda_{r,s,\varepsilon}$. Moreover, each $x \in \Lambda_T$ belongs to some $\Lambda_{r,s,\varepsilon}$.

Now, like in $(2.6)_I$–$(2.8)_I$ we define for $x \in \Lambda_{r,s,\varepsilon}$:

$$\lambda'(x) = \lambda(x)\exp(3\varepsilon),\quad \mu'(x) = \mu(x)\exp(-3\varepsilon),\quad \nu'(x) = \exp(-\varepsilon). \tag{12.6}$$

For $x \in \Lambda_T$, let us define the finite dimensional subspace $W_0(x)$ as the orthogonal complement in E of the subspace $V_0(x)$ (cf. (12.3)).

From (12.1.3) it follows now that for every $m,n \ge 0$ and every $v \in V_n(x), x \in \Lambda_{r,s,\varepsilon}$, one has (cf. $(2.1.1)_I$)

$$\|d\Phi^m_{\Phi^n x} v\| \le \|v\|\alpha_\varepsilon(x) r^m \exp(\varepsilon(n+m)) , \tag{12.7}$$

where ∂_ε is some positive measurable function defined on $\Lambda_{r,s,\varepsilon}$.

Moreover, from (12.1.4) it follows that there exists a measurable positive function β_ε defined on Λ_T, such that (cf. $(2.1.3)_I$) for $x \in \Lambda_T$ and every $n \ge 0$

$$(\beta_\varepsilon(x))^{-1}\exp(-\varepsilon n) \le \gamma(W_n(x), V_n(x)), \tag{12.8}$$

where $W_n(x) \overset{\text{def}}{=\!=} d\Phi^n_x(W_0(x))$. Now, let us define $p(x) = \dim W_0(x) = \dim W_n(x)$; $p(x) < +\infty$. For every $n \ge 0$, $p(x)$ is also the codimension in E of the subspace $V_n(x)$.

Now, let us remark that the inequality $(2.1.2)_I$ remains true for $n,m \ge 0$; indeed the proof given in Sec. 1.3 of $[\mathrm{Pes}]_1$ works also in our framework. Thus for every $\varepsilon > 0$ there exists a measurable positive function γ_ε defined on Λ_T such that for all $n,m \geq 0$ and every $v \in W_n(x)$

$$\|d\phi^m_{\phi^n_x} v\| \geq \|v\| \frac{1}{\gamma_\varepsilon(x)}[\mu(x)]^m \exp(-\varepsilon(n+m)). \tag{12.9}$$

Let us fix once and for all a natural number $p \geq \ell$. Let us denote

$$\Lambda_p = \{x \in \Lambda_T;\ p(x) = p\}.$$

It is clear that $\phi(\Lambda_p) \subset \Lambda_p$.

Let $x \in \Lambda_p$. Let us denote by $\lambda_1(x) \geqq \lambda_2(x) \geqq \ldots$ all Lyapunov characteristic exponents at x, where any exponent $\chi_i(x)$ is repeated $k_i(x)$ times ; i.e. the first $k_1(x)$ of them coincides with $\chi_1(x)$, the following $k_2(x)$ with $\chi_2(x)$, etc. Now, let us consider an arbitrary base $f_1(x),\ldots,f_p(x)$ of subspace $W_0(x)$, such that for every $i, 1 \leqq i \leqq p$ $\lim_{n\to\infty} \frac{1}{n} \log \|d\phi^n_x(f_i(x))\| = \lambda_i(x)$

The Theorem T.1. of Appendix 2 applied to the measurable mapping $\Lambda_p \ni x \longrightarrow d\phi_x\big|_{W_0(x)}$, implies that for $p \geq 2$ there exists a ϕ invariant measurable subset $Z \subset \Lambda_p$, $\mu(Z) = \mu(\Lambda_p)$ such that for every $x \in Z$ and for every two non-empty disjoint subsets $P, Q \subset \{1,2,\ldots,p\}$ we have:

$$\lim_{n\to\infty} \frac{1}{n} \log \gamma(d\phi^n_x(E_P),\ d\phi^n_x(E_Q)) = 0 \tag{12.11}$$

where E_P and E_Q denote the linear subspaces of $W_0(x)$ spanned by the vectors $\{f_i(x)\}_{i\in P}$ and $\{f_i(x)\}_{i\in Q}$ respectively.

Thus for every $\varepsilon > 0$ there exists a measurable positive function $C_\varepsilon > 1$ defined on Z such that for every $x \in Z$ and every $n \geqq 0$

$$\gamma(d\phi^n_x(E_P),\ d\phi^n_x(E_Q)) \geqq \frac{1}{C_\varepsilon(x)} \exp\left(-\frac{\varepsilon}{p(x)} n\right) \tag{12.12}$$

Now, (12.12) easily implies that for μ-almost every $x \in Z$, for every $n \geqq 0$ and for every $a_1,\ldots,a_p \in \mathbb{R}$

$$\sum_{i=1}^{p} |a_i|\ \|d\phi^n_x(f_i(x))\| \leqq \eta_\varepsilon(x) \exp(\varepsilon n) \|\, d\phi^n_x(\sum_{i=1}^{p} a_i f_i(x))\| \tag{12.13}$$

where $\eta_\varepsilon(x) = (4C_\varepsilon(x))^p$. Indeed, the simple geometrical consideration implies that for every two vector $u, u \in E$, $u \neq 0$, $v \neq 0$, such that $\gamma(\varepsilon_1 u, \varepsilon_2 v) \geq \frac{1}{q}$, $\varepsilon_1 = \pm 1$, $\varepsilon_2 = \pm 1$, one has $\|u\| + \|v\| \leq 4q\|u+v\|$.

For $p \geq 1$, let us denote

$$\Lambda_{p,r,s,\varepsilon} = \Lambda_p \cap \Lambda_{r,s,\varepsilon}.$$

Clearly, $\Phi(\Lambda_{p,r,s,\varepsilon}) \subset \Lambda_{p,r,s,\varepsilon}$.

For $\ell > 1$, let us define the set

$$\tilde{\Lambda}^{\ell}_{p,r,s,\varepsilon} = Z \cap \{x \in \Lambda_{p,r,s,\varepsilon};\ \alpha_\varepsilon(x) \leq \ell, \quad \beta_\varepsilon(x) \leq \ell,\ \gamma_\varepsilon(x) \leq \ell,\ \eta_\varepsilon(x) \leq \ell\} \tag{12.14}$$

Obviously, if $\ell_1 > \ell_2$ then $\tilde{\Lambda}^{\ell_2}_{p,r,s,\varepsilon} \supset \tilde{\Lambda}^{\ell_2}_{p,r,s,\varepsilon}$ and $\bigcup_{\ell>0} \tilde{\Lambda}^{\ell}_{p,r,s,\varepsilon} = \Lambda_{p,r,s,}$.

Now, let us denote $\tilde{\Lambda}^{\ell}_{p,r,s} = \tilde{\Lambda}^{\ell}_{p,r,s,\delta}$, where $\delta = \frac{\varepsilon_{r,s}}{2p}$.

12.5. We pass now to the heart of the matter, the counterpart of Theorem 2.2_I; i.e., to the construction of the Lyapunov norm (scalar product) along the semi orbit $\{\Phi^n(x)\}_{n\geq 0}$, for $x \in \tilde{\Lambda}^{\ell}_{p,r,s}$.

<u>THEOREM 12.2. For every</u> $x \in \tilde{\Lambda}^{\ell}_{p,r,s}$ <u>and every</u> $n \geq 0$, <u>there exists a scalar product</u> $\langle\cdot,\cdot\rangle'_{x,n}$ <u>on the space</u> E, <u>which depends on</u> x <u>in a measurable way, and positive numbers</u> $A = A(\ell,p,r,s)$, $B = B(\ell,p,r,s)$ <u>such that</u>

(12.2.1) <u>The associated norm</u> $\|\cdot\|'_{x,n}$ <u>is related to the Hilbert norm</u> $\|\cdot\|$ <u>on</u> E <u>by the following inequalities</u>

$$B \exp\left(-\frac{n\varepsilon_{r,s}}{p}\right)\|\cdot\| \leq \|\cdot\|'_{x,n} \leq A \exp\left(\frac{n\varepsilon_{r,s}}{p}\right)\|\cdot\|$$

$$(12.2.2)\quad \begin{cases} \| d\Phi_{\Phi^n(x)}\Big|_{V_n(x)} \|' \le \lambda'(x), \\ \\ (\| [d\Phi_{\Phi^n(x)}\Big|_{W_n(x)}]^{-1}\|')^{-1} \ge \mu'(x), \end{cases}$$

where $\|A\|'$ denotes the usual operator norm of an operator A acting between two spaces (or their subspaces) endowed with norms $\|\cdot\|'_{y,n}$ generated by the scalar products $<\cdot,\cdot>'_{y,n}$.

(12.2.3) The subspaces $V_n(x)$ and $W_n(x)$ are orthogonal with respect to the scalar product $<\cdot,\cdot>'_{x,n}$.

Let us underline that two main differences between theorems 2.2_I and 12.2. First, for Theorem 12.2 the invertibility of Φ is not required. Secondly, in Theorem 12.2 we do not claim that the scalar product $<\cdot,\cdot>'_{x,n}$ and $<\cdot,\cdot>'_{\Phi^n(x),0}$ coincide, even if $\Phi^n x = x$. The fact that in (12.2.1) instead of $\frac{1}{\sqrt{2}}$ one has $B\exp(-\frac{n\varepsilon}{p})$ is not a substantial one.

Proof of Theorem 12.2. The proof follows the same idea as the proof of Theroem 1.5.1 in $[\text{Pes}]_1$. First we construct independently scalar product we need on $V_n(x)$ and on $W_n(x) = d\Phi^n_x(W_0(x))$. Afterwards, as the subspaces $V_n(x)$ and $W_n(x)$ are othogonal with respect to the scalar product $<\cdot,\cdot>'_{x,n}$, we obtain the scalar product $<\cdot,\cdot>'_{x,n}$ one whole space E.

Let $x \in \tilde{\Lambda}^{\ell}_{p,r,s}$ and let $n \ge 0$. For $u,v \in V_n(x)$ let us define following $[\text{Pes}]_1$, the following scalar product

$$<u,v>'_{x,n} = \sum_{m\ge 0} (\lambda'(x))^{-2m} < d\Phi^m_{\Phi^n x}(u),\ d\Phi^m_{\Phi^n x}(v)>. \qquad (12.15)$$

As $\exp(\chi(\Phi^n(x),u)) \le \lambda(x)$ and $\exp(\chi(\Phi^n(x),v)) \le \lambda(x)$, then from Schwarz inequality, one easily obtains that the series in (12.15) converges.

We will now verify that for some constant $\tilde{A} > 0$ the norm $\|\cdot\|'_{x,n}$ verifies

$$\|\cdot\| \le \|\cdot\|'_{n,x} \le \tilde{A}\exp(n\delta)\|\cdot\| \qquad (12.16)$$

and the first inequality of (12.2.2).

Indeed, from (12.15), from Schwarz inequality and from (12.7) one obtains that for $u \in V_n(x)$

$$\begin{cases} \|u\|_{x,n} = \sqrt{\sum_{m\ge 0} (\lambda'(x))^{-2m} \|d\Phi^m_{\Phi^n x}(u)\|^2} \le \\ \le \alpha_\delta(x)\|u\|\exp(n\delta)\sqrt{\sum_{m\ge 0}(\lambda'(x))^{-2m}(r\exp\delta)^{2m}} = \\ = \widetilde{A}(x)\exp(n\delta)\|u\|, \end{cases}$$

because from (12.5), $r\exp(2\delta) < \lambda'(x) = \lambda(x)\exp(3\delta) \le r\exp(3\delta)$.
As $x \in \Lambda^\ell_{p,r,s}$ (cf. (12.14)) one obtains

$$\widetilde{A}(x) = \alpha_\delta(x)\sqrt{\sum_{m\ge 0}(\lambda'(x))^{-2m}(r\exp\delta)^{2m}} \le \ell\sqrt{\frac{1}{1-\exp(-\delta)}} \stackrel{\text{def}}{=} \widetilde{A}$$

and thus (12.16) is proved.

Let us prove now the first inequality of (12.2.2). For $u \in V_n(x)$ one has

$$\begin{aligned} \|d\Phi_{\Phi^n x} u\|_{x,n+1} &= \sqrt{\sum_{m\ge 0}(\lambda'(x))^{-2m}\|d\Phi^{m+1}_{\Phi^n(x)}(u)\|^2} = \\ &= \lambda'(x)\sqrt{\sum_{m\ge 0}(\lambda'(x))^{-2(m+1)}\|d\Phi^{m+1}_{\Phi^n(x)}(u)\|^2} \le \\ &\le \lambda'(x)\sqrt{\sum_{m\ge 0}(\lambda'(x))^{-2m}\|d\Phi^m_{\Phi^n(x)}(u)\|^2} = \lambda'(x)\|u\|'_{x,n}. \end{aligned}$$

We pass now to the construction of the scalar product $\langle\cdot,\cdot\rangle'_{x,n}$ on subspace $W_n(x)$. For any vector $u \in W_n(x)$, $u \ne 0$, let us define

$$\widetilde{\|u\|}_{x,n} = \frac{1}{\sum_{m\geq 0} \frac{(\mu'(x))^m}{\|d\Phi^m_{\Phi^n(x)}(u)\|}} \quad . \tag{12.17}$$

Let us recall that $\|d\Phi^m_{\Phi^n(x)}(u)\| \neq 0$. Moreover, as $\exp \chi(x,u) \geq$ $\geq \mu(x)$, the series $\sum_{m\geq 0} \frac{(\mu'(x))^m}{\|d\Phi^m_{\Phi^n(x)}(u)\|}$ converges and thus $\|\tilde{u}\|_{x,n} > 0$ is well defined. $\|\tilde{\cdot}\|_{x,n}$ is not a norm on $W_n(x)$, but $\|\tilde{\cdot}\|_{x,n}$ is homogeneous, i.e. for $\gamma \in \mathbb{R}$, $\gamma \neq 0$, $\widetilde{\|\gamma u\|}_{x,n} = |\gamma| \cdot \|\tilde{u}\|_{x,n}$. Let us note that $\|u\|_{x,n} \leq \|u\|$.

Now, let us fix once at for all a measurable family of bases $f_1(x),\ldots,f_p(x)$ of subspaces $W_0(x)$, such that (12.10) is satisfied.

Then, for every $n \geq 0$, the vectors

$$f_{n,i} = \frac{d\Phi^n_x f_i}{\widetilde{\|d\Phi^n_x f_i\|}_{x,n}}, \quad 1 \leq i \leq p,$$

form a basis of $W_n(x)$.

We will define now the scalar product $\langle\cdot,\cdot\rangle'_{x,n}$ on $W_n(x)$ as follows: if $u = \sum_{i=1}^{p} a_i f_{n,i}$ and $v = \sum_{i=1}^{p} b_i f_{n,i}$, then

$$\langle u,v\rangle'_{x,n} = \sum_{i=1}^{p} a_i b_i, \tag{12.18}$$

i.e. with respect to this scalar product, $\widetilde{\{f_{n,i}\}}_{1\leq i\leq p}$ is an orthonormal basis of $W_n(x)$; in particular $\|f_{n,i}\|'_{x,n} = \|f_{n,i}\|_{x,n} = 1$.

Now, we shall verify that on subspace $W_n(x)$, for some constant $\tilde{B} > 0$, the norm $\|\cdot\|'_{x,n}$ verifies

$$\tilde{B}\exp(-n\delta)\|\cdot\| \leq \|\cdot\|'_{x,n} \leq \ell\,\exp(n\delta)\ \|\cdot\| \tag{12.19}$$

and the second inequality of (12.2.2).

First, let us prove the left side of the inequality

12.19. From (12.9), (12.14) and (12.17), one obtains that for every vector $f_{n,i}$, $1 \le i \le p$,

$$\begin{cases} \sum_{m\ge 0} \dfrac{(\mu'(x))^m}{\|d\Phi^m_{\Phi^n x}(f_{n,i})\|} \le \dfrac{\widetilde{\|d\Phi^n_x f_i\|_{x,n}}}{\|d\Phi^n_x f_i\|} \gamma_\delta(x)\exp(\delta n) \sum_{m\ge 0} \exp(-2\delta m) \le \\ \qquad \le \dfrac{\ell}{1-\exp(-2\delta)} \exp(\delta n) \dfrac{1}{\|f_{n,i}\|}, \end{cases}$$

i.e.

$$\|f_{n,i}\|'_{x,n} \ge \frac{1-\exp(-2\delta)}{\ell} \exp(-\delta n)\|f_{n,i}\|.$$

Consequently, for every $v \in W_n(x)$, the left side inequality of (12.19) is true with $\widetilde{B} = \dfrac{1-\exp(-2\delta)}{\ell\sqrt{p}}$. The right side inequality of (12.19) is evident, because for every i, $1 \le i \le p$, $\|f_{n,i}\|'_{x,n} \le \|f_{n,i}\|$ and because the inequality (12.13) is true with $\eta\delta(x) \le \ell$.

Let us prove now the second inequality of (12.2.2). Let $v \in W_n(x)$. The second inequality of (12.2.2) is equivalent to the inequality.

$$\|d\Phi_{\Phi^n(x)}(v)\|'_{x,n+1} \ge \mu'(x)\|v\|'_{x,n} . \tag{12.20}$$

Like in the proof of the first inequality of (12.2.2), using (12.17), one easily verifies (12.20) for $v = f_{n,i}$, $1 \le i \le p$. Now (12.20) follow from the fact that the vectors $\{d\Phi_{\Phi^n x}(f_{n,i})\}_{1\le i\le p}$ are orthogonal in $W_{n+1}(x)$ with respect to the scalar product $\langle\cdot,\cdot\rangle'_{x,n+1}$ (cf. (12.18)).

To complete the proof of Theorem 12.2, we have only to prove (12.2.1) for every $n \ge 0$ and every vector $w \in E = V_n(x) \oplus W_n(x)$, i.e., that

$$B\exp(-2\delta n)\|w\| \le \|w\|_{x,n} \le A\exp(2\delta n)\|w\|. \tag{12.21}$$

Let $w \in E$, $w = u + v$, $u \in V_n(x)$, $v \in W_n(x)$. As $\ell > 1$, one immediately deduces that $\widetilde{B} \le 1$ and thus from the left side inequality of (12.16) one obtains that for every $u \in V_n(x)$:

$$\widetilde{B} \exp(-\delta n) \|u\| \le \|u\|'_{x,n}. \tag{12.22}$$

Now from (12.19), as the subspaces $V_n(x)$ and $W_n(x)$ are orthogonal with respect to the scalar product $\langle \cdot,\cdot \rangle_{x,n}$, one finally obtains the left side of (12.21) with $B = \frac{\widetilde{B}}{2}$

Let us go now to the right side inequality of (12.21). As $x \in \widetilde{\Lambda}^{\ell}_{p,r,s}$, then from (12.8) and (12.10),

$$\gamma(V_n(x), W_n(x)) \ge \frac{1}{\ell} \exp(-\delta n).$$

The simple geometric considerations imply that for every $u \in V_n(x)$ and $v \in W_n(x)$

$$\|u\| + \|v\| \le 4\ell \exp(\delta n) \|u+v\|.$$

Consequently, from (12.16) and (12.19) one obtains

$$\|u+v\|_{x,n} \le \|u\|_{x,n} + \|v\|_{x,n} \le \widetilde{A} \exp(\delta n) \|u\| + \ell \exp(n\delta) \|v\| \le$$

$$\le (\widetilde{A}+\ell) \exp(\delta n) (\|u\|+\|v\|) \le 4\ell(\widetilde{A}+\ell) \exp(2\delta n) \|u+v\|,$$

i.e. one obtains (12.21) with $A = 4\ell(\widetilde{A}+\ell)$. ∎

As usual let us denote by ℓ^2 the Hilbert space of all real square summing sequences.

Like in Sec. 2.3_I, for $x \in \widetilde{\Lambda}^{\ell}_{p,r,s}$ and $n \ge 0$, we can define a measurable family of invertible linear transformations $\{\tau'_{x,n}\}$,

$$\begin{cases} \tau'_{x,n}: E \to \ell^2 \times \mathbb{R}^p, \ \tau'_{x,n} = \tau'_{1x,n} \times \tau'_{2x,n}, \\ \\ \tau'_{1x,n}: V_n(x) \to \ell^2, \ \tau'_{2x,n}: W_n(x) \to \mathbb{R}^p, \end{cases}$$

such that τ'_x transforms the scalar product $\langle\cdot,\cdot\rangle'_{x,n}$ into the standard scalar product in $\ell^2 \times \mathbb{R}^p$.

12.6. Let us suppose that besides all our preceding assumptions, $\Phi: U \to E$ be a C^2 mapping with bounded first and second derivatives, i.e. for some constant $L > 0$ and every $x \in U$

$$\|d\Phi_x\| \le L \quad \text{and} \quad \|d^2\Phi_x\| \le L. \tag{12.23}$$

In fact it is sufficient to suppose that Φ is a C^1 mapping with bounded derivative satisfying the Hölder condition, but we do not enter into this easy refinement, leaving the details to the reader.

Now let us note that Pesin's Abstract Invariant Manifold Theorem (cf. Sec. 5_I), together with his proof given in Sec. 2 of $[\text{Pes}]_1$ remains true without any change if one replaces the spaces $\mathbb{R}^k$ and $\mathbb{R}^{m-k}$ by arbitrary Banach spaces.

Consequently, using Theorem 12.2 instead of Theorem 2.2_I, we are able to prove for Φ the counterpart of all results of Part I, up to the Theorem $7.1.3_I$, if instead of the sets $\Lambda^\ell_{k,r,s,\alpha,\gamma}$ one considers the sets $\tilde{\Lambda}^\ell_{p,r,s}$. All proofs remain valid with only some minor changes. In fact as Φ is defined on a subset of a linear space and as the singularities are absent, the proofs are even simpler. Indeed, the exponential mapping is unnecessary and it can now be identified with the identity mapping. Thus we can completely forget and in the further text consider Φ instead of Φ_{jw} etc. Moreover in absence of singularities, we drop α and γ among the indices.

Moreover on the sets $\tilde{\Lambda}^\ell_{p,r,s}$, the estimation $(6.1.5)_I$ is uniformly satisfied, because

$$\sup_{x \in \tilde{\Lambda}^\ell_{p,r,s}} G(x)A^2(x) < \infty.$$

In the other words, for $x \in \tilde{\Lambda}^\ell_{p,r,s}$, the mappings ϕ_x which defines TLSM are uniformly Lispchitz.

12.7. Now we will define the measurable subsets $\Lambda^\ell_{p,r,s} \subset \tilde{\Lambda}^\ell_{p,r,s}$ such that $\mu(\Lambda_{p,r,s} \setminus \bigcup_{\ell>0} \Lambda^\ell_{p,r,s}) = 0$ and that if instead of the sets $\Lambda^\ell_{k,r,s,\alpha,\gamma}$ one considers the sets $\Lambda^\ell_{p,r,s}$, the Corollary 7.1_I remains valid.

The space of all p dimensional subspaces of a separable Hilbert space is separable if $p < \infty$. The space of all scalar products on finite dimensional vector space is also separable. Thus from Luzin's theorem (see exercise II. 7.3 from J. Neveu, Bases mathematiques du calcul des probabilités, Paris, Masson ed., 1964) one can find a measurable subset $\Lambda^{\ell}_{p,r,s} \subset \tilde{\Lambda}^{\ell}_{p,r,s}$ such that $\mu(\tilde{\Lambda}^{\ell}_{p,r,s} \setminus \Lambda^{\ell}_{p,r,s}) \le \frac{1}{\ell}$, that the subspace $W_0(x)$ and the scalar product $\langle\cdot,\cdot\rangle'_{x,0}$ on it depends continuously on $x \in \Lambda^{\ell}_{p,r,s}$ and that the function $\lambda = \lambda(x)$ is continuous on $\Lambda^{\ell}_{p,r,s}$.

As $V_0(x) = [W_0(x)]^{\perp}$, then the subspaces $V_0(x)$ depends continuously on $x \in \Lambda^{\ell}_{p,r,s}$. Moreover, from the definition (12.15) and from (12.1.3) it is easy to see that the scalar product $\langle\cdot,\cdot\rangle'_{x,0}$ on $V_0(x)$ depends continuously on $x \in \Lambda^{\ell}_{p,r,s}$. Consequently the scalar product $\langle\cdot,\cdot\rangle'_{x,0}$ depends continuously on $x \in \Lambda^{\ell}_{p,r,s}$.

Now, from the above properties of the sets $\Lambda^{\ell}_{p,r,s}$ together with the last remark of Sec. 12.6 one easily follows that on the sets $\Lambda^{\ell}_{p,r,s}$, Corollary 7.1_I remains valid.

12.8. Let us suppose now that $\phi: U \longrightarrow E$ is an injective mapping.

The Absolute Continuity Theorem 4.1 also remains valid in our infinite dimensional framework. As for the proof, there is only one point, the proof of Lemma 9.2 which needs more substantial changes. The remaining parts of proof need only minor, inessential changes.

The point is that in the proof of Lemma 9.2 one uses $d\Phi^{-1}$ and $d^2\Phi^{-1}$ which are presently not defined. Let us note by the way that the mapping $d\Phi^{-n}(a_n)\Big|_{F^1_n(a)}$ etc., used in the formulation of Lemma 9.2 are nevertheless well defined in our new framework as the inverse of the invertible mapping $d\Phi^n(a) : F^1_0(a) \to F^1_n(a)$ (cf. remark following (12.1.4)).

We will now describe in detail how to adapt the proof of Lemma 9.2 to our infinite dimensional framework.

The unique point in the proof of Lemma 9.2 where $d\Phi^{-1}$ and $d^2\Phi^{-1}$ are used is the estimate of the quantity $L_n(a,b)$ from (9.6) In fact, one proves that when the transversal submanifolds $\tilde{W}^1$ and $\tilde{W}^2$ coincide (case (i)) and when $a \in V(w)$, $b \in \tilde{W}^1$ then

$$\begin{cases} R_j(a,b) \stackrel{\text{def}}{=} \dfrac{\left|\, \left| d\Phi^{-1}(a_j)\big|_{F_j^1(a)} \right| - \left| d\Phi^{-1}(b_j)\big|_{F_j^1(b)} \right| \,\right|}{\left| d\Phi^{-1}(b_j)\big|_{F_j^1(b)} \right|} \le \\ \\ \le C(\ell,p,r,s)\,\theta^j \end{cases} \tag{12.24}$$

for appropriate $\theta = \theta(\ell,p,r,s)$, $0 < \theta < 1$. The analogous estimations in the case (ii) when $a,b \in V(w)$ have similar proof and we leave it to the reader.

We will give now a proof of (12.24), which assuming (12.1), avoids completely the use of $d\Phi^{-1}$ and $d^2\Phi^{-1}$.

It is clear that

$$\begin{cases} \left| d\Phi^{-1}(a_j)\big|_{F_j^1(a)} \right| = \left| d\Phi(a_{j-1})\big|_{F_{j-1}^1(a)} \right|^{-1}, \\ \\ \left| d\Phi^{-1}(b_j)\big|_{F_j^1(b)} \right| = \left| d\Phi(b_{j-1})\big|_{F_{j-1}^1(b)} \right|^{-1}. \end{cases} \tag{12.25}$$

Consequently,

$$\begin{cases} \left|\, \left| d\Phi^{-1}(a_j)\big|_{F_j^1(a)} \right| - \left| d\Phi^{-1}(b_j)\big|_{F_j^1(b)} \right| \,\right| = \\ \\ = \left|\, \left| d\Phi(a_{j-1})\big|_{F_{j-1}^1(a)} \right|^{-1} - \left| d\Phi(b_{j-1})\big|_{F_{j-1}^1(b)} \right|^{-1} \,\right| = \\ \\ = \dfrac{\left|\, \left| d\Phi(a_{j-1})\big|_{F_{j-1}^1(a)} \right| - \left| d\Phi(b_{j-1})\big|_{F_{j-1}^1(b)} \right| \,\right|}{\left| d\Phi(a_{j-1})\big|_{F_{j-1}^1(a)} \right| \cdot \left| d\Phi(b_{j-1})\big|_{F_{j-1}^1(b)} \right|}. \end{cases} \tag{12.26}$$

Substituting (12.26) into (12.24) one obtains

$$R_j(a,b) = \frac{\left|\, \left| d\Phi(a_{j-1})\big|_{F_{j-1}^1(a)} \right| - \left| d\Phi(b_{j-1})\big|_{F_{j-1}^1(b)} \right| \,\right|}{\left| d\Phi(a_{j-1})\big|_{F_{j-1}^1(a)} \right|}. \tag{12.27}$$

Now, using Lemma 3.2, the Mean Value Theorem and (12.20) we

estimate the last numerator (cf. (12.23))

$$\begin{cases} \left| |d\Phi(a_{j-1})|_{F^1_{j-1}(a)}| - |d\Phi(b_{j-1})|_{F^1_{j-1}(b)}| \right| \le \\ \le C_1 L^p [\| d\Phi(a_{j-1}) - d\Phi(b_{j-1}) \| + \Gamma(F^1_{j-1}(a), F^1_{j-1}(b))] \le \\ \le C_1 L^p [L \| a_{j-1} - b_{j-1} \| + \Gamma(F^1_{j-1}(a), F^1_{j-1}(b))] \le \\ \le C_1 (\max(L,1))^{p+1} [\operatorname{diam}(\widetilde{W}^1_{j-1}) + \Gamma(F^1_{j-1}(a), F^1_{j-1}(b))]. \end{cases} \tag{12.28}$$

The estimation of $\operatorname{diam}(\widetilde{W}^1_{j-1})$ and of $\Gamma(F^1_{j-1}(a), F^1_{j-1}(b))$ remains unchanged (cf. (9.10) and (9.11) respectively). Thus, as in the proof of Lemma 9.2 from (12.28) one obtains that

$$\begin{cases} \left| |d\Phi(a_{j-1})|_{F^1_{j-1}(a)}| - |d\Phi(b_{j-1})|_{F^1_{j-1}(b)}| \right| \le \\ \le C(\ell,r,s,L)(\delta_{10} + C)\exp(-5j\varepsilon_{r,s}). \end{cases} \tag{12.29}$$

We will now estimate the denominator of (12.26). Like in (12.28), but using the fact that $a \in V(w)$, applying $(6.1.1)_I$ and the inequality $\kappa_{r,s} < \eta$ one obtains

$$\begin{cases} \left| |d\Phi(a_{j-1})|_{F^1_{j-1}(a)}| - |d\Phi(\Phi^{j-1}(w))|_{W_{j-1}(w)}| \right| \le \\ \le C_1 L^p [\| d\Phi(a_{j-1}) - d\Phi(\Phi^{j-1}(w)) \| + \Gamma(F^1_{j-1}(a), W_{j-1}(w))] \le \\ \le C_1 (\max(L,1))^{p+1} [\| a_{j-1} - \Phi^{j-1}(w) \| + \Gamma(F^1_{j-1}(a), W_{j-1}(w))] \le \\ \le C_1(\ell,p,r,s,L)(\|a-w\|(\kappa_{r,s})^{j-1} + C\eta^{j-1}\exp(2(j-1)\varepsilon_{r,s})) \le \\ \le C_2(\ell,p,r,s,L)q(2,C)((\kappa_{r,s})^{j-1} + \eta^{j-1}\exp(2(j-1)\varepsilon_{r,s})) \le \\ \le C_2(\ell,p,r,s,L)q(2,C)\exp(-5j\varepsilon_{r,s}), \end{cases}$$

where recall, $q(2,C)$ is defined by (7.12)-(7.14). Consequently, one obtains

$$\begin{cases} |d\Phi(a_{j-1})\Big|_{F^1_{j-1}(a)}| \geq |d\Phi(\Phi^{j-1}(w))\Big|_{W_{j-1}(w)}| + \\ \\ \qquad\qquad - C_2(\ell,p,r,s,L)q(2,C)\exp(-5j\varepsilon_{r,s}). \end{cases} \tag{12.30}$$

To estimate from below $|d\Phi(\Phi^{j-1}(w)\Big|_{W_{j-1}(w)}|$ we proceed as follows.

From (12.1.1) and (12.2.2) one immediately obtains that for any vector $h \in W_{j-1}(w)$

$$\|d\Phi(\Phi^{j-1}(w))(h)\| \geq C_3(\ell,p,r,s)\ \exp(-\frac{2j\ \ r,s}{p})\ \|h\|. \tag{12.31}$$

Now, from (2.16) and (12.31) one deduces easily that

$$|d\Phi(\Phi^{j-1}(w))\Big|_{W_{j-1}(w)}| \geq C_3(\ell,p,r,s)\exp(-4j\varepsilon_{r,s}).$$

This, together with (12.30) implies that

$$\begin{cases} |d\Phi(a_{j-1})\Big|_{F^1_{j-1}(a)}| \geq C_3(\ell,p,r,s)\exp(-4j\varepsilon_{r,s}) + \\ \\ \qquad\qquad -C_2(\ell,p,r,s,L)q(2,C)\exp(-5j\varepsilon_{r,s}). \end{cases} \tag{12.32}$$

Thus if,

$$q(2,C) \leq \frac{C_3(\ell,p,r,s)}{2C_2(\ell,p,r,s,L)} \quad , \tag{12.33}$$

then (12.32) implies that for all $j \geq 1$

$$|d\Phi(a_{j-1})|_{F^1_{j-1}(a)}| \geq \frac{C_3(\ell,p,r,s)}{2} \exp(-4\varepsilon_{r,s}). \tag{12.34}$$

Finally, putting together (12.27), (12.29) and (12.34) one obtains that if $q(2,C)$ satisfies (12.33) then

$$R_j(a,b) \leq C(\ell,p,r,s)(\delta_{10} + C)\exp(-\ j\varepsilon_{r,s}),$$

i.e. (12.24) is proved.

The rest of the proof remains unchanged.

12.9. Let us compare now the assumption that $\phi\colon U \longrightarrow E$ is an injective mapping with those of [Str], where the announcement of Absolute Continuity Theorem in infinite dimensional case was published.

On page 101 of [Str] the set U has to be supposed convex. This together with the assumption of injectivity of $d\phi_x$ for $x \in U$ (and not $x \in K$ as written on page 104 of [Str]), implies that $\phi\colon U \longrightarrow E$ is an injective mapping.

13. FINAL REMARKS

13.1. Once proved the existence of local stable manifolds and the absolute continuity of their families, one can rewrite almost without changes the content of Sec. 7 of $[\text{Pes}]_3$. The same is true for the content of $[\text{Kat}]_2$, which relies only on the facts from Sec. 1_I-4_I. We will formulate now these two sets of results in our framework.

13.2. Let M,N,μ,Φ satisfy conditions (A), (B) and (1.1) - (1.3) from Sec. 1_I. Let

$$\{x \in \Lambda;\ \chi_i(x) \neq 0 \quad \text{for every} \quad i, 1 \leq i \leq s(x)\} \overset{\text{def}}{=\!=} \Lambda'$$

(cf. Sec. 2.2_I). Let us suppose that $\mu(\Lambda') > 0$.

The following theorem repeats word by word Theorems 7.2 and 7.9 from $[\text{Pes}]_3$. Corollary 12.1 below corresponds to Corollary 7.2 from $[\text{Pes}]_3$.

THEOREM 13.1. *There exist subsets* $\Lambda_i \subset \Lambda'$, $i = 0,1,2,\ldots$ *such that*

(13.1.1) $\Lambda_i \cap \Lambda_j = \emptyset$ if $i \neq j$ *and* $\bigcup_{i=0}^{\infty} \Lambda_i = \Lambda'$,

(13.1.2) $\mu(\Lambda_0) = 0$, $\mu(\Lambda_i) > 0$ *for* $i > 0$,

(13.1.3) $\Phi(\Lambda_i) = \Lambda_i$ for $i \geq 0$

(13.1.4) *for every* $i > 0$, *the mapping* $\Phi|_{\Lambda_i}$ *is ergodic with respect to the measure* μ,

(13.1.5) *for every* $i \geq 1$ *there exist a decomposition*

$\Lambda_i = \bigcup_{j=1}^{n_i} \Lambda_i^j$, $n_i \in \mathbb{Z}^+$, $\Lambda_i^{j_1} \cap \Lambda_i^{j_2} = \emptyset$ if $j_1 \neq j_2$, *such that*

$\Phi(\Lambda_i^j) = \Lambda_i^{j+1}$ *for* $1 \leq j \leq n_i - 1$, $f(\Lambda_i^{n_i}) = \Lambda_i^1$ *and the mapping*

$\Phi^{n_i}|_{\Lambda_i^1}$ *is a* K *authomorphism*.

COROLLARY 13.1. *If for some* $i > 0$, *the mapping* $\Phi|_{\Lambda_i}$ *has a continuous spectrum* (*see Sec*. 2 *of* [Roc]$_3$), *then* Φ *restricted to* Λ_i *is a* K *authomorphism*.

In fact there is no doubt that the mapping $\Phi^{n_i}|_{\Lambda_i^1}$ is Bernoulli (cf. Theorem 8.1 from [Pes]$_3$). It seems that to prove this result we need an estimation of Jacobian $J(p)$, stronger that the estimation given in Theorem 4.3.

13.3. Let us formulate now the counterpart of the principal results from [Kat]$_2$ (Theorems (4.1)-(4.3)) which do not involve topological entropy.

Let M, N, μ, Φ satisfy conditions (A), (B) and (1.1)-(1.3) from Sec. 1_I. Let us denote by $P_n(\Phi)$ the number of periodic points of Φ of period n.

THEOREM 13.2. *Let us suppose that the measure* μ *has nonzero Lyapunov exponents*. *Then*

(13.2.1) _periodic points_ of Φ _are dense in the support of measure_ μ,

(13.2.2) $$\max\left(0, \limsup_{n\to\infty} \frac{\log P_n(\Phi)}{n}\right) \geq h_\mu(\Phi),$$

(13.2.3) _if in addition the measure_ μ _is ergodic and not concentrated on a single trajectory, then the periodic hyperbolic points of_ Φ _having a transversal homoclinic point are dense in the support of measure_ μ.

PART III

THE ESTIMATION OF ENTROPY FROM BELOW THROUGH LYAPUNOV CHARACTERISTIC EXPONENTS*

F. Ledrappier and J.-M. Strelcyn

1. INTRODUCTION AND FORMULATION OF THE RESULTS

1.1. The celebrated Pesin entropy formula asserts that when M is a compact Riemannian manifold, when $\Phi \in \mathrm{Diff}^{1+\varepsilon}(M)$ and when μ is an absolutely continuous (i.e. absolutely continuous with respect to the measure on M induced by the Riemannian metric) invariant probability measure, then

$$h_\mu(\Phi) = \int_M \Big(\sum_{\chi_i(x)>0} k_i(x)\chi_i(x)\Big) d\mu(x), \tag{1.1}$$

where $\chi_1(x) < \chi_2(x) < \ldots < \chi_{s(x)}(x)$ denote the Lyapunov characteristic exponents of Φ at x, $\{k_i(x)\}_{1\le i\le s(x)}$ their multiplicities respectively. (Cf. Sec. 2.2_I) and $h_\mu(\Phi)$ the metric entropy of the system (M,μ,Φ).

The aim of this and of the next part is to give a proof of formula (1.1) which remains valid if one considers maps with singularities as well as some class of measures somewhat larger than the class of absolutely continuous measures (see Sec. 1.3 below).

The fact that the Pesin entropy formula holds for the class of measures considered here was well known to Pesin himself, although not clearly stated. Furthermore, R. Mañé [Mañ]$_1$ recently gave a very ingenious simple proof of the estimation of the entropy from below in formula (1.1). It seems that Mañé's proof can also be extended to the more general framework we are considering. Nevertheless, the original proof given in [Pes]$_2$ whose idea goes back to Ja. G. Sinaï (see [Sin]$_1$) is also beautiful, natural and, if one admits only the Pesin theory of invariant manifolds, this proof is in fact rather simple. It allows us also to identify the Pinsker algebra of the system, that is the σ-algebra of sets A such that the mean entropy $h(\Phi,\gamma_A)$ of the partition $\gamma_A = \{A, M\setminus A\}$ is zero. The proof given here follows Pesin. One of our aims is to make it understandable by a large readership. Except for the knowledge of results stated in

---

* With minor modifications, this part reproduces the paper A proof of the estimation from below in Pesin's entropy formula. Ergod. Th. & Dynam. Sys. (1982), 2, 203-219.

Part I (Propositions 1.1 and 3.3 below) and for elementary entropy theory (see $[\mathrm{Roh}]_3$) this part is self-contained.

1.2. The subset $\Lambda \subset M$ is defined in Sec. 2_I. For $x \in \Lambda$,

$$0 \neq v \in T_x N,$$

let us denote by

$$\chi^+(x,v) = \lim_{n\to\infty} \frac{1}{n} \log \| d\phi_x^n(v) \|$$

the Lyapunov characteristic exponent of vector v.

Like in Sec. 11.4_{II} let us define

$$\Lambda_{r,s,\alpha} = \bigcup_{\substack{\ell>2 \\ \gamma>0 \\ 0\le k\le m}} \Lambda^{\ell}_{k,r,s,\alpha,\gamma}$$

(cf. Secs. 3.5_I) where $\alpha = \alpha(r,s)$, $0 < \alpha < 1$, satisfies $(2.5)_I$, $(6.5)_I$, $(2.8)_{II}$ and thus all results of Parts I and II are applicable.

We call Λ^+ the following subset of Λ:

$$\Lambda^+ = \{x \in \Lambda; \chi^+(x,v) > 0 \text{ for some } 0 \neq v \in T_x N\}.$$

Let us denote also

$$\Lambda^+_{r,s,\alpha} = \Lambda^+ \cap \Lambda_{r,s,\alpha} \quad \text{and} \quad P^+ = \bigcup_{r<1,s\ge 1} \Lambda^+_{r,s,\alpha}.$$

The local unstable manifold of Φ (i.e. the local stable manifold for Φ^{-1}) passing through $x \in \tilde{P}^+$ will be denoted here by $V_{loc}(x)$. Let us denote

$$V(x) = \begin{cases} \bigcup_{-\infty}^{\infty} \Phi^n(V_{loc}(\Phi^{-n}x)) & \text{for } x \in P^+, \\ \{x\} & \text{for } x \text{ elsewhere.} \end{cases}$$

We call $V(x)$ the global unstable manifold passing through x. Let us also note that $\mu(\Lambda^+) = \mu(P^+)$ and that $\tilde{P}^+$ is Φ invariant.

Theorem 1.1 is proved in our framework exactly in the same way as in $[\mathrm{Pes}]_3$ or in $[\mathrm{Rue}]_2$.

THEOREM 1.1. Let $x,y \in P^+$. Then the following are true:

(1.1.1) if $y \notin V(x)$ then $V(x) \cap V(y) = \emptyset$;

(1.1.2) if $y \in V(x)$ then $V(x) = V(y)$;

(1.1.3) if $K(x) = \dim N - \dim\{v \in T_xN, \chi^+(x,v) \leq 0\}$ then $V(x)$ is a $K(x)$ dimensional open submanifold of N which is generally not connected;

(1.1.4) $\Phi(V(x)) = V(\Phi(x))$;

(1.1.5) if $y \in V(x)$ then $\lim_{n\to\infty} \rho_{V(\Phi^{-n}x)}(\Phi^{-n}x, \Phi^{-n}y) = 0$;

where $\rho_{V(z)}$ denotes the distance induced on the connected components of the submanifold $V(z)$ by the Riemannian metric on N.

Let us note that the proof of (1.1.5) is easy if derived from the estimates obtained in the proof of Theorem 6.1_I.

Let us remark also that the partition of N into global unstable manifolds can also be defined (mod 0) as the classes of the following equivalence relation between the points x and y

$$\limsup_{n\to\infty} \frac{1}{n} \log \rho(\Phi^{-n}(x), \Phi^{-n}(y)) < 0$$

(see Theorem 6.1_I and the very definition of unstable manifolds.) We will call this partition the global unstable foliation and we will denote it by V.

1.3. In what follows, we shall use the notions of measurable partitions and conditional measures (see $[\text{Roh}]_3$).

Let ξ be a measurable partition of M. For $x \in M$, we denote $C_\xi(x)$ the element of the partition ξ which contains the point x and $\{\mu_{C_\xi(x)}\}_{x\in M}$ the set of conditional measures defined by μ on the elements $\{C_\xi(x)\}_{x\in M}$ of the partition ξ. If μ is another measurable partition, then $\xi \leq \mu$ means that for μ almost every $x \in M$ one has $C_\mu(x) \subset C_\xi(x)$. By $\mathcal{M}$ we denote the σ-algebra of all measurable subsets of M.

The global unstable foliation defined in §1.2 is generally not a measurable partition. But we may consider the σ-algebra $\mathcal{M}_V$ consisting of the measurable subsets of M which are the union of unstable manifolds ($A \in \mathcal{M}_V$ iff $A \in \mathcal{M}$ and $x \in A$ implies $V(x) \subset A$).

We shall also consider measures which are in some sense well disintegrated by the global unstable foliation. For if W is a manifold imbedded in N, W inherits from N a Riemannian structure and hence a Riemannian measure that we shall call the induced measure on W. We denote μ_x the induced measure on $V(x)$.

Definition 1. A measure μ on M will be called absolutely continuous with respect to the global unstable foliation of Φ iff for any measurable partition ξ of M such that $C_\xi(x) \subset V(x)$ and $\mu_x(C_\xi(x)) > 0$ for μ almost every $x \in P_\xi^+$, the conditional measures $\mu_{C_\xi(x)}$ are absolutely continuous with respect to the induced measure μ_x.

This property could also be called quasi-invariance by the global unstable foliation (think of the foliation by the orbits of a flow, for instance).

A short name ought to be Sinaï measures. In fact Sinaï did stress the importance of that property in $[\text{Sin}]_1$ and in a lot of examples, the existence of such a measure is the key property for explaining 'stochastic' properties of the system (see $[\text{Ano}]_{1,2}$, $[\text{Bow}]_{1,2}$, $[\text{Bun}]_5$, $[\text{Pes}]_{2,3}$, $[\text{Rue}]_4$). One is led to believe (see $[\text{Rue}]_5$) that measures absolutely continuous with respect to the unstable foliation play the same role and have the same importance for multidimensional systems as absolutely continuous invariant measures do in the theory of maps of an interval.

The absolute continuity theorem proved for diffeomorphisms of compact manifolds in $[\text{Pes}]_1$ and for the mappings with singularities in Sec. 11_{II} asserts that if μ is an absolutely continuous measure on N, then μ is absolutely continuous with respect to both global unstable and global stable foliation of Φ.

It is interesting to note that the measures which are absolutely continuous with respect to global unstable (or stable) foliation of Φ but which are not absolutely continuous occur very frequently in Anosov and related dynamical systems (see $[\text{Sin}]_3$ and $[\text{Bow}]_1$.

Let M, N, Φ and μ satisfy conditions (A), (B) and (1.1)-(1-3) from Sec. 1.1_I.

Let us now recall Conditions $(1.1)_I$ and $(1.2)_I$ concerning the Φ-invariant measure μ.

Condition 1.1_I. There exist positive constants C_1 and a such that for every positive ε,

$$\mu(U_\varepsilon(A)) \le C_1 \varepsilon^a.$$

Here A denotes the singular set of mapping Φ and U_ε its ε neighborhood.

Condition 1.2_I. $\int_M \log^+ \| d\Phi_x^{\pm 1} \| d\mu(x) < +\infty$ where $\log^+ x = \max(\log x, 0)$.

Let us note that when Φ is a diffeomorphism of a smooth compact manifold these conditions are automatically satisfied.

THEOREM 1.2. If the Φ-invariant probability measure μ satisfying the above conditions $(1.1)_I$ and (1.2) is absolutely continuous with respect to the global unstable foliation of Φ, then

$$h_\mu(\Phi) \geq \int_M \Big(\sum_{\chi_i(x)>0} k_i(x)\chi_i(x)\Big) d\mu(x) \tag{1.2}$$

and the Pinsker σ-algebra of the system coincides (mod 0) with $\mathcal{M}_V$ (i.e. for any subset A in the Pinsker σ-algebra there is a $B \in \mathcal{M}_V$ with

$$\mu(A \backslash B) = \mu(B \backslash A) = 0.$$

If moreover the conditions (C) and (1.4) from Sec. 1.1_I are satisfied then in the inequality (1.2) one has an equality.

The above characterization of Pinsker σ-algebra was given by Ya. B. Pesin in $[\text{Pes}]_3$.

We will prove here only the estimation of the entropy through Lyapunov characteristic exponents from below. Indeed, the above estimate is completely independent. It is proved in the next part.

COROLLARY 1.1. Let M, N, Φ and μ satisfy all the conditions from Sec. 1.1_I, i.e. the conditions (A)-(C), $(2.1)_I$-$(2.4)_I$ and $\Phi : N \to \Phi(N)$ is a diffeomorphic imbedding. Moreover let us suppose that the μ is an absolutely continuous measure on N. Then in the inequality (1.2) one has an equality.

This corollary follows immediately from above Theorem 1.2 and from Theorem 11.1_{II} which implies (by application to Φ^{-1}) that μ is a Sinaï measure.

Before proving Theorem 1.2, we want to emphasize that you can exchange the role of Φ and Φ^{-1} in all that is written here, thus replacing in every place unstable by stable and vice versa. The characterization we give is more 'natural' because it relates entropy and the coefficient of expansion of the volume.

If A is a linear operator between two Euclidean spaces of the same finite dimension m, we denote $A^{\wedge k}$ the k'th exterior power of A. Let us denote

$$\|A^{\wedge}\| = 1 + \sum_{K=1}^{m} \|A^{\wedge k}\|.$$

With this notation, it follows from [Rue]$_1$ (see also Appendix 2) that μ almost everywhere

$$\sum_{\chi_i(x)>0} k_i(x)\chi_i(x) = \lim_{n\to\infty} \frac{1}{n} \log\|(d\Phi_x^n)^{\wedge}\| \tag{1.3}$$

and

$$\int_M \Big(\sum_{\chi_i(x)>0} k_i(x)\chi_i(x)\Big) d\mu(x) = \lim_{n\to\infty} \frac{1}{n} \int_M \log\|(d\Phi_x^n)^{\wedge}\| d\mu(x).$$

Thus, the Pesin entropy formula may be also written as follows:

$$h_\mu(\Phi) = \lim_{n\to\infty} \frac{1}{n} \int_M \log\|(d\Phi_x^n)^{\wedge}\| d\mu(x). \tag{1.4}$$

In their important work [Led]$_6$, F. Ledrappier and L. S. Young prove that for the diffeomorphisms of class C^2 of smooth compact manifolds, the property (1.4), or equivalently the equality in (1.2) is a characteristic property of Sinaï measures. Earlier Ledrappier (Led]$_{4,5}$ proved this fact for measures with non-zero Lyapunov exponents. It seems that the proofs from [Led]$_{4,5}$ can be adapted to the case of maps with singularities studied in this book.

Let us note that inequality (1.2) remains true for the class of mappings formed by all finite composition $\Phi_K \circ \ldots \circ \Phi_1$ of mappings satisfying the conditions (A), (B) and (2.1)-(2.3) from Sec. 1_I. (See Secs. 1.2_I and 4.2_I). We leave the details to the reader.

2. PRELIMINARIES

2.1. If H is a finite dimensional Euclidean space, we denote Vol_H the volume on H. Let E and F be two Euclidean spaces of the same dimension, $E_1 \subset E$ be a linear subspace of E and $A: E \to F$ a linear mapping. Let us define

$$|A|_{E_1}| = \frac{\mathrm{Vol}_{F_1}(A(U))}{\mathrm{Vol}_{E_1}(U)}$$

where U is an arbitrary open and bounded subset of E_1, F_1 is an arbitrary linear subspace of F of the same dimension as E_1 and $A(U) \subset F_1$. We also denote

$$|\det A| = |A|_E| .$$

Let X and Y be two Riemannian manifolds of the same finite dimension and $T : X \to Y$ be a C^1 diffeomorphism. Riemannian measures (i.e. the measures induced by the Riemannian metrics) on X and Y will be denoted ν_x and ν_Y respectively. We recall the change of variables formula.

PROPOSITION 2.1. If $f \in L^1(Y,\nu_Y)$, then

$$\int_X (f \circ T)|\det dT| d\nu_X = \int_Y f d\nu_Y .$$

2.2. Let $(M,\mathcal{M},\mu)$ be a measure space of finite measure and $T : M \to M$ a measurable measure preserving map. We shall use the following well-known result.

PROPOSITION 2.2. Let g be a positive finite measurable function defined on M such that

$$\log^- \frac{g \circ T}{g} \in L^1(M,\mu), \quad \text{where} \quad \log^- a = \min(\log a, 0).$$

Then

$$\lim_{n\to\infty} \frac{1}{n} \log g(T^n x) = 0 \quad \mu\text{-almost everywhere,} \tag{2.1}$$

$$\int_M \log \frac{g \circ T}{g} d\mu = 0. \tag{2.2}$$

Proof. Let us first note that when $\log g \in L^1(M,\mu) \geq$ (2.2) is immediate and (2.1) follows from the Birkhoff ergodic theorem applied to the function

$$\log \frac{g \circ T}{g}.$$

Let us also note that the Birkhoff ergodic theorem is still true when applied to a function h, $h = h_+ - h_-$, with $h_+ \geq 0$, $h_- \geq 0$ and $h_- \in L^1(M,\mu)$, but in general the limit can be infinite. As

$$\log^{-} \frac{g \circ T}{g} \in L^1(M,\mu),$$

this shows that the following limit exists μ almost everywhere

$$\lim_{n\to\infty} \frac{1}{n} \sum_{i=0}^{n-1} \log \frac{g \circ T^{i+1}}{g \circ T^i} = \lim_{n\to\infty} \frac{1}{n} \log \frac{g \circ T^n}{g} \overset{\text{def}}{=} K$$

and moreover

$$\int_M K d\mu = \int_M \log \frac{g \circ T}{g} d\mu,$$

where both sides may be equal $+\infty$.

As

$$\frac{1}{n} \log g \to 0 \quad \mu \text{ almost everywhere}$$

we have therefore

$$K = \lim_{n\to\infty} \frac{1}{n} \log(g \circ T^n) \quad \mu \text{ almost everywhere.}$$

On the other hand, we know, as $0 < g < \infty$ μ almost everywhere, that

$$\frac{1}{n} \log(g \circ T^n)$$

converges to 0 in measure, because T preserves the measure μ. Thus there is a sequence $n_i \nearrow +\infty$ such that

$$\lim_{i\to\infty} \frac{1}{n_i} \log(g \circ T^{n_i}) = 0 \quad \mu \text{ almost everywhere.}$$

This implies $K(x) = 0$ μ almost everywhere and proves (2.1) and (2.2). □

Let M_1 be some sub σ-algebra of M. We will denote $L^2(M,M_1,\mu)$ the space of all M_1 measurable functions in $L^2(M,\mu)$. If α is a measurable partition, we denote M_α the σ-algebra generated by α and call a function f α-measurable iff f is M_α measurable.

2.3. Let Φ be an invertible measurable and measure preserving map on a probability space (M,M,μ). We recall two results of entropy theory.

PROPOSITION 2.3. *Let* α *be a measurable partition of* M *such that* $\alpha \le \Phi^{-1}\alpha$. *Then*

$$h_\mu(\Phi) \ge h(\Phi^{-1},\alpha) \ge H(\Phi^{-1}\alpha|\alpha).$$

Proof. Proposition 2.3 follows immediately from the very definition of entropy of Φ^{-1} (see Sec. 7.1 of $[\text{Roh}]_3$). □

PROPOSITION 2.4. *Let* α *be a measurable partition of* M *such that* $\alpha \le \Phi^{-1}\alpha$,

$$h_\mu(\Phi) = H(\Phi^{-1}\alpha|\alpha) < +\infty$$

and that the partitions $\{\Phi^n\alpha\}_{n\in\mathbb{Z}}$ *generate* $\mathcal{M}$. *Then the Pinsker* σ*-algebra of the system coincides* (mod 0) *with* $\bigwedge_{n\in\mathbb{Z}} \mathcal{M}_{\Phi^n\alpha}$.

Proof Proposition 2.4 is only a reformulation in terms of σ-algebra of theorems 12.1 and 12.3 of $[\text{Roh}]_3$. □

2.4. From now on $\Phi : M \to M$ will be a map with singularities as in the statement of the theorem we want to prove.

For $x \in N$ let us recall that T_xN is the tangent space to N at the point x. For $x \in \Lambda$, T_xN decomposes in

$$T_xN = E^u_x \oplus E^0_x \oplus E^s_x$$

where E^u_x, E^0_x, and E^s_x are linear subspaces corresponding respectively to positive, zero and negative Lyapunov characteristic exponents of Φ at x. This decomposition is invariant in the sense that

$$d\Phi_x(E^u_x) = E^u_{\Phi(x)}, d\Phi_x(E^0_x) = E^0_{\Phi(x)} \text{ and } d\Phi_x(E^s_x) = E^s_{\Phi(x)}.$$

Let us note that for $x \in \Lambda^+$, E^u_x is a non-trivial subspace of T_xN.

Let us consider (see Sec. 2.1)

$$\mathcal{T}^u(x) = |d\Phi_x|_{E^u_x}|.$$

PROPOSITION 2.5. $\log \mathcal{T}^u \in L^1(M,\mu)$ *and*

$$\int_M \Big(\sum_{\chi_i(x)>0} k_i(x)\chi_i(x) \Big) d\mu(x) = \int_M \log T^u(x)\, d\mu(x). \tag{2.3}$$

Proof. The proof is based on the following fact implicitly contained in [Ose] but explicitly in $[\text{Rue}]_1$ (compare with (1.3)). If $x \in \Lambda$ then

$$\sum_{\chi_i(x)>0} k_i(x)\chi_i(x) = \lim_{n\to\infty} \frac{1}{n} \log |d\Phi_x^n|_{E_x^u}|. \tag{2.4}$$

From the Hadamard inequality we have

$$\frac{1}{\| d\Phi^{-1}_{\Phi(x)} \|^m} \le \frac{1}{|d\Phi^{-1}_{\Phi(x)}|_{E^u_{\Phi(x)}}|} \le |d\Phi_x|_{E_x^u}| = T^u(x) \le \| d\Phi_x \|^m,$$

where $m = \dim N$. Thus condition (2.2) and the Φ invariance of the measure μ, implies that $\log T^u \in L^1(M,\mu)$. Now (2.3) follows from (2.4) and the Birkhoff ergodic theorem, because

$$\log |d\Phi_x^n|_{E_x^u}| = \sum_{i=0}^{n-1} T^u(\Phi^i(x)).$$ □

2.5. Let us now formulate a classical remark due to E. Hopf (see $[\text{Ano}]_2$ Theorem 4.4). We write down the proof here for the sake of completeness.

PROPOSITION 2.6. Let M_I be the σ-algebra of the Φ invariant subsets of M. Then $M_I \subset M_V$.

Proof. We shall prove an equivalent inclusion

$$L^2(M, M_I, \mu) \subset L^2(M, M_V, \mu).$$

We first consider a dense set F of continuous functions in $L^2(M, M, \mu)$. By the Birkhoff ergodic theorem, for any $f \in F$, the limit

$$f^-(x) = \lim_{n\to\infty} \frac{1}{n} \sum_{i=0}^{n} f(\Phi^{-i}(x))$$

exists on a set $M_f, M_f \in M_I$, $\mu(M_f) = 1$.

Let P be the orthogonal projection of $L^2(M,\mu)$ on $L^2(M,M_I,\mu)$. By mean ergodic theorem, f^- represents Pf. The set of Pf where $f \in F$ is dense in $L^2(M,M_I,\mu)$. But for any two points $x, y \in M_f$ which are V equivalent, i.e. such that there exists some $z \in P^+$ with $x, y \in V(z)$ we have by (1.1.5) $\lim_{n\to\infty} \rho(\Phi^{-n}x, \Phi^{-n}y) = 0$ and thus $f^-(x) = f^-(y)$. Therefore any function in $\{Pf; f \in F\}$ is constant along the global unstable manifolds restricted to M_f and so coincides (mod 0) with an M_V measurable function. □

2.6. Suppose now $\mu(P^+) = 1$. Using Propositions 2.3, 2.4 and 2.5 we shall prove Theorem 1.2 by constructing a measurable partition η of P^+ such that

$$\left.\begin{array}{l} \eta \leq \Phi^{-1}, \ \bigwedge_{n\in\mathbb{Z}} M_{\Phi^n\eta} = M_V. \\ \{\Phi^n\eta\}_{n\in\mathbb{Z}} \ \text{generate} \ M \ \text{(with respect to} \ \mu) \end{array}\right\} \tag{2.5}$$

and

$$H(\Phi^{-1}\eta|\eta) = \int_M \log T^u(x)\,d\mu(x). \tag{2.6}$$

This will at once prove the estimation of entropy from below and hence formula (1.2). The characterization of the Pinsker σ-algebra comes by applying Proposition 2.4 also to the partition η.

Let us remark also that there is nothing to prove in Theorem 1.2 if $\mu(\widetilde{P}^+) = 0$. The Pinsker σ-algebra is thus M and as $V(x) = x$ μ almost everywhere, M_V also coincides (mod 0) with M.

In general, when $0 < \mu(P^+) < 1$, Theorem 1.2 follows clearly because μ can be written as

$$\mu = \mu(P^+)\mu_1 + (1-\mu(P^+))\mu_2,$$

with $\mu_1(P^+) = 1$ and $\mu_2(P^+) = 0$.

This shows that to finish the proof we have only to construct a partition η satisfying (2.5) and (2.6) in the case when $\mu(P^+) = 1$, and this is what the rest of the paper is devoted to.

3. CONSTRUCTION OF THE PARTITION η

3.1. In this section we construct a measurable partition satisfying (2.5) for which we compute the entropy in §4. More precisely we want to prove

PROPOSITION 3.1. *Let* μ *be a* Φ*-invariant probability measure such that* $\mu(P^+) = 1$. *Then there exists a measurable partition* η *of* N *such that*

(3.1.1) $\eta \le \Phi^{-1}\eta$;

(3.1.2) *for* μ*-almost every point* $x \in N$, $C_\eta(x) \subset V(x)$ *contains a* V(x) *neighborhood of* x *and*

(3.1.3) $\bigcup_n \Phi^n C_\eta(\Phi^{-n}x) = V(x)$ μ*-almost everywhere and* $\bigcap_n M_{\Phi^n\eta} = M_V \pmod 0$;

(3.1.4) *the partitions* $\{\Phi^n\eta\}_{n\in\mathbb{Z}}$ *generate* M;

(3.1.5) *for any Borel subset* $B \subset M$ *the function* $\psi(x) = \mu_x(C_\eta(x) \cap B)$ *is measurable and* μ*-almost everywhere finite.*

Let us emphasize before proving Proposition 3.1 that the absolute continuity of the measure μ with respect to the global unstable foliation does not play any role in this section.

We want also to point out one meaning of (3.1.3); we have already said that μ being absolutely continuous with respect to the global unstable foliation means that μ is quasi-invariant for the equivalence relation $V : x \sim y$ iff there exists z such that $x, y \in V(z)$. In that framework, it follows easily from (3.1.3) that the measured equivalence relation $(M,\mu;V)$ is hyperfinite (see [Sch]).

The second property in (3.1.3) is a reformation of the first one. Note that

$$\Phi^n C_\eta(\Phi^{-n}x) = C_{\Phi^n\eta}(x).$$

Therefore

$$\bigcup_n C_{\Phi^n\eta}(x) = V(x)$$

implies clearly that $M_V \subset M_{\Phi^n\eta}$ for all n and that any measurable set which is in all $M_{\Phi^n\eta}$ belongs to M_V.

3.2. We first prove a general lemma from measure theory (see also 3.1 of $[\text{Led}]_3$).

PROPOSITION 3.2. *Let* $r_0 > 0$ *and* ν *be a finite non-negative Borel measure on* $\mathbb{R}$, *concentrated on* $[0,r_0]$, $0 < a < 1$. *Then the Lebesgue*

<u>measure of the set</u> L_a,

$$L_a = \{r; 0 \le r \le r_0, \sum_{k=0}^{\infty} \nu([r-a^k, r+a^k]) < +\infty\}$$

<u>is equal to</u> r_0.

<u>Proof</u>. Let us define

$$N_{a,k} = \{r; 0 \le r \le r_0, \nu([r-a^k, r+a^k]) > \frac{\nu([0,r_0])}{k^2}\}.$$

Let 'bad interval' denote an interval of length $2a^k$ with its centre in a point of $N_{a,k}$. It is easy to see that $N_{a,k}$ can be covered by bad intervals $C_{i,k}, 1 \le i \le S(k)$, so that any point meets at most two bad intervals. We have

$$\frac{S(k)\nu([0,r_0])}{k^2} \le \sum_{i=1}^{S(k)} \nu(C_{i,k}) \le 2\nu([0,r_0])$$

and $|N_{a,k}| \le 2S(K)a^k$, where $|W|$ denotes the Lebesgue measure of the set W. We get $|N_{a,k}| \le 4a^k k^2$ and therefore, by the Borel-Cantelli lemma, Lebesgue almost any point r belongs only to a finite number of $N_{a,k}$ and thus the series

$$\sum_{k=0}^{\infty} \nu([r-a^k, r+a^k])$$

converges. □

3.3. Let us recall some facts from Secs. 5_I and 6_I. Let $B(x,r)$ denote the closed ball in M with centre x and radius r. Let us define the distance ρ_V:

$$\rho_V(x,y) = \begin{cases} \rho_{V(w)}(x,y) & \text{if } x,y \in V(w) \text{ for some } w \in P^+, \\ +\infty & \text{otherwise.} \end{cases}$$

The distance $\rho_V(x,y)$ does not depend on the choice of the point w.

<u>PROPOSITION 3.3. There exists an increasing sequence</u> $\{\Lambda_\ell\}_{\ell \ge 1}$ <u>of closed subsets of</u> P^+, $\mu(P^+ \setminus \bigcup_\ell \Lambda_\ell) = 0$ <u>and positive numbers</u> r_ℓ, A_ℓ, B_ℓ <u>and</u> C_ℓ <u>such that</u>

(3.3.1) *for each* $y \in \Lambda_\ell$, $B(y,3r_\ell) \subset N$;

(3.3.2) *for each* $x \in \Lambda_\ell$ *there exists* $\varepsilon(\ell)$, $0 < \varepsilon(\ell) < 1$, *such that for any*

$$y \in \Lambda_\ell \cap B(x,\varepsilon(\ell)r),\ 0 < r \le r_\ell,$$

the local unstable manifold $V_{loc}(y)$ *is such that* $V_{loc}(y) \cap B(x,r)$ *is connected*;

(3.3.3) *the map* $y \to V_{loc}(y) \cap B(x,r_\ell)$ *is continuous from* $B(x,\varepsilon(\ell)r_\ell) \cap \Lambda_\ell$ *into the space of subsets of* $B(x,r_\ell)$ *(endowed with the Hausdorff topology)*;

(3.3.4) *for each* $y \in \Lambda_\ell$ $V_{loc}(y)$ *contains the closed ball of centre* y *and* ρ_V *radius* A_ℓ *in* $V(y)$ *and*

(3.3.5) *if* $z \in V_{loc}(y)$, *then for every* $n \ge 0$

$$\rho_V(\Phi^{-n}y,\Phi^{-n}z) \le B_\ell e^{-nC_\ell}\rho_V(y,z);$$

(3.3.6) *for any* $r, 0 < r \le r_\ell$, *the two points* z_1 *and* z_2 *belong to*

$$S(x,r) = \bigcup_{y\in\Lambda_\ell\cap B(x,\varepsilon(\ell)r)} V_{loc}(y) \cap B(x,r)$$

and are not in the same local leaf $V_{loc}(y) \cap B(x,r)$ *for some* y *belonging to* $\Lambda_\ell \cap B(x,\varepsilon(\ell)r)$, *then* $\rho_V(z_1,z_2) > 2r$.

This proposition follows from results (and from the proofs) of Secs. 6_I and 7_I. If one takes as Λ_ℓ the set

$$\bigcup_{k=0}^{m-1} \bigcup_{j=2}^{\ell+1} \Lambda^j_{k,\frac{1}{j},1+\frac{1}{j},\alpha_j,\frac{1}{j}}$$

where $\alpha_j = \alpha(\frac{1}{j},\ell+\frac{1}{j})$, i.e. $r = \frac{1}{j}$, $s = 1+\frac{1}{j}$, $\alpha = \alpha_j$, $\gamma = \frac{1}{j}$ in $\Lambda^\ell_{k,r,s,\alpha,\gamma}$.

3.4. We will now prove Proposition 3.1 in a particular case. We choose ℓ and $x \in \Lambda_\ell$ such that $\mu(S(x,r)) > 0$ for all r, $0 < r \le r_\ell$. This is possible by choosing first ℓ such that $\mu(\Lambda_\ell) > 0$ and then x in the support of the trace of the measure μ on Λ_ℓ. For any r, $0 < r \le r_\ell$, we consider the partition ξ_r of M defined by all the sets

$$V_{loc}(y) \cap B(x,r)$$

for $y \in \Lambda_\ell \cap B(x,\varepsilon(\ell)r)$ and the set $M\backslash S(x,r)$. From (3.3.3) it follows clearly that ξ_r is a measurable partition of M.

We define $\eta_r = \bigvee_{n=0}^{\infty} \Phi^n\xi_r$. The partition η of Lemma 3.1 will be a partition η_r for some r, $0 < r \le r_\ell$ that we choose later. Let us define

$$S_r = \bigcup_{n\ge 0} \Phi^n S(x,r).$$

We now prove properties (3.1.1)-(3.1.5) when $\mu(S_r) = 1$.

(3.1.1) This property is clear from the definition of η_r.

(3.1.2) It is clear that for $z \in S_r$ and for some $n \ge 0$.

$$C_{\eta_r}(z) \subset \Phi^n V_{loc}(\Phi^{-n}z) \subset V(z).$$

On the other hand, we claim that there exists a function β_r, $\beta_r \ge 0$, such that $y \in V(z)$, $\rho_V(y,z) \le \beta_r(z)$ implies $y \in C_{\eta_r}(z)$.

The proof of (3.1.2) consists in constructing such a β_r and then choosing r such that $\beta_r > 0$ μ almost everywhere.

We define β_r only on $\bigcup_\ell \Lambda_\ell$. For $z \in \bigcup_\ell \Lambda_\ell$, put

$$\ell(z) = \inf\{\ell';z \in \Lambda_{\ell'}\}$$

and

$$\beta_r(z) = \inf_{n\ge 0}\{A_{\ell(z)}, \frac{1}{2B_{\ell(z)}} \rho(\Phi^{-n}z,\partial B(x,r))e^{nC_{\ell(z)}}, \frac{r}{B_{\ell(z)}}\}.$$

Let us first prove our claim. Let $z \in \bigcup_\ell \Lambda_\ell$, $y \in V(z)$ and $\rho_V(y,z) \le \beta_r(z)$. We have to check that for any $n \ge 0$

$$C_{\xi_r}(\Phi^{-n}y) = C_{\xi_r}(\Phi^{-n}z). \tag{3.1}$$

First we know by (3.3.4) as $y \in V(z)$ and $\rho_V(y,z) \le A_{\ell(z)}$ that $y \in V_{loc}(z)$ and that (3.3.5) applies. Therefore we have for any $n \ge 0$

$$\rho_V(\Phi^{-n}y,\Phi^{-n}z) \le B_{\ell(z)}e^{-nC_{\ell(z)}}\rho_V(y,z) \le \frac{1}{2}\rho(\Phi^{-n}z,\partial B(x,r)) \tag{3.2}$$

and

$$\rho_V(\Phi^{-n}y,\Phi^{-n}z) \le B_{\ell(z)} e^{-nC_{\ell(z)}} \rho_V(y,z) \le r. \tag{3.3}$$

We have four cases to consider.

(i) If $\Phi^{-n}y$ and $\Phi^{-n}z$ both belong to $S(x,r)$, we have (3.1) by (3.3.6) and (3.3).

(ii) If neither $\Phi^{-n}y$ nor $\Phi^{-n}z$ belong to $S(x,r)$, we have (3.1) by the definition of ξ_r.

(iii)-(iv) If $\Phi^{-n}y$ belongs to $S(x,r)$ but not $\Phi^{-n}z$, or vice versa, we should have

$$\rho_V(\Phi^{-n}y,\partial(x,r)) \le \rho_V(\Phi^{-n}y,\Phi^{-n}z)$$

which would contradict (3.2). Thus only (i) and (ii) occur, which proves the claim.

We will now choose r, $0 < r \le r_\ell$ such that $\beta_r > 0$ μ almost everywhere. We will even show that for Lebesgue almost every choice of r, $0 < r \le r_\ell$ $\beta_r > 0$ μ almost everywhere.

Let $x \in M$. Let ν be the finite non-negative measure on $[0,r_\ell]$ defined by

$$\nu(A) = \mu(\{y \in M; \rho(x,y) \in A\}).$$

and let p be an integer, $p \ge 1$. We get by Proposition 3.2, applied to $a = e^{-C_p}$, that $|K_p| = r_\ell$, where

$$K_p = \{r; 0 \le r \le r_\ell, \sum_{k=0}^{\infty} \mu(\{y \in M; |\rho(x,y)-r| < e^{-kC_p}\}) < +\infty\}.$$

As Φ preserves the measure μ, we have also

$$K_p = \{r; 0 \le r \le r_\ell, \sum_{k=0}^{\infty} \mu(\{y \in M; |\rho(x,\Phi^{-k}y)-r| < e^{-kC_p}\}) < +\infty\}.$$

Note that from the uniformity of the Riemannian metric on $B(x,r_\ell)$ (see Proposition 3.3.1) there is a constant $D > 0$ such that

$$\rho(z, \partial B(x,r_1)) \le \tau$$

implies

$$|\rho(x,z) - r_1| \le D\tau$$

for r_1 and τ such that $0 < \tau \leq r_1 \leq r_\ell$.

Thus we have for r in K_p

$$\sum_{k=0}^{\infty} \mu(\{y \in M; \rho(\Phi^{-k}y, \partial B(x,r)) \leq \frac{e^{-kC_p}}{D}\}) < +\infty.$$

This implies by the Borel-Cantelli lemma that for μ almost every y there exist only a finite number of k with

$$\rho(\Phi^{-k}y, \partial B(x,r)) \leq \frac{e^{-kC_p}}{D}.$$

The set

$$T = \{r; \mu(\bigcup_n \Phi^n(\partial B(x,r))) > 0\}$$

is at most countable and for r in $(\bigcap_{p\geq 1} K_p)\setminus T$ we have clearly $\beta_r > 0$ μ almost everywhere. This completes the proof of (3.12).

Fix $r \in (\bigcap_{p\geq 1} K_p)\setminus T$ and omit the subscript r in β_r, ξ_r, y_r etc., except S_r.

(3.13) Firstly it is clear that for all n and z in S_r, $\Phi^n C_\eta(\Phi^{-n}z)$ are contained in $V(z)$.

On the other hand, let y be in $V(z)$. By (1.1.5) we have

$$\lim_{n\to\infty} \rho_V(\Phi^{-n}y, \Phi^{-n}z) = 0.$$

The invariance of μ implies

$$\lim_n \frac{1}{n} \sum_{i=1}^{n} \beta(\Phi^{-i}z) > 0 \quad \mu \text{ almost everywhere}$$

and hence there will be some n as large as we want such that

$$\beta(\Phi^{-n}z) > \rho_V(\Phi^{-n}y, \Phi^{-n}z).$$

By the proof of (3.1.2), we have $\Phi^{-n}y \in C_\eta(\Phi^{-n}z)$. Therefore for any $y \in V(z)$ there exists some n such that

$$y \in \Phi^n C_\eta(\Phi^{-n}z).$$

(3.1.3) is proved.

(3.1.4). To show that the partitions $\{\Phi^n\eta\}_{n\in\mathbb{Z}}$ generate M, we will show that any two different points y and z of a set of full measure in S_r are separated by some partition $\Phi^{-n}\eta$, $n \geq 1$. Suppose y belongs to infinitely many $\Phi^{-n}S(x,r)$ and that

$$\Phi^{-n}C\eta(z) = \Phi^{-n}C_\mu(z) \quad \text{for all } n.$$

$\Phi^n z$ and $\Phi^n y$ are infinitely often in the same local unstable manifold and by (3.3.5)

$$\rho_V(y,z) \leq 2B_\ell r e^{-nC_\ell}.$$

Therefore $\rho_V(y,z) = 0$ and they coincide.

(3.1.5) Let B be a Borel subset of M. By (3.3.3) the function

$$y \to \mu_y(C_\xi(y) \cap B)$$

is measurable and finite on the set $S(x,r)$, possibly not in the complement.

It follows clearly that the function

$$f_n(y) = \mu_y(C_{\bigvee_{i=0}^{n-1}\Phi^i\xi}(y) \cap B)$$

is measurable and finite on the set $\bigcup_{i=0}^{n-1} \Phi^{-i}S(x,r)$, though possibly not on the complement. Moreover $f_n \geq f_{n+1}$. Therefore

$$\mu_y(C_\eta(y) \cap B) = \lim_{n\to\infty} f_n(y)$$

is measurable and finite μ almost everywhere on S_r and this completes the proof of Proposition 3.1 on S_r.

3.5. The proof of Proposition 3.1 is completed if $\mu(S_r) = 1$ and in particular if μ is ergodic. If $\mu(S_r) < 1$, let us remark that by Proposition 2.6 any invariant set A is a union of global unstable manifolds.

So if $\mu(A) < 1$, let us consider for some ℓ a point x in the support of the restriction of the measure μ to $(M\backslash A) \cap \Lambda_\ell$. If we take the trace on $M\backslash A$ of the partition constructed in the same way as in Sec.3.4 starting from a small neighborhood of x, it clearly satisfies the properties (3.1.1)-(3.1.5) on a new invariant subset of $M\setminus A$, of positive measure.

Therefore we can construct inductively a partition satisfying (3.1.1)-(3.1.5) almost everywhere. As this will be done in a countable number of steps, all measurability properties are preserved and we thus complete the proof of Proposition 3.1. □

4. COMPUTATION OF THE ENTROPY

4.1. We have only to prove (2.6), i.e.

$$H(\Phi^{-1}\eta|\eta) = \int_M \log T^u(x)d\mu(x). \qquad (2.6)$$

Let us define the Borel measure ν on M by

$$\nu(K) = \int_M \mu_y(C_\eta(y) \cap K)d\mu(y) \qquad (4.1)$$

for all Borel subsets K of M. By (3.1.5) ν is a σ-finite measure on M.

Let us recall that by definition of the conditional measures $\{\mu_{C_\eta(y)}\}_{y\in M}$ we have

$$\mu(K) = \int_M \mu_{C_\eta(y)}(C_\eta(y) \cap K)d\mu(y) \qquad (4.2)$$

for all Borel Subsets K of M. In future we will use the short form $\mu_{C_\eta(y)}(K)$ instead of $\mu_{C_\eta(y)}(C_\eta(y) \cap K)$.

As the measure μ is absolutely continuous with respect to the global unstable foliation (we now use this assumption for the first time) we have that μ is absolutely continuous with respect to ν. We call g the Radon-Nikodym derivative $d\mu/d\nu$; $g(r) \geq 0$ μ almost everywhere on M. The next result is a well-known measure theoretic statement.

PROPOSITION 4.1. *For* μ *almost every* $y \in M$

$$g = \frac{d\mu_{C_\eta(y)}}{d\mu_y}$$

μ_y *almost everywhere on* $C_\eta(y)$.

Proof. Let $A \in \mathcal{M}_\eta$, $B \in \mathcal{M}$ be two arbitrary sets. As

$$\int_{A\cap B} g d\nu = \int_{A\cap B} d\mu,$$

from (4.1) and (4.2) one obtains that

$$\left.\begin{aligned} \int_{A\cap B} g d\nu &= \int_A \left(\int_{B\cap C_\eta(y)} g(z) d\mu_y(z)\right) d\mu(y) = \\ &= \int_{A\cap B} d\mu = \int_A \mu_{C_\eta(y)}(B) d\mu(y). \end{aligned}\right\} \tag{4.3}$$

The measure space $(M, \mathcal{M}, \mu)$ is separable as a Lebesgue space (see [Roh]$_3$). Let $\{B_i\}_{j=1}^\infty \subset \mathcal{M}$ be a dense subset in $\mathcal{M}$ (with respect to the standard metric

$$d(C,D) = \mu(C \setminus D) + \mu(D \setminus C)).$$

Fix j, $1 \le j < \infty$, and apply (4.3) to an arbitrary set $A \subset \mathcal{M}_\eta$ and to $B = B_j$. As A is arbitrary, (4.3) implies that there exists a measurable subset Z_j of $\mathcal{M}$, $\mu(Z_j) = 1$ such that for every $y \in Z_j$ one has

$$\int_{B_j \cap C_\eta(y)} g(z) d\mu_y(z) = \mu_{C_\eta(y)}(B_j).$$

This implies our assertion. □

We may therefore write for any Borel subset B of M

$$\int_{B\cap C_\eta(y)} g(z) d\mu_y(z) = \mu_{C_\eta(y)}(B) \tag{4.4}$$

μ almost everywhere. In particular

$$\int_{C_\eta(y)} g(z) d\mu_\eta(z) = 1 \tag{4.5}$$

μ almost everywhere .

4.2. We now compute the entropy. We denote

$$I(\Phi^{-1}\eta|\eta)(y) = -\log \mu_{C_\eta}(y)(C_{\Phi^{-1}\eta}(y)).$$

By (4.4) as $C_{\Phi^{-1}\eta}(y) \subset C_\eta(y)$, we have

$$I(\Phi^{-1}\eta|\eta)(y) = -\log \int_{C_{\Phi^{-1}\eta}(y)} g(z)\,d\mu_y(z).$$

Using now that $C_{\Phi^{-1}\eta}(y) = \Phi^{-1}(C_\eta(\Phi y))$ and Proposition 2.1, we have

$$\int_{C_{\Phi^{-1}\eta}(y)} g(z)\,d\mu_y(z) = \int_{\Phi^{-1}[C_\eta(\Phi y)]} g(z)\,d\mu_y(z) =$$

$$= \int_{C_\eta(\Phi_y)} g(\Phi^{-1}z)\,\frac{1}{T^u(\Phi^{-1}z)}\,d\mu_{\Phi(y)}(z).$$

The following proposition is the key point in our computation. We will prove it in Sec. 4.3.

PROPOSITION 4.2. *The function*

$$L(z) = \frac{g(z)T^u(\Phi^{-1}z)}{g(\Phi^{-1}z)}$$

is η *measurable*.

We have now

$$I(\Phi^{-1}\eta|\eta)(y) = -\log \int_{C_\eta(\Phi y)} \frac{g(z)}{L(z)}\,d\mu_{\Phi(y)}(z).$$

But by Proposition 4.2, the function L is constant on μ almost all elements of the partition η. Therefore on $C_\eta(\Phi y)$ we have $L(z) = L(\Phi y)$ for μ almost every y.

Consequently by (4.5)

$$\int_{C_\eta(\Phi y)} \frac{g(z)}{L(z)}\,d\mu_{\Phi(y)}(z) = \frac{1}{L(\Phi y)} \int_{C_\eta(\Phi y)} g(z)\,d\mu_{\Phi(y)}(z) = \frac{1}{L(\Phi y)}$$

and finally

$$I(\Phi^{-1}\eta|\eta)(y) = \log L(\Phi y) = \log T^u(y) + \log \frac{g(\Phi y)}{g(y)}. \tag{4.6}$$

Now $I(\Phi^{-1}\eta|\eta) \geq 0$ and by Proposition 2.5 $\log J^u \in L^1(M,\mu)$. Consequently (4.6) implies

$$\log^- \frac{g\circ\Phi}{g} \in L^1(M,\mu).$$

But the entropy $H(\Phi^{-1}\eta|\eta)$ is given by

$$H(\Phi^{-1}\eta|\eta) = \int_M I(\Phi^{-1}\eta|\eta)(y)\,d\mu(y)$$

and (2.6) follows immediately from (4.6) and from Proposition 2.2.

4.3. We give now the proof of Proposition 4.2.

Let $Z(y)$ denote the $M_{\Phi^{-1}\eta}$ measurable function defined by

$$Z(y) = \mu_{C_\eta(y)}(C_{\Phi^{-1}\eta}(y)).$$

We shall write the family of conditional measures with respect to $\Phi^{-1}\eta$ in two ways.

Firstly as $\Phi^{-1}\eta \geq \eta$, we can write for any Borel subset K of M

$$\mu_{C_{\Phi^{-1}\eta}(y)}(K) = \frac{\mu_{C_\eta(y)}(K \cap C_{\Phi^{-1}\eta}(y))}{\mu_{C_\eta(y)}(C_{\Phi^{-1}\eta}(y))} = \frac{1}{Z(y)}\int_{C_{\Phi^{-1}\eta}(y)\cap K} g(z)\,d\mu_y(z). \tag{4.7}$$

Secondly by invariance of μ we also have for any Borel subset K of M

$$\mu_{C_{\Phi^{-1}\eta}(y)}(K) = \mu_{C_\eta(\Phi y)}(\Phi(K)). \tag{4.8}$$

Therefore we get from (4.7) and (4.8) for any Borel subset K of M that

$$\frac{1}{Z(y)}\int_{C_{\Phi^{-1}\eta}(y)\cap K} g(z)\,d\mu_y(z) = \int_{C_\eta(\Phi y)\cap\Phi(K)} g(z)\,d\mu_{\Phi(y)}(z) =$$

$$= \int_{\Phi(C_{\Phi^{-1}\eta}(y)\cap K)} g(z)\,d\mu_{\Phi(y)}(z) = \int_{C_{\Phi^{-1}\eta}(y)\cap K} g(\Phi z)\,T^u(z)\,d\mu_y(z).$$

The last equality follows from Proposition 2.1.

Consequently we get for μ almost every $y \in M$

$$\frac{1}{Z(y)}\,g(z) = g(\Phi z)\,T^u(z)$$

for μ_y almost every z in $C_{\Phi^{-1}\eta}(y)$. Thus the function

$$L \circ \Phi = \frac{1}{Z}$$

is $\Phi^{-1}\eta$ measurable and proves Proposition 4.2. This achieves the proof of Theorem 1.2. □

4.4. Let us finally point out that this last Proposition 4.2 can be written as follows:

PROPOSITION 4.2 (bis). *Let* η *be a measurable partition with the properties from Proposition* 3.1. *Then there exists a strictly positive, measurable function* h *such that the function*

$$T^u \frac{h \circ \Phi}{h}$$

is η *measurable.*

This result was proved in a completely different way for Anosov systems by A. N. Livsic and Ya. G. Sinaï (see [Liv and [Bow]$_1$). It turns out that it is true in a more general setting.

PART IV

THE ESTIMATION OF ENTORPY FROM ABOVE THROUGH LYAPUNOV CHARACTERISTIC NUMBERS

A. Katok and J.-M. Strelcyn

1. INTRODUCTION AND FORMULATION OF THE RESULTS

1.1 In the present part we prove for the broad class of smooth maps with singualrities the inequality which gives the estimation from above of metric entropy of such mapping through the integral of the sum of its positive Lyapunov Characteristic Exponents.

THEOREM 1.1. Let M,N and μ satisfy the conditions (A),(Ba),(Bb), (C) and (1.1) from Sec. 1.1_I. Let $\Phi:N \to V$ be a C^2 mapping preserving the measure μ, satisfying the conditions $(1.3)_I$, $(1.4)_I$ and such that

$$\int_M \log^+ \|d\Phi_x\| d\mu(x) < +\infty. \tag{1.1}$$

Then

$$h_\mu(\Phi) \leq \int_M \Big(\sum_{\chi_i(x)>0} k_i(x)\chi_i(x)\Big) d\mu < +\infty, \tag{1.2}$$

where $-\infty \leq \chi_1(x) <\ldots< \chi_{s(x)}(x)$ denote the Lyapunov characteristic exponents of Φ at x, $\{k_i(x)\}_{1\leq i\leq s(x)}$ their multiplicaties respectively (see Appendix 2 and Sec. 2.2_I).

1.2 This inequality was first found by G.A. Margulis around 1967-68. He considered diffeomorphisms of compact smooth manifolds and absolutely continuous invariant measures. Margulis never published his proof, although it is mentioned in [Mil] and in Sec. 5 of $[\text{Pes}]_2$. In 1978 D. Ruelle published a proof of this inequality for a more general situation, namely for an arbitrary Borel invariant measure of a C^1 (not necessary invertible) map of a compact smooth manifold into itself. Our proof is an adaptation of Ruelle's proof to the framework of smooth maps with singularities. But in order to overcome the presence of singularities, we apply a trick which can be useful also in other circumstances. Let us note that unlike the compact manifold case, the finiteness of the integral in (1.2) is not obvious.

In Appendix 1, using a variation of Ruelle arguments, we describe the inequality giving the estimation from above of the metric entropy of skew product through its "vertical Lyapunov Characteristic Numbers". This answers a question of H. Furstenberg.

The content of this part is in fact independent from the rest of the book. We will only use the definitions from Sec. 1.1_I and the Multiplicative Ergodic Theorem (see Appendix 2 and Sec. 2.2_I). Necessary elementary facts concerning entropy can be found in [Roc].

2. PRELIMINARIES

2.1. The following inequality is basic for our proof.

PROPOSITION 2.1. *Let $p \geq 2$. Let X_i, $1 \leq i \leq p+1$ be the Euclidean spaces and let $U_i \subset X_i$ be an open subset of X_i, $1 \leq i \leq p$. Let $H_i: U_i \to X_{i+1}$, be a map of class C^2, $1 \leq i \leq p$. Let us note $G_i = H_i \circ \dots \circ H_2 \circ H_1$, $1 \leq i \leq p$ and by G_0 let us denote the identity mapping. Let $u \in U_1$ be such a point that $G_p(u)$ is well defined. Then*

$$\left.\begin{aligned}\|d^2G_p(u)\| \leq p \cdot \max_{0\leq i\leq p-1} \|d^2H_{i+1}(G_i(u))\| \cdot \\ \cdot [\max(\max_{0\leq i\leq p-1} \|dH_{i+1}(G_i(u))\|, 1)]^{2(p-1)}\end{aligned}\right\}.$$

Proof. Let $U \subset \mathbb{R}^m$ and $V \subset \mathbb{R}^n$ be open subsets. Let $f: U \to \mathbb{R}^n$ be such mapping that $f(U) \subset V$ and that f is twice differentiable at some point $a \in U$. Let $g: V \to \mathbb{R}^p$ be a mapping twice differentiable in $b = f(a)$. Then the mapping $h = g \circ f: U \to \mathbb{R}^p$ is twice differentiable at a and for any $x_1, x_2 \in \mathbb{R}^m$

$$d^2h(a)(x_1,x_2) = dg(b)(d^2f(a)(x_1,x_2)) + d^2g(b)(df(a)x_1, df(a)x_2)$$

(see [Car], formula (7.5.1)).

Consequently one obtains the inequality

$$\|d^2h(a)\| \leq \|dg(b)\| \cdot \|d^2f(a)\| + \|d^2g(b)\| \cdot \|df(a)\|^2.$$

One deduces now by induction that

$$\left.\begin{aligned}\|d^2G_p(u)\| \leq \sum_{r=1}^{p} \|dH_p(G_{p-1}(u))\| \cdot \dots \cdot \|dH_{r+1}(G_r(u))\| \cdot \|d^2H_r(G_{r-1}(u))\| \cdot \\ \cdot (\|dH_{r-1}(G_{r-2}(u))\| \cdot \dots \cdot \|dH_1(u)\|)^2,\end{aligned}\right\}$$

where for $r = 1$ and $r = p$ we read the terms in the above formula respectively as follows

$$\|d^2H_1(u)\| \cdot \prod_{j=2}^{p} \|dH_j(G_{j-1}(u))\| \text{ and } \|d^2H_p(G_{p-1}(u))\| \cdot \prod_{j=1}^{p-1} \|dH_j(G_{j-1}(u))\|.$$

Our proposition follows now immediately. □

2.2 Let $T: M \to M$ be a measurable mapping preserving the measure μ. The following Proposition is an easy consequence of results from Sec. 9.1 and 8.6 of [Roh].

PROPOSITION 2.1. *Let* $\xi_i = (A_o^i, A_1^i, \ldots, A_{p(i)}^i)$, $i = 1,2\ldots.$ *be a sequence of finite measurable partitions of space* M *satisfying the following conditions:*

(2.1) $\lim_{i\to\infty} \mu(A_o^i) = 0,$

(2.2) $\lim_{i\to\infty} \left(\sup_{1\le k\le p(i)} \operatorname{diam}(A_k^i)\right) = 0.$

Then $h_\mu(T) = \lim_{i\to\infty} h(T,\xi_i)$.

PROPOSITION 2.2. *Let* ξ *be a finite measurable partition of space* M. *Then*

$$h(T,\xi) = \lim_{k\to\infty} H(T^{-k}\xi \mid \xi \vee \ldots \vee T^{-(k-1)}\xi).$$

Proof. First let us recall that if α and β are finite measurable partitions of M, then $H(\alpha\vee\beta) = H(\alpha|\beta) + H(\beta)$. Applying this equality to $\alpha = T^{-1}\xi$ and $\beta = \xi \vee \ldots T^{-(k-1)}\xi$ one obtains that

$$\left.\begin{aligned} s_k &\overset{\text{def}}{=} H(T^{-1}\xi \mid \xi \vee \ldots \vee T^{-(k-1)}\xi) = \\ &= H(\xi \vee \ldots \vee T^{-k}\xi) - H(\xi \vee \ldots \vee T^{-(k-1)}\xi) \end{aligned}\right\} \quad (2.3)$$

On the other hand, it is easy to see that for every $k \ge 1$ one has $s_k \ge s_{k+1}$. Thus from (2.3) one has

$$\lim_{k\to\infty} s_k = \lim_{k\to\infty} \frac{s_1 + \ldots + s_k}{k} =$$

$$= \lim_{k\to\infty} \frac{H(\xi \vee T^{-1}\xi \vee \ldots \vee T^{-k}\xi) - H(\xi)}{k} = h(T,\xi). \quad □$$

As $s_1 \geq s_k \geq s_{k+1}$, Proposition 2.2 implies immediately the following Corollary which is at the basis of the proof of Theorem 1.1.

COROLLARY 2.1. *For every finite measurable partition* ξ *of* M *one has*

$$h(T,\xi) \leq H(T^{-1}\xi|\xi)$$

3. CONSTRUCTION OF PARTITIONS $\{\xi_t\}_{t\geq 1}$

3.1 For $x \in N$ and $r > 0$ by $B(x,r)$ one denotes the closed ball in N of radius r and center at x.

Let us note $N(r) = N\backslash U_r(A)$, where $U_r(A)$ is the open r-neighborhood of A.

PROPOSITION 3.1. *For any* $r > 0$ *there exists a finite measurable partition* $\rho_r = (C_0, C_1, \ldots, C_{p(r)})$ *of the space* M *such that*

$$A \subset C_0 \subset U_{2r}(A) \tag{3.1.1}$$

and for all i, $1 \leq i \leq p(r)$,

$$C_i \cap U_r(A) = \emptyset, \tag{3.1.2}$$

$$\operatorname{diam} C_i \leq 2r \tag{3.1.3}$$

$$C_i \text{ contains an open ball in } N \text{ of radius } \tfrac{r}{6}. \tag{3.1.4}$$

Proof. As $N(r)$ is compact, one can find its finite covering by closed balls of radius $\frac{r}{6}$, centered at points of $N(r)$. Let $B_1 = B_1(z_1,\frac{r}{6}),\ldots,B_{\ell(r)} = B(z_{\ell(r)},\frac{r}{6})$ be a covering of $N(r)$ of minimal cardinality by such closed balls.

One can assume that the balls are ordered in such a way that $B_1,\ldots,B_s$ are pairwise disjoint and that for any j, $s+1 \leq j \leq \ell(r)$

$B_j \cap (\bigcup_{k=1}^{s} B_k) \neq 0$. Let us denote by $B_{j1},\ldots,B_{jq_j}$ all the balls among the balls $\{B_i\}_{s+1\leq i\leq \ell(r)}$ which intersect the ball B_j.

Now, let us define the disjoint sets $\tilde{C}_1,\ldots,\tilde{C}_q$, as follows

$\tilde{C}_1 = B_1 \cap \bigcup_{k=1}^{q_1} B_{1k}$. Let j_2, $2 \leq j_2 \leq s$ be the smallest number such that $\widetilde{C}_1 \cap B_{j_2} = \emptyset$. We define $\widetilde{C}_2 = (B_{j_2} \cap (\bigcup_{k=1}^{q_{j_2}} B_{j_2 k}))\backslash\widetilde{C}_1$. Now, let j_3, $j_2+1 \le j_3 \le s$ be the smallest member such that $(\widetilde{C}_1 \cup \widetilde{C}_2) \cap B_{j_3} = \emptyset$. We define $\widetilde{C}_3 = (B_{j_3} \cap (\bigcup_{k=1}^{q_{j_3}} B_{j_3 k}))\backslash(\widetilde{C}_1 \cup \widetilde{C}_2)$ etc. After some step $q(r)$, $1 \le q(r) \le s$, this procedure can not be continued. At this step any ball $\{B_i\}_{1\le i\le \ell(r)}$ from our covering is either contained in the set $\bigcup_{j=1}^{q(r)} \widetilde{C}_j$ or intersect some of the set $\{\widetilde{C}_j\}_{1\le j\le q(r)}$. Let us denote by $B_j^1,\ldots,B_j^{t_j}$ all the ball among the balls $\{B_i\}_{s+1\le i\le \ell(r)}$ which intersects the set $\widetilde{C}_j$, $1 \le j \le q(r)$.

Now the definition of the sets $C_1,\ldots,C_{q(r)}$ goes as follows:

$$C_1 = \widetilde{C}_1 \cap \bigcup_{k=1}^{t_1} B_1^k \quad \text{and} \quad C_j = \widetilde{C}_j \cap (\bigcup_{k=1}^{t_j} B_j^k)\backslash C_{j-1} \quad \text{for} \quad 2 \le j \le q(r).$$

Finally we define $C_o = M\backslash \bigcup_{k=1}^{q(r)} C_k$.

It is evident that for the partition $\rho_r = (C_o,\ldots,C_{q(r)})$ conditions (3.1.1)-(3.1.3) are satisfied. □

Let us remark that in general the elements of the partition ρ_r are not connected.

For $t \ge 1$, the partition $\xi_t = \{A_o,A_1,\ldots,A_{p(t)}\}$ is defined as follows: $\xi_t = \rho_{\frac{r}{2t}}$ where R comes from condition (Ba) of Sec. 1.1_I. It is clear that the sequence of partitions $\{\xi_t\}_{t\ge 1}$ satisfies all the assumptions of Proposition 2.2.

Since the set $N(r)$ is a compact subset of N its capacity does not exceed m (cf. condition (C) of Sec. 1.1_I). This implies that for every $\beta > 0$

$$\lim_{t\to\infty} \frac{\log p(t)}{t^\beta} = 0. \tag{3.1}$$

4. THE GOOD AND BAD ELEMENTS OF PARTITIONS $\{\xi_t\}_{t\ge 1}$

4.1 From now on n and t always denote the positive integers. The constants C, g, c_3, a and d come from condition (Ba) from Sec. 1.1_I, conditions $(1.1)_I$ and $(1.4)_I$ respectively.

Let us define

$$\alpha(n) = \frac{1}{2(nd^2+g)}, \tag{4.1}$$

and

$$H_t^n = \{x \in M;\ \rho(x,A) \le \frac{2}{t^{\alpha(n)}}\}.$$

For a fixed n, let us define the good (noted $G(n,t)$) and the bad (noted $B(n,t)$) elements of partition ξ_t as follows:

$$\left.\begin{aligned} G(n,t) &= \{S \in \xi_t;\ \text{for every } k,\ 0 \le k \le n,\ \Phi^k(S) \not\subset H_t^n\}, \\ B(n,t) &= \{S \in \xi_t;\ \text{there exists } k,\ 0 \le k \le n,\ \Phi^k(S) \subset H_t^n\} \end{aligned}\right\} \tag{4.2}$$

Now, the strategy of the proof of Theorem 1.1 is as follows. Instead of estimating directly the entropy $h_\mu(T) = \lim_{t\to\infty} h(T,\xi_t)$, we will estimate (cf. Corollary 2.1 applied to Φ^n)

$$H(\Phi^{-n}\xi_t|\xi_t) = \sum_{A\in\xi_t} \mu(A)\Big[-\sum_{B\in\xi_t} \frac{\mu(\Phi^{-n}B\cap A)}{\mu(A)} \log \frac{\mu(\Phi^{-n}B\cap A)}{\mu(A)}\Big].$$

In order to evaluate this sum, we divide it into two parts: $\sum_{A\in G(n,t)} + \sum_{A\in B(n,t)}$. The first sum will be estimated in a fine manner, using the counterpart of Ruelle's Main Lemma in our framework (see Secs. 5 and 6). The second sum will be estimated very crudely using (3.1) and the fact that for any fixed n,

$$\mu\Big(\bigcup_{A\in B(n,t)} A\Big) \le 2^a (n+1) \frac{1}{t^{a\alpha(n)}} \tag{4.3}$$

where a comes from condition $(1.1)_I$.

Indeed, for any k, $0 \le k \le n$, let us define

$$B_0(n,t) = \{S \in \xi_t;\ S \subset H_t^n\}$$

and for $1 \le k \le n$

$$B_k(n,t) = \{S \in \xi_t;\ \Phi^i(S) \not\subset H_t^n \text{ for } 0 \le i \le k-1 \text{ and } \Phi^k(S) \subset H_t^n\}.$$

Clearly $B(n,t) = \bigcup_{k=0}^{n} B_k(n,t)$ and $B_k(n,t) \cap B_l(n,t) = \emptyset$ for $k \ne l$. By the definition of $\{B_k(n,t)\}_{0\le k\le n}$ one obtains that $B_k(n,t) \subset \Phi^{-k}(H_t^n)$

and thus as μ is an Φ invariant measure one obtains that $\mu(B_k(n,t)) \le \mu(H_t^n)$. Thus

$$\mu(B(n,t)) = \sum_{k=0}^{n} \mu(B_k(n,t)) \le (n+1)\mu(H_t^n) = (n+1)\mu(U_{\frac{2}{t^{\alpha(n)}}}(A)).$$

Now the condition $(1.1)_I$ immediately implies (4.3).

4.2 The first step to the estimation of the $\sum_{A \in G(n,t)}$ is the proof of the following Lemma.

LEMMA 4.1. *For every* $n \ge 1$ *there exists a number* $t(n)$ *such that for every* $t \ge t(n)$, *every* $S \in G(n,t)$, *every* $z \in \Phi^k(S)$ *and every* k, $0 \le k \le n$, *one has*

$(4.1.1)_k$ $\quad \Phi^k(S) \subset \{x \in M;\ \rho(x,A) \ge \frac{1}{t^{\alpha(n)}}\}$,

$(4.1.2)_k$ $\quad \operatorname{diam} \Phi^k(S) \le (2c_3^d)^k \frac{1}{t^{1-(n+k)\alpha(n)d}}$,

$(4.1.3)_k$ $\quad U_{\frac{1}{t}}(\Phi^k(S)) \subset B(z,(k+2)(2^d c_3)^k \frac{1}{t^{1-(n+k)\alpha(n)d}}) \subset$

$$\subset B(z,(2^d c_3)^k \frac{1}{t^{1-(n+k)\alpha(n)d^2}}) \subset B(z,R(z,N)),$$

where, recall, $R(z,N)$ *is defined by* $(3.12)_I$ *and*

$(4.1.4)_k$ $\quad (B(z,(2^d c_3)^k \frac{1}{t^{1-(n+k)\alpha(n)d^2}}) \subset$

$$\subset B(\Phi(z),(2^d c_3)^{k+1} \frac{1}{t^{1-(n+k+1)\alpha(n)d^2}}).$$

Proof For a fixed $n \ge 1$, we will now prove $(4.1.1)_n$-$(4.1.4)_n$ by induction. Indeed, we will prove $(4.1.1)_0$-$(4.1.4)_0$ and that if $(4.1.1)_k$-$(4.1.4)_k$ are true for $t \ge t_k(n)$, then $(4.1.1)_{k+1}$-$(4.1.4)_{k+1}$ are true for $t \ge t_{k+1}(n)$, $0 \le k \le n$.

Let us fix $S \in G(n,t)$.

$(4.1.1)_0$ From the definition of the partition ξ_t it follows that $\operatorname{diam}(S) \le \frac{1}{t}$. Let us fix $w \in S \setminus H_t^n$. Thus for every $z \in S$ and every

$t \geq 1$ one has

$$\rho(z,A) \geq \rho(w,A) - \rho(z,w_0) \geq \frac{2}{t^{a(n)}} - \frac{1}{t} \geq \frac{1}{t^{a(n)}}$$

because $0 < \alpha(n) < 1$. Thus $(4.1.1)_0$ is proved.

$(4.1.2)_0$ follows from the fact that for every $t \geq 1$

$$\text{diam}\,(S) \leq \frac{1}{t} \leq \frac{1}{t^{1-n\alpha(n)d}}$$

because $c_3 \geq 1$.

$(4.1.3)_0$ Let $z \in S$. As $\text{diam}\,(S) \leq \frac{1}{t}$ and as $n\alpha(n)d < 1$ then for every $t \geq 1$ and every $y \in U_{\frac{1}{t}}(S)$ one has

$$\rho(y,z) \leq \frac{2}{t} < \frac{2}{t^{1-n\alpha(n)d}}; \text{ i.e. } U_{\frac{1}{t}}(S) \subset B(z, \frac{2}{t^{1-n\alpha(n)d}}).$$

To finish, it remains only to see that there exists $t_0^1(n)$ such that for $t \geq t_0^1(n)$ one has (cf. $(3.12)_I$)

$$\frac{2}{t^{1-n\alpha(n)d^2}} \leq R(z,N) = \min(\rho(z,A),R(z,V)) = \min(CR,C[\rho(z,A)]^g).$$

Indeed, as $S \in G(n,t)$, for t big enough $\frac{2}{t^{1-n\alpha(n)d^2}} \leq CR$ and

$$\frac{2}{t^{1-n\alpha(n)d^2}} \leq \frac{C}{t^{g\alpha(n)}} \leq C[\rho(z,A)]^g.$$

$(4.1.4)_0$ Let $z \in S$. Let us fix $w \in S$ such that $\Phi(w) \notin H_t^n$. Then since $\text{diam}\,(S) \leq \frac{1}{t}$ and $n\alpha(n)d < 1$ we have

$$B(z, \frac{1}{t^{1-n\alpha(n)d^2}}) \subset B(w, \frac{2}{t^{1-n\alpha(n)d^2}}).$$

From $(4.1.3)_0$, $B(w, \frac{2}{t^{1-n\alpha(n)d^2}}) \subset B(w,R(w,N))$. Moreover (4.1) and $(4.1.1)_0$ imply that

$$B(w, \frac{2}{t^{1-n\alpha(n)d^2}}) \subset \{x \in M;\ \rho(z,A) \geq \frac{1}{2t^{\alpha(n)}}\}. \tag{4.4}$$

Let $y \in B(w, \frac{2}{t^{1-n\alpha(n)d^2}})$ be an arbitrary point. Let us denote by Γ the shortest geodesic joining w and y; $\Gamma \subset B(w,R_N(w,N))$ and $\ell(\Gamma) \le \frac{2}{t^{1-n\alpha(n)d^2}}$, where $\ell(\Gamma)$ denote the length of Γ. Now, by Mean Value Theorem, using condition $(1.4)_I$ and (4.4), one obtains that for $t \ge t_0(n) \ge t_0^1(n)$

$$\left.\begin{aligned} &\rho(\Phi(w),\Phi(y)) \le \ell(\Gamma) \sup_{z\in\Gamma} \|d\Phi_z\| \le \\ &\le \frac{2}{t^{1-n\alpha(n)d^2}} \cdot c_3 2^d t^{\alpha(n)d} < 2^d c_3 \frac{1}{t^{1-(n+1)\alpha(n)d^2}} \end{aligned}\right\},$$

i.e. $(4.1.4)_0$ is true.

Let us suppose that $(4.1.1)_k$-$(4.1.4)_k$ is true for $t \ge t_k(n)$, $0 \le k \le n-1$.

$(4.1.2)_{k+1}$ Let us fix $w \in \Phi^k(S)$ such that $\Phi(w) \notin H_t^n$. From $(4.1.1)_k$ and $(4.1.2)_k$ one easily sees that for $t \ge t'_{k+1}(n)$

$$\Phi^k(S) \subset B(w,2(2^d c_3)^k \frac{1}{t^{1-(n+k)\alpha(n)d}}) \subset \{x \in M;\ \rho(x,A) > \frac{1}{2t^{\alpha(n)}}\}. \tag{4.5}$$

Let $y \in \Phi^k(S)$ be an arbitrary point. From $(4.1.3)_k$ it follows that $\Phi^k(S) \subset B(w,R(w,N))$. Let us denote by Γ the shortest geodesic joining w and y, $\Gamma \subset B(w,R(w,N))$. As follows from (4.5), $\ell(\Gamma) \le 2(2^d c_3)^k \frac{1}{t^{1-(n+k)\alpha(n)d}}$.

Now, by Mean Value Theorem, from condition $(1.4)_I$ and from (4.5) one obtains that for $t \ge t''_{k+1}(n) \ge t'_{k+1}(n)$,

$$\left.\begin{aligned} &\rho(\Phi(w),\Phi(y)) \le \ell(\Gamma) \sup_{z\in\Gamma} \|d\Phi_z\| < \\ &(2^d c_3)^k \frac{1}{t^{1-(n+k)\alpha(n)d}} \cdot c_3 2^d t^{\alpha(n)d} = (2^d c_3)^{k+1} \frac{1}{t^{1-(n+k+1)\alpha(n)d}} \end{aligned}\right\},$$

$(4.1.2)_{k+1}$ thus follows.

$(4.1.1)_{k+1}$ As $\Phi(w) \notin H_t^n$, then from $(4.1.2)_{k+1}$ one obtains that for $t \ge t'''_{k+1}(n) \ge t''_{k+1}(n)$

$$\Phi^{k+1}(S) \subset \{x \in M;\ \rho(x,A) \geq \frac{2}{t^{\alpha(n)}} - 2(2^d c_3)^n \frac{1}{t^{1-2n\alpha(n)d}}\} \subset$$

$$\subset \{x \in M;\ \rho(x,A) \geq \frac{1}{t^{\alpha(n)}}\}.$$

Indeed, for t big enough $\frac{1}{t^{\alpha(n)}} > (2^d c_3)^n \frac{1}{t^{1-2n\alpha(n)d}}$. Thus $(4.1.1)_{k+1}$ is proved.

$(4.1.3)_{k+1}$ From $(4.1.2)_{k+1}$ and (4.1) it follows that for every $z \in \Phi^{k+1}(S)$ one has

$$U_{\frac{1}{t}}(\Phi^{k+1}(S)) \subset B(z,(k+2)(2^d c_3)^{k+1} \frac{1}{t^{1-(n+k+1)\alpha(n)d}} + \frac{1}{t})$$

and thus the first inclusion of $(4.1.3)_{k+1}$ is proved. The second one follows, for $t \geq t^{IV}_{k+1} \geq t'''_{k+1}$, from the fact that $d > 1$.

The third inclusion of $(4.1.3)_{k+1}$ follows from (4.1). Indeed, there exists $t_{k+1}(n) \geq t^{IV}_{k+1}(n)$ such that for $t \geq t_{k+1}(n)$ one has

$$(2^d c_3)^{k+1} \frac{1}{t^{1-(n+k+1)\alpha(n)d}} < \frac{C}{t^{g\alpha(n)}} \leq \min(CR, \frac{C}{t^{g\alpha(n)}}) \leq$$

$$\leq \min(CR, C[\rho(z,A)]^g) = R(z,N).$$

$(4.1.4)_{k+1}$ The proof is along exactly the same line as the proof of $(4.1.4)_0$. We leave it to the reader. □

5. THE MAIN LEMMA

5.1 When E is a linear space, by $L(E,E)$ we denote as usual the space of all linear mappings of E into itself.

Let $\|\cdot\|$ denote the standard Euclidean norm of $\mathbb{R}^m$. For $T \in L(\mathbb{R}^m, \mathbb{R}^m)$ by $T^{\wedge k}$ we denote the k-th exterior power of T. Let us define $T^\wedge = Id \oplus (\bigoplus_{k=1}^{m} T^{\wedge k})$, where Id denotes the identity mapping in $\mathbb{R}^m$. Let us define the norm of $T^\wedge$ by $\|T^\wedge\| = 1 + \sum_{k=1}^{m} \|T^{\wedge k}\|$, where $\|T^{\wedge k}\|$ denotes the operator norm of $T^{\wedge k}$ induced by the norm $\|\cdot\|$.

We know from Sec. 2_I that there exists a measurable family of linear isometries

$$\{\tau_x : (T_xV, \|\cdot\|_x) \to (\mathbb{R}^m, \|\cdot\|)\}_{x \in V}.$$

For $x \in N$ and $n \geq 0$, let us note

$$\widetilde{d\Phi}^n_x = \tau^{-1}_{\Phi^n(x)} \circ d\Phi^n_x \circ \tau_x. \tag{5.1}$$

5.2 The following "Main Lemma" is in our framework the exact counterpart of the key remark used in the Ruelle's proof (see remark (a) from Sec. 2 of $[\text{Rue}]_1$).

LEMMA 5.1. *There exists a number* $K > 0$ *such that for every* $n \geq 2$ *there exists a number* $\tilde{t}(n)$ *such that for every* $t \geq \tilde{t}(n)$ *and for every* $S \in G(n,t)$ *and every* $x \in S$

$$\#\{A \in \xi_t;\ A \cap \Phi^n(S) \neq \emptyset\} \leq K\|(\widetilde{d\Phi^n_x})^{\wedge}\|. \tag{5.2}$$

As it will be clear from the proof, this lemma is a nonlinear variation of the following geometrical fact.

PROPOSITION 5.1. *For any* $m \geq 1$, $a \geq 0$ *and* $r > 0$, *there exists* $C_1 = C_1(m,a,r)$ *such that for every* $T \in L(\mathbb{R}^m, \mathbb{R}^m)$ *one has*

$$\mathrm{Vol}(U_a[T(B(0,r))]) \leq C_1\|T^{\wedge}\|$$

where Vol *denotes the Lebesgue measure on* $\mathbb{R}^m$.

Proof Let us note by $s_1 \geq \ldots \geq s_m$ all eigenvalues of the linear operator $(T^*T)^{1/2}$. It is well known that the lengths of principal axis of the ellipsoid $T(B(0,r))$ are equal to $rs_1,\ldots,rs_m$. Moreover $\|T^{\wedge k}\| \geq \prod_{i=1}^{k} s_i$ for $1 \leq k \leq m$. Consequently

$$\mathrm{Vol}(U_a[T(B(0,r))]) \leq \prod_{i=1}^{m} (rs_i + a) \leq$$

$$\leq C_1(m,r,a)[1 + \sum_{k=1}^{m} \prod_{i=1}^{k} s_i] \leq C_1(m,r,a)\|T^{\wedge}\|. \qquad \square$$

5.3 We pass now to the proof of Lemma 5.1. Let us introduce the following notations for $w \in S$ and $n \geq 0$ (do not confuse with $(3.13)_I$)

$$\Phi'_{kw} = \tau_{\Phi^{k+1}(w)} \circ \exp^{-1}_{\Phi^{k+1}(w)} \circ \Phi \circ \exp_{\Phi^k(w)} \circ \tau^{-1}_{\Phi^k w}, \; 0 \le k \le n-1,$$

$$\psi_{nw} = \Phi'_{(n-1)w} \circ \dots \circ \Phi'_{0w} = \tau_{\Phi^n(w)} \circ \exp^{-1}_{\Phi^n(w)} \circ \Phi^n \circ \exp_w \circ \tau_w^{-1}.$$

The mappings $\{\Phi'_{kw}\}_{0\le k\le n-1}$ and ψ_{nw} are defined in some neighborhoods of the origin in $\mathbb{R}^m$ and are all of class C^2. The symbols $d\Phi'_{kw}(y)$, $d^2\Phi'_{kw}(y)$ etc., always assume the differentiation with respect to y. As $\{\tau_w\}$ are the isometries, then the estimations of $\|d\Phi'_{kw}(h)\|$ and of $\|d^2\Phi'_{kw}(h)\|$ for $\|h\| \le R(\Phi^k(w),N)$ are given by the right hand sides of $(1.4)_I$ and $(1.3)_I$ respectively.

When $Z \subset Q(\Phi^k(w), R_N(\Phi^k(w))$, by $\hat{Z}^k$ we will note the subset $(\tau_{\Phi^k(w)} \circ \exp^{-1}_{\Phi^k(w)})(Z)$ of $\mathbb{R}^m$. As follows from Lemma 4.1, the mapping $\psi_{nw}(z)$ is well defined for all $z \in \mathbb{R}^m$, $\|z\| \le \frac{1}{t^{1-n\alpha(n)d}}$.

Now, for $t \ge t(n)$, we will estimate for $x, w \in S$ the quantity

$$J_n(w,x) = \|\psi_{nw}(\hat{x}^\circ) - d\psi_{nw}(0)\hat{x}^\circ\|.$$

Let us note by the way that $\psi_{nw}(0) = 0$.

Let us denote $v = \hat{x}^\circ - \hat{w}^\circ = \hat{x}^\circ$. As $x, w \in S$, then from condition (Bb) from Sec. 1.1_I, from $(4.1.2)_0$ and (4.1) it follows that

$$\|v\| \le \operatorname{diam}(S^\circ) \le q \operatorname{diam}(S) \le \frac{q}{t^{1-n\alpha(n)d}} \le \frac{q}{t^{3/4}}.$$

Now we arrive to the only point in this part when the second derivative of Φ plays a substantial role.

As ψ_{nw} is well defined for all $z \in \mathbb{R}^m$, $\|z\| \le \frac{1}{t^{1-n\alpha(n)d}}$, then using the Taylor Formula one obtains that $J_n(w,x) \le \|d^2\psi_{nw}(z)\| \cdot \|v\|^2$ where z is some point belonging to the interval joining v and 0. Now, using Proposition 2.1, conditions $(1.3)_I$, $(1.4)_I$ and (5.1), one obtains that for $t \ge t(n)$

$$\left.\begin{aligned} &J_n(w,x) \le \|d^2\psi_{nw}(z)\|\cdot\|v\|^2 \le \\ &\le n \cdot \max_{0\le i\le n-1} \|d^2\Phi'_{iw}(\psi_{iw}(z))\| \cdot \\ &\cdot \max\Big(\max_{0\le i\le n-1} \|d\Phi'_{iw}(\psi_{iw}(z)\|,1\Big)^{2(n-1)} \|v\|^2 \le \end{aligned}\right\} \quad (5.3)$$

$$\leq nc_2 c_3^{2(n-1)} \cdot 2^{b+2d(n-1)} \cdot t^{[b+2d(n-1)]\alpha(n)} \cdot \frac{q^2}{t^{1/2}} =$$

$$= C(n) t^{[b+2d(n-1)]\alpha(n)} \cdot \frac{1}{t^{1/2}} \leq \frac{C(n)}{t^{1/4}},$$

where $C(n) = q^2 nc_2 c_3^{2(n-1)} 2^{b+2d(n-1)}$.

Thus one obtains that for all $t \geq \tilde{t}(n) \overset{\text{def}}{=} \max(t(n), [C(n)]^4)$ and $w, x \in S$ one has

$$\|\psi_{nw}(\hat{x}^\circ) - d\psi_{nw}(0)\hat{x}^\circ\| \leq \frac{1}{t}.$$

Now we can finish the proof. As $A \in \xi_t$ and $A \cap \Phi^n(S) \neq \emptyset$, as $\operatorname{diam}(A) \leq \frac{1}{t}$, then from $(4.1.3)_n$ one obtains that $A \subset Q(\Phi^n(w), R(\Phi^n(w), N))$. Thus for such A, $\hat{A}^n = \exp^{-1}_{\Phi^n(w)}(A)$ is well defined. Consequently

$$\#\{A \in \xi_t;\ A \cap \Phi^n(S) \neq \emptyset\} \leq$$

$$\leq \#\{A \in \xi_t;\ \hat{A}^n \cap U_{\frac{1}{t}}[d\psi_{nw}(0)(\hat{S}^n)] \neq \emptyset\} \overset{\text{def}}{=} L_{nw}(S).$$

From the condition (Bb) from Sec. 1.1_I it follows that $\operatorname{diam}(\hat{A}^n) \leq \frac{q}{t}$. Moreover from the Proposition 3.1.3 it follows that $\hat{A}^n$ contains an open ball of radius $\frac{R}{12qt}$. As $\operatorname{diam}(\hat{S}^n) \leq \frac{q}{t}$, then $\hat{S}^n \subset B(0,\frac{q}{t})$ and thus

$$L_{nw}(S) \leq \#\{B \in \xi_t;\ \hat{A}^n \subset U_{\frac{1+q}{t}}[d\psi_{nw}(0)(\hat{S}^n)] \neq \emptyset\} \leq$$

$$\leq \frac{\operatorname{Vol}(U_{\frac{1+q}{t}}[d\psi_{nw}(0)(B(0,\frac{q}{t}))])}{\operatorname{Vol}(B(0,\frac{R}{12qt}))} =$$

$$= \frac{\operatorname{Vol}(U_{1+q}[d\psi_{nw}(0)(B(0,q))])}{\operatorname{Vol}(B(0,\frac{R}{12q}))} \leq$$

$$\leq C_2 \|(d\psi_{nw}(0))^\wedge\| \leq K\|(\widetilde{d\Phi_w})^\wedge\|,$$

where the constants C_2 and K do not depend on t, but only on q and R. Indeed, the second inequality follows form Proposition 5.1 and the last one from condition (Bb) of Sec. 1.1_I. □

6. THE ESTIMATION OF ENTROPY

6.1 We can now finish the proof of Theorem 1.1. Let us fix $n \geq 1$. From Propositon 2.2 and Corollary 2.1 it follows that for every $\varepsilon > 0$ there exists $t(n,\varepsilon) \geq \tilde{t}(n)$ such that for all $t \geq t(n,\varepsilon)$

$$nh_\mu(\Phi) = h_\mu(\Phi^n) \leq H(\Phi^n,\xi_t) + \varepsilon \leq H(\Phi^{-n}\xi_t|\xi_t) + \varepsilon. \tag{6.1}$$

Let us denote $\xi_t = \{A_0,A_1,\ldots,A_{p(t)}\}$ where A_0 is the "singular" element of ξ_t containing the set A. Let us denote also $z(x) = -x \log x$ for $0 < x < 1$ and $z(0) = z(1) = 0$. Clearly $z(x) \geq 0$ for $0 \leq x \leq 1$. If $a_i > 0$, $1 \leq i \leq q$ and $\sum_{i=1}^{q} a_i = 1$ then $\sum_{i=1}^{q} z(a_i) \leq \log q$.

Let us proceed now to the estimation of $H(\Phi^{-n}\xi_t|\xi_t)$. Let us consider $t \geq t(n,\varepsilon)$

$$\left.\begin{aligned}
&H(\Phi^{-n}\xi_t|\xi_t) = \\
&= \sum_{i=0}^{p(t)} \mu(A_i)\Big[-\sum_{j=1}^{p(t)} \frac{\mu(\Phi^{-n}A_j \cap A_i)}{\mu(A_i)} \log \frac{\mu(\Phi^{-n}A_j \cap A_i)}{\mu(A_i)}\Big] = \\
&= \sum_{i=0}^{p(t)} \mu(A_i)\Big[\sum_{j=0}^{p(t)} z(\Phi^{-n}A_j \cap A_i)\Big] = \\
&= \mu(A_0)\Big[\sum_{j=0}^{p(t)} z(\Phi^{-n}A_j \cap A_0)\Big] + \\
&+ \sum_{i=1}^{p(t)} \mu(A_i)\Big[\sum_{j=0}^{p(t)} z(\Phi^{-n}A_j \cap A_i)\Big] = H_B(n,t) + H_G(n,t).
\end{aligned}\right\} \tag{6.2}$$

We will estimate separately $H_B(n,t)$ and $H_G(n,t)$.

$$H_B(n,t) \leq \log p(t) \sum_{A_i \in B(n,t)} \mu(A_i).$$

Thus, from (4.3) one obtains

$$H_B(n,t) \leq 2^a(n+1)\,\frac{\log p(t)}{t^{a\alpha(n)}}. \tag{6.3}$$

Let us estimate $H_G(n,t)$.

$$H_G(n,t) \le \sum_{A_i \in G(n,t)} \mu(A_i)\cdot \log \#\{j\ ;\ \Phi^{-n}A_j \cap A_i \ne \emptyset\}.$$

Now $\#\{j;\Phi^{-n}A_j \cap A_i \ne \emptyset\} \le \#\{j;A_j \cap \Phi^n A_i \ne \emptyset\}$. Hence one obtains that

$$H_G(n,t) \le \sum_{A \in G(n,t)} \mu(A_i)\log \#\{j;\Phi^{-n}A_j \cap A_i \ne \emptyset\} \le$$

$$\le \int_{G(n,t)} \log \#\{j;A_j \cap \Phi^n A(x) \ne \emptyset\}d\mu(x),$$

where by $A(x)$ we denote the element of partition ξ_t which contains x.

Thus using the Main Lemma 5.1 one obtains:

$$\left.\begin{aligned} H_G(n,t) &\le \log K + \int_{M\setminus A_0} \log \|\widetilde{(d\Phi_x^n)}^\wedge\| d\mu(x) \le \\ &\le \log K + \int_M \log \|\widetilde{(d\Phi_x^n)}^\wedge\| d\mu(x). \end{aligned}\right\} \qquad (6.4)$$

Now, (6.1)-(6.4), (4.3) and (3.1) imple that for any $n \ge 1$, when $t \to \infty$, one has

$$h_\mu(\Phi) \le \frac{\log K}{n} + \frac{1}{n}\int_M \log \|\widetilde{(d\Phi_x^n)}^\wedge\| d\mu(x).$$

To finish the proof it is sufficient to prove that

$$\lim_{n\to\infty} \frac{1}{n}\int_M \log\|\widetilde{(d\Phi_x^n)}^\wedge\| d\mu(x) = \int_M \Big(\sum_{\chi_i(x)>0} k_i(x)\chi_i(x)\Big)d\mu(x) < +\infty. \qquad (6.5)$$

Now, as noted by Ruelle in $[\text{Rue}]_1$ (see also Appendix 2) μ almost everywhere on M, one has the following equality:

$$\lim_{n\to\infty} \frac{1}{n} \log\|\widetilde{(d\Phi_x^n)}^\wedge\| = \sum_{\chi_i(x)>0} k_i(x)\chi_i(x).$$

Indeed, this follows easily from Oseledec Multiplicative Ergodic Theorem in the form given in $[\text{Rue}]_1$ (see also $[\text{Rue}]_2$).

Thus the equality from (6.5) is equivalen to

$$\lim_{n\to\infty} \frac{1}{n}\int_M \log\|\widetilde{(d\Phi_x^n)}^\wedge\| d\mu(x) = \int_M \lim_{n\to\infty} \Big[\frac{1}{n} \log\|\widetilde{(d\Phi_x^n)}^\wedge\|\Big]d\mu(x) \qquad (6.6)$$

But (6.6) is a direct consequence of Kingman's Subadditive Ergodic Theorem (see Theorem A.1. of $[\mathrm{Rue}]_2$ and [Der], see also Theorem 1.8 of [Kin]) applied to the multiplicative cocycle $(x,n) \to \widetilde{(d\Phi_x^n)^\wedge}$, $n \geq 0$.

Taking into account (6.6), in order to prove the finiteness of (6.5), it suffices to prove that

$$\int_M \lim_{n\to\infty} [\frac{1}{n} \log\|\widetilde{(d\Phi_x^n)^\wedge}\|] d\mu(x) < +\infty. \tag{6.7}$$

This turns out to be true because $\|\widetilde{(d\Phi_x^n)^\wedge}\| \leq m(1 + \|d\Phi_x^n\|)^m$ and thus

$$\log \|\widetilde{(d\Phi_x^n)^\wedge}\| \leq \log m + m \log(1 + \|d\Phi_x^n\|) \leq$$

$$\leq \log m + m \log(1 + \|d\Phi_x\| \cdot \|d\Phi_{\Phi(x)}\| \cdot \ldots \cdot \|d\Phi_{\Phi^{n-1}(x)}\|) \leq$$

$$\leq \log m + m \sum_{k=0}^{n-1} \log(1 + \|d\Phi_{\Phi^k(x)}\|).$$

Consequently

$$\frac{1}{n} \log\|\widetilde{(d\Phi_x^n)^\wedge}\| \leq \log m + \frac{m}{n} \sum_{k=0}^{n-1} \log(1 + \|d\Phi_{\Phi^k(x)}\|). \tag{6.8}$$

Now, (6.7) follows from (6.8) and from Birkhoff Ergodic Theorem applied to Φ and μ, because condition (1.1) implies that the function $\psi(x) = \log(1 + \|d\Phi_x\|)$ belongs to $L^1(M,\mu)$. □

6.2 Now, let Φ be a C^1 mpapping of smooth compact manifold M into itself. For such Φ, inequality (1.2) remains true (see $[\mathrm{Rue}]_1$).

Indeed, in this case the validity of (4.4) follows from the simple uniform continuity argument concerning $d\Phi$.

In fact, in this case the whole demonstration is considerably simplified because $B(n,t) = \emptyset$. In particular, when Φ is a diffeomorphism of a smooth compact manifold into itself, then for every $n \geq 1$

$$1 \leq \|\widetilde{(d\Phi_x^n)^\wedge}\| \leq m(1 + \max_{m\in M}\|d\Phi_x\|)^{nm}.$$

Thus in this case, (6.6) follows directly form the Lebesgue Dominated Convergence Theorem.

APPENDIX 1. ESTIMATION OF ENTROPY OF SKEW PRODUCT FROM ABOVE THROUGH VERTICAL LYAPUNOV CHARACTERISTIC EXPONENTS

Let (X,μ) be a Lebesgue space (cf. Sec. 11.2_{II}), M be a smooth (at least of class C^3) compact Riemannian manifold, ν be a Borel probability measure defined on M. The Riemannian norm in T_m, $m \in M$, will be simply noted by $\|\cdot\|$. As in Sec. 2.2_I let

$$\{\tau_m: (T_mM, \|\cdot\|) \to (\mathbb{R}^{\dim M}, \|\cdot\|)\}_{m\in M}$$

be a measurable family of linear isometries.

Let $\{f_x\}_{x\in X}$ be a measurable family of C^1 differentiable mappings of M into itself, such that each f_x preserves the measure ν and

$$\sup_{x\in X, m\in M} \|df_x(m)\| = L < +\infty \tag{A.1}$$

Let us consider the skew product

$$\Psi(x,m) = (Tx, f_x(m)) \tag{A.2}$$

It is easy to see that Ψ preserves the probability measure $\mu\times\nu$ on $X\times M$.

With the skew product Ψ we associate the multiplicative cocycle defined on $X\times M$ with values in $GL(\dim M, \mathbb{R})$ given by the formula

$$X \times M \times Z^+ \ni (x,m,k) \to (T^k(x), \tau_{f_x^k(m)} \circ df_{xk}(m) \circ \tau_x^{-1}) \in GL(\dim M, \mathbb{R}), \tag{A.3}$$

where Z^+ denotes the non-negative integers and where $f_{x0} = Id$ $f_{xk} = f_{T^{k-1}(x)} \circ \ldots \circ f_{T(x)} \circ f_x, k \geq 1$.

From (A.2) it follows that the Oseledec Noncommutative Ergodic Theorem can be applied to the cocyle (A.3). The Lyapunov Characteristic Exponents of the cocycle (A.3) will be called the vertical Lyapunov Characteristic Exponents of the skew product Ψ. The exponents at the point (x,m) will be noted by $\{\chi_i(x,m)\}$ and their corresponding multiplicities by $\{k_i(x,m)\}$.

Let us cover the manifold M by a finite number of local charts $U_1,\ldots,U_q$. From (A.1) it follows that there exists $\beta > 0$ such that for every $x \in X$ and $m_1,m_2 \in M$, such that $d(m_1,m_2) \leq \beta$, there exists i and j, $1 \leq i, j \leq q$ such that $m_1,m_2 \in U_i$ and $f_x(m_1) \in U_j$ and $f_x(m_2) \in U_j$. The use of local coordinates in U_i

and U_j gives the possibility to define the numbers (the norms) $\|df_x(m_1) - df_x(m_2)\|$ if only $d(m_1,m_2) \le \beta$; we leave the details to the reader.

THEOREM A.1 Let us suppose that there exists a non-negative function ϕ defined on $[0,\beta]$ such that $\phi(0) = 0$, $\lim_{r\to 0} \phi(r) = 0$ and that for every $m_1,m_2 \in M$ one has

$$\sup_{x\in X} \|df_x(m_1) - df_x(m_2)\| \le \phi(d(m_1,m_2)) \tag{A.4}$$

if $d(m_1,m_2) \le \beta$. Then

$$h_{\mu\times\nu}(\Psi) \le h_\mu(T) + \int_{X\times M} \Big(\sum_{\chi_i(x,m)>0} k_i(x,m)\chi_i(x,m)\Big) d\mu d\nu \tag{A.5}$$

From condition (A.1) it easily follows that for all i, $\sup_{x\in X, m\in M} |\chi_i(x,m)| \le L < +\infty$. Thus from Theorem A.1 one obtains directly

COROLLARY A.1. If the assumptions of Theorem A.1 are satisfied, then $h_{\mu\times\nu}(\Psi) < +\infty$ iff $h_\mu(T) < +\infty$.

The proof of Theorem A.1 goes exactly along the same lines as the proof of Theorem 2.1 in absence of singularities (i.e. as the Ruelle's proof), if only one uses the Abramov-Rohlin-Adler formula for the entropy of a skew product (see [Abr], [Adl], [Pet]). For the sake of completeness we report this formula.

Let us consider the skew product (A.2). For every measurable partition α of M let us denote by α_n the partition $\bigvee_{k=0}^{n-1} (f_{xk})^{-1}\alpha$. One proves that if $H_\nu(\alpha) < +\infty$ then the limit

$$\lim_{n\to\infty} \frac{H_\nu(\alpha_n)}{n} \stackrel{\text{def}}{=} h(x,\alpha)$$

exists almost everywhere and is finite. Here by $H_\nu(\beta)$ we denote the entropy of partition β computed with respect to the measure ν.

The entropy of the skew product (A.2) is given by

$$h_{\mu\times\nu}(\Psi) = h_\mu(T) + \int_M h(\{f_x\})d\mu(x),$$

where

$$h(\{f_x\}) = \sup\{h(x,\alpha): \alpha \text{ is a measurable partition of } M,\ H_\nu(\alpha)<\infty\}.$$

This formulas allows to reduce (A.5) to the following inequality

$$h(\{f_x\}) \le \int_{X\times M} \Big(\sum_{\chi_i(x,m)>0} k_i(x,m)\chi_i(x,m)\Big)d\mu d\nu \tag{A.6}$$

Now for the entropy $h(\{f_x\})$ one can prove the exact counterparts of results from Sec. 2.2. Thus if instead of Φ^n one considers F_{nx} the proof of Theorem 1.1 can be rewritten with minor modifications in such a way that one obtains the proof of (A.6). We leave the details to the reader.

Let us note that it is also possible to formulate a counterpart of Theorem A.1 and of Corollary A.1 for the case when the mappings $\{f_x\}_{x\in X}$ are the mappings with singularities in the sense of Sec. 2.

The following easy Proposition gives sufficient conditions for the fulfillment of (A.4).

PROPOSITION A.2. *In every one of the two following cases, the condition* (A.4) *is fulfilled.*

(A.2.1) *All mappings* $\{f_x\}_{x\in X}$ *are of class* C^2 *and*

$$\sup_{x\in X, m\in M} \|d^2 f_{0x}(m)\| < +\infty, \text{ where } f_{0x}(m) = \exp_{f_x(m)}\circ f_x\circ \exp_m.$$

(A.2.2) X *is a compact metric space and the mapping* $(x,m)\to df_x(m)$ *is continuous with respect to both variables* $x\in X$ *and* $m\in M$.

When the condition (A.2.1) is satisfied and when the measure is absolutely continuous with respect to the Lebesgue measure on M, then in the estimation (A.6) and consequently in the estimation (A.5) we have an equality, i.e. the Pesin Entropy Formula for the skew product holds. The proof of the fact can be obtained along the same lines as the proof of estimation from below in Pesin formula given in Part III.

PART V

PLANE BILLIARDS AS SMOOTH DYNAMICAL SYSTEMS WITH SINGULARITIES

J.-M. Strelcyn

1. INTRODUCTION

1.1. The aim of this part is to provide full and detailed proofs of the results which allow the consideration of the large class of plane billiards in the framework of the smooth mappings with singularities in the sense explained in Parts I-IV. The class of billiards to which one can apply all the results of the present part contains in particular all the billiards whose boundary is the union of a finite number of real analytic curves (i.e. of real-analytic images of closed intervals) as well as all the billiards considered in $[\text{Sin}]_2$ and $[\text{Bun}]_{1-4}$. Consequently, one obtains for such billiards the validity of all results proved in Parts I-IV, in particular the validity of the Pesin entropy formula.

In this part we consider only the plane bounded billiards. Nevertheless one can apply our results also to the unbounded horizon case. The method is explained by an example in Sec. 10.

Although the billiards considered here represent a very large class of plane billiards, there exist many to which the method developed in Parts I-IV cannot be directly applied. The limitations of this method seem to be very difficult to overcome at present.

The multidimensional billiards and more general "Riemannian billiards" are not considered here. Nevertheless there is no doubt that similar results are true for them.

1.2. We will now indicate the exact places where the conditions from Sec. 1.1_I are verified for our billiards.

The space M is defined in Sec. 3.2. From its definition the validity of conditions (A) - (C) follows directly.

The invariant measure μ is defined in Sec. 4.4.

The singular set A is defined in Sec. 6.1.

The conditions (1.1) - (1.4) follows from Theorem 6.3, Theorem 5.1 and Theorem 7.2 respectively.

One should note that we usually give the weakest but still reasonable hypotheses which assure the validity of our assertions. Consequently, the billiard classes considered in the various sections are not the same but they become more and more restrictive.

2. TERMINOLOGY AND NOTATION

2.1. An interval means always a straight line interval. For two different points $A, B \in \mathbb{R}^2$ one denotes by $[A,B]$ (A,B), $(A,B]$ and $[A,B)$ respectively the closed, open, half-open on the left and on the right intervals connecting A and B.

By differentiability of mappings defined on closed (resp. half closed) intervals, we understand the differentiability in the interior of the interval and the existance of the right and (resp. or) left derivatives at the ends of interval.

An arc (resp. closed or open) means a homeomorphic image of an interval (resp. closed or open).

A closed curve means a Jordan curve, i.e. an homeomorphic image of a unit circle S^1.

A closed arc of class C^k means a C^k diffeomorphic image of a closed interval, $1 \leq k \leq \infty$.

Let Γ be the boundary of an open plane region Ω.

A ramification point of Γ is a point $z \in \Gamma$ in whose neighborhood Γ is not an arc containing z or beginning at z.

By points of nondifferentiability of the boundary Γ we understand the points in which Γ has no uniquely defined two sided tangents as well as the points of ramification, and the points belonging to $\partial\Gamma$ - the boundary considered as a one dimensional space. All other points of Γ will be called the points of differentiability.

A smooth boundary piece (resp. boundary piece of class C^k) means any smooth arc (resp. arc of class C^k) belonging to Γ and not containing any ramification point of Γ except perhaps its own ends. The smooth boundary pieces will be also simply called boundary pieces.

The boundary Γ of an open plane region Ω will be called piecewise C^k if it is a union of an at most countable number of boundary pieces of class C^k.

We will call a smooth boundary piece strictly curvilinear if it does not contain any interval.

A boundary piece γ is of the first kind if for any point $z \in \gamma$ any plane neighborhood of this point contains an open subset of Ω and of $\mathbb{R}^2 \backslash \Omega$. Otherwise γ is of the second kind.

A strictly curvilinear boundary piece of the first kind will be called strictly convex (resp. strictly concave) if the interval connecting any two points of this boundary piece belong to Ω (resp. to $\mathbb{R}^2\setminus\Omega$), except its ends. For the boundary pieces of the second kind the notion of convexity or concavity are meaningless. Thus, for convenience, we will call a strictly curvilinear boundary piece of the second kind strictly convex if it is visible as strictly convex from some points of Ω.

Let L be a smooth oriented arc of class C^2 and let $\{x(\phi),y(\phi)\}_{a\leq\phi\leq b}$ be the parametrisation by the arc length which agrees with the orientation of L. The curvature of L at the point $z = (x(\phi),y(\phi)) \in L$ is given by the formula

$$k(\phi) = \det \begin{pmatrix} x'(\phi), & y'(\phi) \\ x''(\phi), & y''(\phi) \end{pmatrix} \tag{2.1.}$$

$|\gamma|$ means the length of arc γ (or of closed curve γ respectively). If $\Gamma = \bigcup_{i\in I} \gamma_i$, where all $\{\gamma_i\}_{i\in I}$ are arcs or closed curves such that $\gamma_r \cap \gamma_s$ have at most one common point for all $r, s \in I$ where $r \neq s$, then, by definition, the length $|\Gamma|$ of Γ is equal $|\Gamma| = \sum_{i\leq I} |\gamma_i|$.

3. THE PLANE BILLIARDS. GENERALITIES

3.1. Consider a bounded open connected subset $\Omega \in \mathbb{R}^2$ with piecewise C^1 boundary Γ. The region Ω will be called billiard.

Denote, as usual, by $\bar{\Omega}$ the closure of Ω. Let X be the set of all tangent vectors of length one at the points of $\bar{\Omega}$, which is naturally identified with $\bar{\Omega} \times S^1$. We proceed now to the definition of the billiard flow $\{T^t\}$ in Ω. If $v \in X$ is a unit tangent vector at point $z \in \bar{\Omega}$ and directed inside Ω, and if $t \geq 0$, then $T^t v \in X$ is obtained by moving v along the straight line with unit speed during time t with elastic collision at the boundary Γ. More precisely, when the vector v reaches the boundary Γ of the region at time s, an instantaneous reflection from Γ takes place according to the rule "the angle of incidence is equal to the angle of reflection", i.e. the component of the vector tangential to Γ is preserved and the normal component changes its sign; then the motion inside $\bar{\Omega}$ continues along a straight line until a new collision with the boundary Γ. By definition, at the time s one has $T^s v = \tilde{v}$, where $\tilde{v}$ is the vector obtained at the moment of collision by the reflection described above. Moreover, if $z \in \Omega$, for all $t \leq 0$ one defines $T^t v = T^{-t}\bar{v}$, where $\bar{w} = -w$ is the tangent vector at the same point as vector w

with opposite direction. When $z \in \Gamma$, for all $t \leq 0$ by definition $T^t v = T^{-(t+r)}(T^r v)$, where $r > 0$ is sufficiently small.

While trying to define correctly the billiard trajectory $\{T^t v\}$ of v for any time $t \in \mathbb{R}$ the following three obstructions may occur:

A. For some $t \in \mathbb{R}$, $T^t v$ reaches a point of nondifferentiability of the boundary;

B. The billiard trajectory of v contains an interval belonging to Γ which is a boundary piece of the first kind as presented on Fig. 1. This interval can be reduced to a point of inflection of the border.

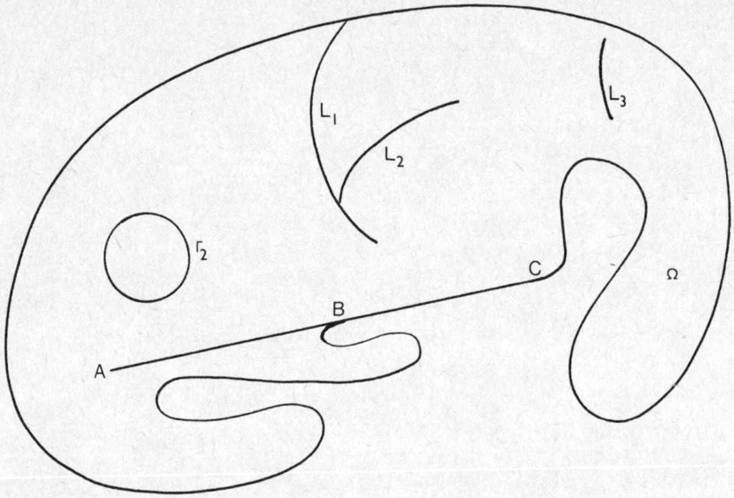

Fig. 1. Points A, B and C belong to a straight line. For $s > d(A,C)$, $T^s v$ is not defined in spite of the fact that Γ_1 is a smooth curve.

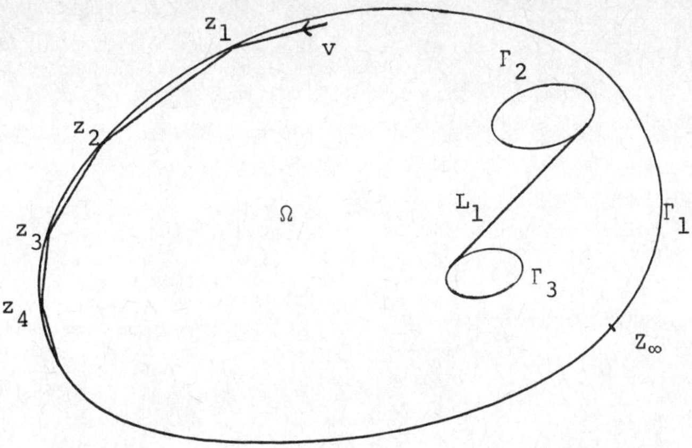

Fig. 2. $\{z_i\}_{i\geq 1}$ denotes a sequence of collisions of a trajectory of v with the boundary. $\lim_{i\to\infty} z_i = z_\infty$, and the total length of the positive half-trajectory $\{T^t v\}$ is finite in spite of the fact that Γ_1 is a smooth curve.

C. Either the positive half-trajectory $\{T^t v\}_{t\geq 0}$ or the negative half-trajectory $\{T^{-t} v\}_{t\geq 0}$ is not defined for $t \geq t(v) > 0$, in spite of the fact that all collisions with the boundary occur at the point of differentiability of Γ (Fig. 2). See [Hal] where this phenomenon was first described.

Denote by Z the set of all vectors $v \in X$ for which one of the three possibilities A - C described above occurs.

Let $Y = X\backslash Z$. It is clear that for all vectors $v \in Y$ and all $t \in \mathbb{R}$, $T^t v$ is well defined.

It follows from the definition of Y that for any $t \in \mathbb{R}$, the mapping $T^t : Y \to Y$ is bijective. It is easy to see that the Lebesque measure in $X = \overline{\Omega} \times S^1$ is invariant for T^t. Then we have on Y a flow $\{T^t\}_{t\in\mathbb{R}}$ which preserves the Lebesque measure. This flow is called the billiard flow in Ω or briefly the billiard in Ω. As it will be proved in Corollary 4.4., the Lebesque measure of Z is always equal to zero for classes of billiards studied in this paper.

It is important to note that for a fixed t, the mapping T^t may be nondifferentiable at some point $v \in X$ at which it is continuous. Indeed, this situation occurs when the straight line passing through v is tangent to the concave boundary piece.

Instead of studying directly the billiard flow $\{T^t\}$ in its phase space $X = \overline{\Omega} \times S^1$, we will apply the method of "section" which is quite natural for this problem. It is well known that the study of many ergodic properties of a flow can be reduced to the study of the same properties of the transformation induced by that flow on the section. In the case when Γ is a convex smooth closed curve, the method of section is widely used in Chap. VI of [Bir]. Nevertheless, as in our case Γ could be much more complicated, the correct description of the section and of the corresponding induced transformation is somewhat more involved. These descriptions are dealt with in the following two subsections respectively.

3.2. It follows from the definition of Ω, that Γ is a union of at most countable number of smooth closed curves $\Gamma_1, \Gamma_2, \ldots$ and of at most countable number of smooth closed arcs $L_1, L_2, \ldots$ without points of ramification except perhaps for their ends. Any two of these curves or arcs have at most one common point (see Figs. 1 and 2).

Note that if Γ contains a nondifferentiable closed curve, then we consider this curve as a union either finite or infinite number of smooth closed arcs. Moreover, the representation $\Gamma = (\bigcup_{i\geq 1} L_i) \cup (\bigcup_{j\geq 1} \Gamma_j)$ is never unique. Dividing the closed curves Γ_i, we can eliminate completely the closed curves from this representation of Γ. Furthermore, in the definition of L_i there is no uniqueness at all. The representation $\Gamma = (\bigcup_{i\geq 1} L_i) \cup (\bigcup_{j\geq 1} \Gamma_j)$ plays in our considerations a purely secondary role and the preference given to one of these representations of Γ rather than to another one depends only on the particularities of the problem under consideration.

In particular, when we consider the billiards with piecewise C^k boundary, then we will always assume that all curves $\{\Gamma_i\}_{i\geq 1}$ and all arcs $\{L_j\}_{j\geq 1}$ are of class C^k. If the boundary Γ is a union of a finite number of closed arcs of class C^k and of closed curves of class C^k, then we will consider only the representation $\Gamma = (\bigcup_{i=1}^{p} L_i) \cup (\bigcup_{j-1}^{r} \Gamma_j)$ with a finite number of arcs and curves, where all L_i and all Γ_j are of class C^k.

We will say that Γ has a fixed orientation if the orientation of any L_i and Γ_j is fixed. Let us fix an orientation of Γ. Let z belong to some L_i or to some Γ_j. Consider first the case when $z \in L_i$. As L_i is an oriented arc, one can consider the oriented tangent to L_i at z, denoted here ℓ_z. Notice that when z is an end of L_i, one considers the one-sided tangent. Let $0 \leq \theta \leq \pi$ if L_i is of the first kind and let $\theta \in S^1 = \mathbb{R}/2\pi\mathbb{Z}$ if L_i is of the second kind. In the latter case we will consider θ as a number belonging to the interval $[0,2\pi)$. We denote by $(z,\theta)_{L_i}$ the unit tangent vector at z which has an oriented angle θ with the oriented straight line ℓ_z. In a similar manner, for $z \in \Gamma_j$ and $0 \le \theta < 2\pi$, we define the vector $(z,\theta)_{\Gamma_j}$. From now on we will always suppose that the orientation of Γ satisfies the following condition: the vectors $(z,\theta)_{L_i}$ and $(z,\theta)_{\Gamma_j}$ defined above are directed inside Ω. This condition uniquely defines the orientation of all closed arcs L_i of the first kind and all closed curves Γ_j (see Sec. 2). If L_i is of the second kind, then for all $0 \le \theta < 2\pi$ and for all $z \in L_i$ the corresponding vector $(z,\theta)_{L_i}$ is directed inside Ω. To simplify the notations, from now on we will omit the indices L_i and Γ_j but this will never lead to any ambiguity. Notice that the angle with oriented normal is used in [Sin]$_2$ and [Bun]$_{1-4}$ instead of the angle with an oriented tangent also used in [Bir].

Define $M_{\Gamma_j} = \Gamma_j \times [0,\pi]$ and $M_{L_i} = L_i \times [0,\pi]$ or $M_{L_i} = L_i \times S^1$ depending on whether L_i is of the first or of the second kind respectively. M_{L_i} and M_{Γ_j} can be naturally considered as subset of $T_1\mathbb{R}^2$. Define $M = (\bigcup_{i\ge 1} M_{L_i}) \cup (\bigcup_{j\ge 1} M_{\Gamma_j})$; if $p = (z,\theta)$ and $q = (v,\Psi)$ belong to two different elements of this summs and if p and q correspond to the same vector $T_1\mathbb{R}^2$, then we identify them in M. All M_{L_i} (resp. M_{Γ_j}) are metric spaces with the metric ρ given by the formula $\rho(p,q) = \sqrt{[d(z,v)]^2 + \alpha^2}$ where $d(z,v)$ denotes the distance between z and v on L_i (resp. Γ_j) and where $\alpha = \min(|\theta-\Psi|, 2\pi-|\theta-\Psi|)$. This leads directly to the metric ρ defined on any connected component of M, by considering the shortest path between two points of such a component. The metric ρ can be extended to the whole space M if one defines $\rho(p,q) = 1$ for any $p, q \in M$ and not belonging to the same connected component of M. Notice that M is connected iff Γ is connected.

We stress that the space M depends not only on the region Ω but also on the representation $\Gamma = (\bigcup_{i\ge 1} L_i) \cup (\bigcup_{j\ge 1} \Gamma_j)$.

Generally speaking, M is not a manifold with a boundary but a set of an at most countable number of smooth compact manifolds with boundary glued together along arcs belonging to some M_{L_i} and M_{Γ_j}, $i, j \ge 1$.

Let us note that when M is a union of finite number of two dimensional compact manifolds with boundary, then capacity of M is equal two (cf. Sec. 1.1_I).

Clearly, M is a global section for the billiard flow in Ω, i.e. any trajectory of the billiard flow in Ω meets M and, after finite time, always returns to M.

3.3. Let us describe now the transformation induced on M by the billiard flow $\{T^t\}$, i.e. the transformation defined by the time of the first return to M along trajectories of the flow $\{T^t\}$. If $w = (z,\theta) \in M$ then by $L_w = L_{(z,\theta)}$ we will denote the half straight line beginning at point z and parallel to the vector $w = (z,\theta)$. By $z_1 = z_1(z,\theta)$ we will denote the nearest point of Γ belonging to the half straight line $L_{(z,\theta)}$ such that $z \neq z_1$ and such that the interval $[z,z_1]$ is entirely contained in $\bar{\Omega}$. Note that such a point z_1 may not exist at all.

Let $w = (z,\theta) \in M$ and $z_1 = z_1(w)$ be a point of differentiability of Γ. Let $s = d(z,z_1)$ and $T^s w = (z_1,\theta_1)$. We define $\Phi(z,\theta) = (z_1,\theta_1)$. This means that Φ associates to the collision (z,θ) of a billiard trajectory with boundary Γ the subsequent collision of the same trajectory with Γ (see Fig. 3).

Let us denote by S the set of ends of all arcs $\{L_i\}_{i\geq 1}$. Clearly, S contains all the points of nondifferentiability of Γ.

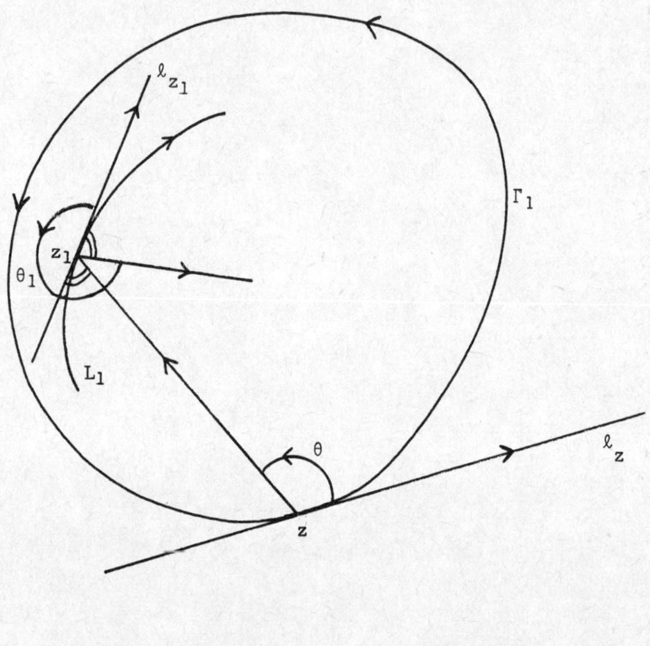

Fig. 3. $\Phi(z,\theta) = (z_1,\theta_1)$. ℓ_z and ℓ_{z_1} are the oriented tangents to L_1 and Γ_1 at points z and z_1 respectively.

We will consider $\Phi(z,\theta)$ not defined for all $w = (z,\theta) \in M$ for which the point $z \in S$ or the point $z_1 = z_1(w)$ is either not defined or is belonging to S. So, the domain of Φ depends on the representation $\Gamma = (\bigcup_{i\geq 1} L_i) \cup (\bigcup_{j\geq 1} \Gamma_j)$. Generally speaking, the domain D_Φ of Φ and

the domain D_{Φ_0} of the transformation Φ_0 induced on M by the billiard flow $\{T^t\}$ in Ω are not exactly the same, but
$\Phi|_{D_\Phi \cap D_{\Phi_0}} = \Phi_0|_{D_\Phi \cap D_{\Phi_0}}$. Thus, although it is formally not correct, we will call Φ a transformation induced on M by the flow $\{T^t\}$.

In general, the mapping $\Phi : D \to M$ is not a continuous one (see Fig. 4). It is easy to see that the mapping Φ is continuous iff Γ is a convex closed curve.

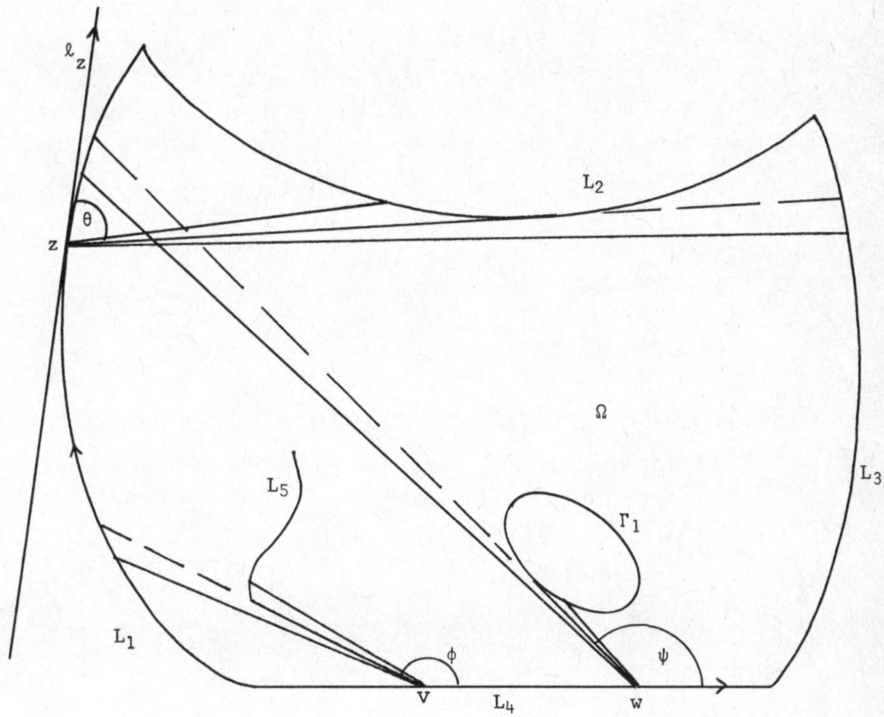

Fig. 4. The points (z,θ), (v,ϕ) and (w,Ψ) are points of discontinuity of mapping $\Phi : D_\Phi \to M$.

Denote by M_1 the subset of all $(z,\theta) \in D_\Phi$ such that $\theta \neq 0,\pi$ and such that $\theta_1 \neq 0,\pi$. Clearly, M_1 is an open subset of M and $\Phi : M_1 \to M$ is a homeomorphic imbedding.

4. THE MAPPING Φ. THE COMPUTATION OF $d\Phi$.

4.1. From now on we will study the mapping Φ. In this and in the next two subsections we deal with the differentiability of Φ and with the computation of the derivative $d\Phi$. This will lead us to the well known Φ invariant absolutely continuous measure discovered by G.D. Birkhoff. All this can be either found explicitely in Chap. VI, Sec. 8 of [Bir] or be easily deduced from it.

THEOREM 4.1. *Let* $(P^0,\theta^0) \in M_1$ *and let* $\Phi(P^0,\theta^0) = (P^1,\theta^1)$. *If* Γ *is of class* C^k, $k \geq 2$, *in some neighborhoods of points* P^0 *and* P^1, *then* Φ *is a local diffeomorphism of class* C^{k-1} *in some neighborhood of* (P^0,θ^0).

Proof. Let ϕ, $a < \phi < b$ be a parameter of class C^k corresponding to the orientation of Γ defined in a neighborhood of point P. Let ϕ_1, $c < \phi_1 < d$, be a similar parameter in a neighborhood of point P^1. Denote by $(F(\phi),G(\phi))$, $\phi \in (a,b)$, F, $G \in C^k(a,b)$ the corresponding parametric representation of Γ in the neighborhood P^0 and by $(F_1(\phi_1),G_1(\phi_1))$. $\phi_1 \in (c,d)$, F_1 $G_1 \in C^k(c,d)$ the corresponding parametric representation of Γ in the neighborhood of P^1. Let $P^0 = (F(\phi^0),G(\phi^0))$ and $P^1 = (F_1(\phi^1),G_1(\phi^1))$.

Let P (resp. P_1) denote a point from a sufficiently small neighborhood of P^0 (resp. P^1). We suppose these neighborhoods disjoint.

We will denote by τ (resp. τ_1) the oriented angle between the x-axis and the oriented tangent to Γ in P (resp. P_1). It is easy to see that $\theta = \alpha - \tau$ and $\theta_1 = \tau_1 - \alpha$ where α is an oriented angle between the x-axis and the trajectory of the billiard going from P to P_1 (see Fig. 5).

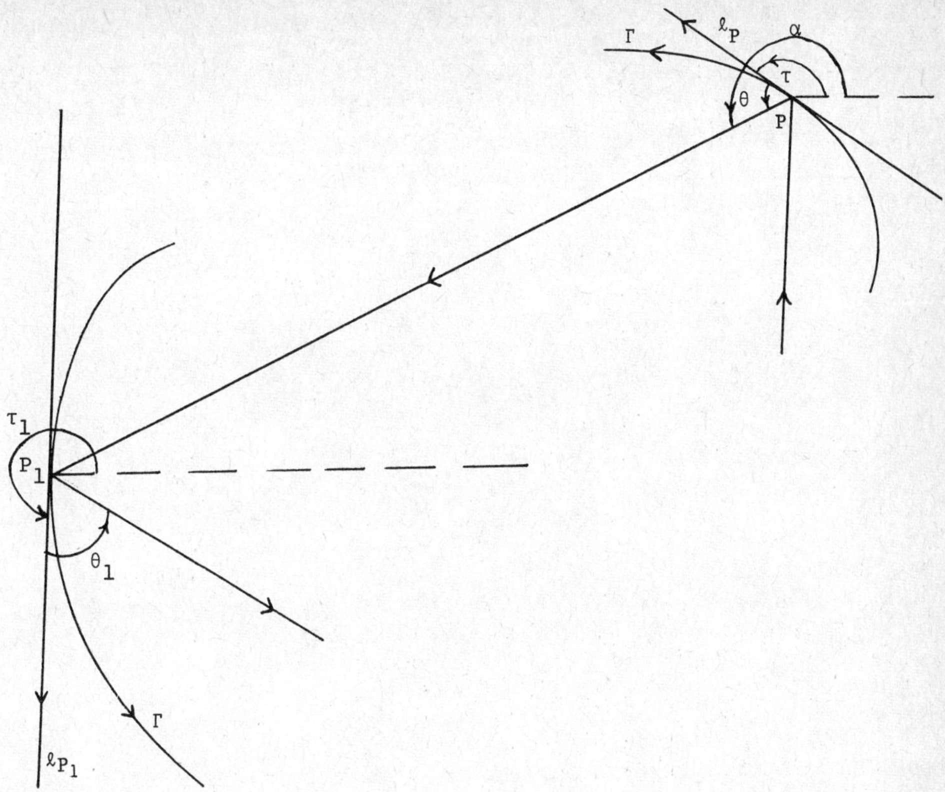

Fig. 5. $\theta = \alpha - \tau$, $\theta_1 = \tau_1 - \alpha$. On this figure the boundary pieces of Γ at the neighborhoods of points P and P_1 are convex. This is of no importance for the proof of Theorem 4.1 where only the C^2 differentiability of Γ in some neighborhoods of points P and P^1 is required.

Rotating, if necessary, the region Ω on the plane, one can always suppose that $F'(\phi^0) \neq 0$ and that $F_1'(\phi') \neq 0$. Without any loss of generality it can be supposed that $F'(\phi) \neq 0$ and $F_1'(\phi_1) \neq 0$ in sufficiently small neighborhoods of ϕ^0 and ϕ^1 respectively.

It is also easy to see that $\tan \tau = \frac{G'(\phi)}{F'(\phi)}$ and that $\tan \tau_1 = \frac{G_1'(\phi_1)}{F_1'(\phi_1)}$. Then:

$$\left.\begin{aligned} \tau &= \text{Arctan}\ \frac{G'(\phi)}{F'(\phi)} \\ \tau_1 &= \text{Arctan}\ \frac{G_1'(\phi_1)}{F_1'(\phi_1)} \end{aligned}\right\} \qquad (4.1.)$$

Thus, taking into account that $\theta = \alpha - \tau$ and $\theta_1 = \tau_1 - \alpha$, we obtain from (4.1.):

$$\begin{cases} \theta = \text{Arctan}\ \dfrac{G_1(\phi_1) - G(\phi)}{F_1(\phi_1) - F(\phi)} - \text{Arctan}\ \dfrac{G'(\phi)}{F'(\phi)} \overset{\text{def}}{=\!=} L(\phi,\phi_1) & (4.2.) \\ \theta_1 = \text{Arctan}\ \dfrac{G_1'(\phi_1)}{F_1'(\phi_1)} - \text{Arctan}\ \dfrac{G_1(\phi_1) - G(\phi)}{F_1(\phi_1) - F(\phi)} \overset{\text{def}}{=\!=} M(\phi,\phi_1) & (4.3.) \end{cases}$$

From (4.2.) and (4.3.) we obtain the following implicite function equations for $\phi_1 = \phi_1(\phi,\theta)$ and $\theta_1 = \theta_1(\phi,\theta)$:

$$\begin{cases} L(\phi,\phi_1) - \theta \overset{\text{def}}{=\!=} A(\phi,\theta,\phi_1,\theta_1) = 0 & (4.4.) \\ M(\phi,\phi_1) - \theta_1 \overset{\text{def}}{=\!=} B(\phi,\theta,\phi_1,\theta_1) = 0 & (4.5.) \end{cases}$$

$$\begin{cases} \phi_1(\phi^0,\theta^0) = \phi^1 \\ \theta_1(\phi^0,\theta^0) = \theta^1 \end{cases} \qquad (4.6.)$$

According to the Implicit Function Theorem, for the existence of functions $\phi_1 = \phi_1(\phi,\theta)$ and $\theta_1 = \theta_1(\phi,\theta)$ of class C^{k-1} satisfying the equations (4.4.) - (4.6.) in some neighborhood of (ϕ^0,θ^0) it is sufficient that:

$$J = \det\begin{pmatrix} A_{\phi_1}(\phi^0,\theta^0) , & A_{\theta_1}(\phi^0,\theta^0) \\ B_{\phi_1}(\phi^0,\theta^0) , & B_{\theta_1}(\phi^0,\theta^0) \end{pmatrix} \neq 0 \qquad (4.7.)$$

where $A_{\phi_1} = \frac{\partial A}{\partial \phi_1}$, $A_{\theta_1} = \frac{\partial A}{\partial \theta_1}$, etc.,

but $A_{\phi_1} = L_{\phi_1}$, $A_{\theta_1} = 0$, $B_{\phi_1} = M_{\phi_1}$ and $B_{\theta_1} = -1$.

Thus, we obtain that:

$$J = \det\begin{pmatrix} L_{\phi_1}(\phi^0,\theta^0), 0 \\ M_{\phi_1}(\phi^0,\theta^0), -1 \end{pmatrix} = -L_{\phi_1}(\phi^0,\theta^0).$$

Let us compute $L_{\phi_1}(\phi^0,\theta^0)$. By differentiation of (4.2.) with respect to ϕ_1 one easily obtains:

$$L_{\phi_1}(\phi^0,\theta^0) = \frac{\det\begin{pmatrix} F_1'(\phi^1) & G_1'(\phi^1) \\ F(\phi^0)-F_1(\phi^1) , & G(\phi^0)-G_1(\phi^1)\end{pmatrix}}{[F(\phi^0)-F_1(\phi^1)]^2 + [G(\phi^0)-G_1(\phi^1)]^2} \qquad (4.8.)$$

Since $P \neq P^1$, the denominator of (4.8.) is never zero. This implies that $L_{\phi_1}(\phi^0,\theta^0) \neq 0$ iff

$$\det\begin{pmatrix} F_1'(\phi^1) & , & G_1'(\phi^1) \\ F(\phi^0)-F_1(\phi^1) & , & G(\phi^0)-G_1(\phi^1)\end{pmatrix} = 0$$

This is equivalent to the fact that the vectors $(F_1'(\phi^1),G_1'(\phi^1))$ and $(F(\phi^0)-F_1(\phi^1),G(\phi^0)-G_1(\phi^1))$ are not parallel. Finally, one obtains that $L_{\phi_1}(\phi^0,\theta^0) \neq 0$ iff the straight line passing through P and P^1 is not tangent to Γ at P^1. This is so because we assume that $(P^0,\theta^0) \in M_1$. Therefore, Φ is a mapping of class C^{k-1} in some neighborhood of (P^0,θ^0).

If, instead of Φ in a neighborhood of (P^0,θ^0), one considers Φ^{-1} in a neighborhood of (P^1,θ^1), then one immediately obtains from what has been proved above, and because $\theta^0 \neq 0,\pi$, that Φ^{-1} is also a mapping of class C^{k-1} in some neighborhood of (P^1,θ^1). This means that Φ is a local diffeomorphism of class C^{k-1} in some neighborhood of (P^0,θ^0). ■

Let now the boundary Γ be a piecewise C^k, $2 \leq k \leq \infty$, then all arcs $\{L_i\}_{i\geq 1}$ and all closed curves $\{\Gamma_j\}_{j\geq 1}$ are also C^k. From Theorem 4.1. one immediately obtains the following corollary.

COROLLARY 4.1. *If the boundary* Γ *is piecewise* C^k, $2 \leq k \leq \infty$, *then* $\Phi : M_1 \to M$ *is a* C^{k-1} *diffeomorphic imbedding*.

4.2. It is quite natural then to find an explicit formula for $d\Phi$.

From now on we will always suppose that the parameters ϕ and ϕ_1 defined in the neighborhoods of points P^0 and P^1 respectively are fixed arc length parameters on Γ, i.e. the parameters given by the length of arcs Γ measured from some fixed points of Γ.

Let us proceed now to the Benettin formula for $d\Phi(P,\theta)$ which is the basis of all future considerations.

From now on we will speak very often without any distinction of points P and P_1, and ϕ and ϕ_1 respectively. Instead of considering $d\Phi(P,\theta)$ we will consider $d\Phi(\phi,\theta)$, etc.

Let us introduce the following notations: $d = \sin\theta$, $d_1 = \sin\theta_1$, $k = k(\phi)$, $k_1 = k(\phi_1)$ where, recall, $k(z)$ denote the curvature of Γ at point z. By $\ell = \ell(P,\theta)$ we will denote the distance between the points P and P_1, i.e. $\ell = \sqrt{[F(\phi)-F(\phi_1)]^2 + [G(\phi)-G(\phi_1)]^2}$ where, as before $\Phi(P,\theta)=(P_1,\theta_1)$.

THEOREM 4.2. <u>Suppose that in some neighborhoods of points</u> ϕ^0 <u>and</u> ϕ^1 <u>respectively</u> Γ <u>is a</u> C^2 <u>curve. Let</u> $(\phi^0,\theta^0) \in M_1$ <u>and let</u> $\Phi(\phi^0,\theta^0) = (\phi^1,\theta^1)$. <u>Then, for</u> (ϕ,θ) <u>belonging to a sufficiently small neighborhood of</u> (ϕ^0,θ^0) <u>one has</u>:

$$d\Phi(\phi,\theta) = \begin{pmatrix} \frac{\partial\phi_1}{\partial\phi}(\phi,\theta) & , & \frac{\partial\phi_1}{\partial\theta}(\phi,\theta) \\ \frac{\partial\theta_1}{\partial\phi}(\phi,\theta) & , & \frac{\partial\theta_1}{\partial\theta}(\phi,\theta) \end{pmatrix} = \tag{4.9.}$$

$$\begin{pmatrix} \varepsilon_1\frac{k\ell - \varepsilon d}{d_1} & , & \varepsilon_1\frac{\ell}{d_1} \\ \varepsilon_1 k_1\frac{k\ell - \varepsilon d}{d_1} - k & , & \varepsilon_1\frac{k_1\ell}{d_1} - 1 \end{pmatrix} \tag{4.10.}$$

<u>where</u> $\varepsilon = \varepsilon(\phi^0,\theta^0) = \pm 1$ <u>and</u> $\varepsilon_1 = \varepsilon_1(\phi^0,\theta^0)$ <u>are the constants depending only on the orientation of</u> Γ. <u>In a particular case when</u> Γ <u>is smoothly parametrised arc or smoothly parametrised closed curve, then</u> $\varepsilon = \varepsilon_1 = 1$.

We precede the proof of Theorem 4.2. by the following lemma:

<u>LEMMA 4.1</u>. <u>Under the conditions of Theorem</u> 4.2. <u>one has</u>:

$$L_\phi = \varepsilon\frac{d}{\ell} - k \tag{4.11.}$$

$$L_{\phi_1} = \varepsilon_1\frac{d_1}{\ell} \tag{4.12.}$$

$$M_\phi = -\varepsilon\frac{d}{\ell} \tag{4.13.}$$

$$M_{\phi_1} = k_1 - \varepsilon_1\frac{d_1}{\ell} \tag{4.14.}$$

<u>where</u> $\varepsilon = \pm 1$ <u>and</u> $\varepsilon_1 = \pm 1$ <u>are absolute constants depending only on the orientation of</u> Γ. <u>In particular case when</u> Γ <u>is smoothly parametrised arc or smoothly parametrised closed curve, then</u> $\varepsilon = \varepsilon_1 = 1$.

<u>Proof</u>. To prove the formulas (4.11.) - (4.14.) one uses the consecutive

differentiation of formulas (4.2.) - (4.3.) and the formula (2.1.). Let us prove for example the formula (4.12.). The other formulas are proved analogously.

We will use the formula (4.3.) which gives the value of $L_{\phi_1}(\phi,\theta)$ in the form of a fraction. The denominator of this formula is equal to ℓ^2. The numerator is equal to the oriented area of the parallelogram spanned by vectors $(F_1'(\phi_1), G_1'(\phi_1))$ and $(F(\phi)-F_1(\phi_1), G(\phi)-G_1(\phi_1))$. The length of the first vector is equal to 1 because ϕ is an arc length parameter on Γ. The length of the other one is equal to 1. Thus, one obtains that:

$$\det\begin{pmatrix} F'(\phi_1) & , G'(\phi_1) \\ F(\phi)-F_1(\phi_1) & , G(\phi)-G_1(\phi_1) \end{pmatrix} = \varepsilon_1 \ell \sin\theta_1 = \varepsilon_1 \ell d_1 \qquad (4.15.)$$

where $\varepsilon_1 = \pm 1$ is a constant depending only on the orientation of Γ. This implies that $L_{\phi_1}(\phi,\theta) = \varepsilon_1 \frac{d}{\ell}$. ∎

<u>Proof of Theorem 4.2.</u> Let us calculate $\frac{\partial\phi_1}{\partial\theta}$, $\frac{\partial\theta}{\partial\phi}$, $\frac{\partial\phi_1}{\partial\phi}$ and $\frac{\partial\theta_1}{\partial\theta}$.

a) Calculation of $\frac{\partial\phi_1}{\partial\theta}$. Differentiating (4.2.) with respect to θ one obtains: $L_{\phi_1}\frac{\partial\phi_1}{\partial\theta} = 1$ and by consequence $\frac{\partial\phi_1}{\partial\theta} = \frac{1}{L_{\phi_1}(\phi,\phi_1)}$. Using (4.15.) one finally obtains:

$$\frac{\partial\phi_1}{\partial\theta}(\phi,\theta) = \varepsilon_1 \frac{\ell}{d_1} \qquad (4.16.)$$

b) Calculation of $\frac{\partial\theta_1}{\partial\phi}$. Differentiating (4.4.) and (4.5.) with respect to ϕ and using Lemma 4.1., one easily obtains:

$$\frac{\partial\theta_1}{\partial\phi} = \frac{1}{L_{\phi_1}} \det\begin{pmatrix} L_{\phi_1} & , M_{\phi_1} \\ L_\phi & M_\phi \end{pmatrix} =$$

$$= \varepsilon_1 \frac{\ell}{d_1} \det\begin{pmatrix} L_{\phi_1} & , M_{\phi_1} \\ L_\phi & , M_\phi \end{pmatrix} = \varepsilon_1 \frac{\ell}{d_1} \det\begin{pmatrix} \varepsilon_1 \frac{d_1}{\ell} & , k_1 - \varepsilon_1 \frac{d_1}{\ell} \\ \varepsilon\frac{d}{\ell} - k & , -\varepsilon\frac{d}{\ell} \end{pmatrix} =$$

$$= \varepsilon_1 k_1 \left(\frac{k\ell - \varepsilon d}{d_1} \right) - k.$$

c) Calculation of $\frac{\partial\phi_1}{\partial\phi}$. Differentiating (4.3.) with respect to ϕ one obtains that $L_\phi + L_{\phi_1}\frac{\partial\phi_1}{\partial\phi} = 0$, and by consequence $\frac{\partial\phi_1}{\partial\phi} = -\frac{L_\phi}{L_{\phi_1}}$.

Thus, from (4.11.) and (4.12.) one obtains: $\frac{\partial \phi_1}{\partial \phi}(\phi_1 \theta) = \varepsilon_1 \frac{k\ell - \varepsilon d}{d_1}$.

d) Calculation of $\frac{\partial \theta_1}{\partial \theta}$. Differentiating (4.4.) with respect to θ one obtains that $\frac{\partial \theta_1}{\partial \theta} = M_{\phi_1} \frac{\partial \phi_1}{\partial \theta}$. Thus, from (4.14.) and (4.16.) one obtains:

$$\frac{\partial \theta_1}{\partial \theta}(\phi_1, \theta) = \varepsilon \frac{k_1 \ell}{d_1} - 1. \qquad \blacksquare$$

4.3. Let us fix once and for all a norm in the space of 2×2 matrice, for example the norm $\left\| \begin{pmatrix} a & , & b \\ c & , & d \end{pmatrix} \right\| = |a| + |b| + |c| + |d|$. From Theorem 4.2. we immediately obtain the following important consequences.

<u>COROLLARY 4.2. Let I and J be two disjoint closed C^2 boundary pieces of Γ. Let $\bar{P}$ be one of the ends of I and let $\bar{P}^1$ be one of the ends of J. Let $\bar{\theta}$ be a fixed angle. Let $(P,\theta) \in M_1$ where $P \in I$; $\theta < \bar{\theta}$ ($\theta > \bar{\theta}$ respectively) and where (P,θ) is sufficiently near $(\bar{P},\bar{\theta})$, so that $P_1 = P_1(P,\theta) \in J$. Let</u>

$$\lim_{\substack{(P,\theta) \to (\bar{P},\bar{\theta}) \\ P \in I,\ (P,\theta) \in M_1 \\ \theta < \bar{\theta}}} P_1(P,\theta) = \bar{P}^1. \quad \text{Then:} \quad \lim_{\substack{(P,\theta) \to (\bar{P},\bar{\theta}) \\ P \in I,\ (P,\theta) \in M_1 \\ \theta < \bar{\theta}}} \| d\Phi(\phi,\theta) \| = +\infty$$

<u>holds iff the straight line passing through $\bar{P}$ and $\bar{P}^1$ is tangent to the boundary piece J at $\bar{P}^1$. A completely analogous assertion is valid in the case of $\theta > \bar{\theta}$. (See Figs. 6a - 6e).</u>

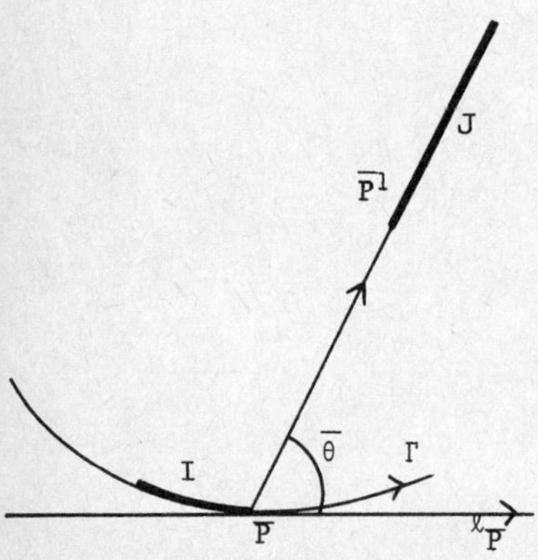

Fig. 6a

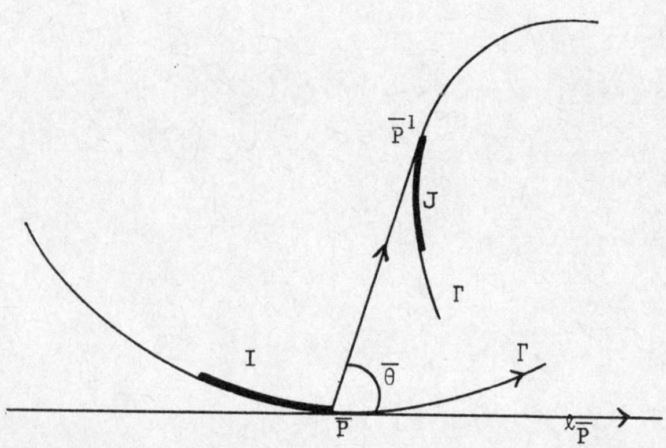

Fig. 6b

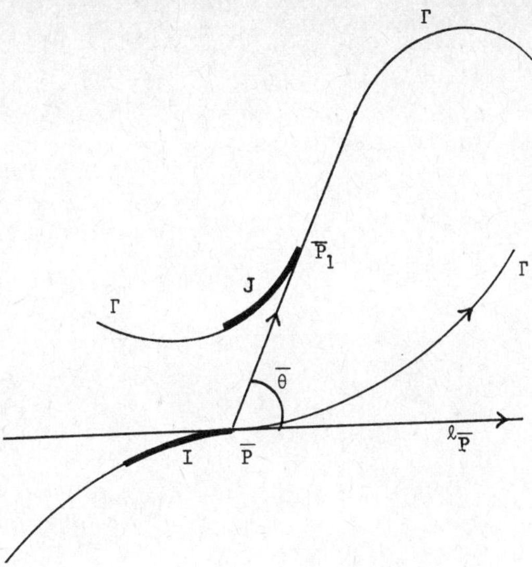

Fib. 6c

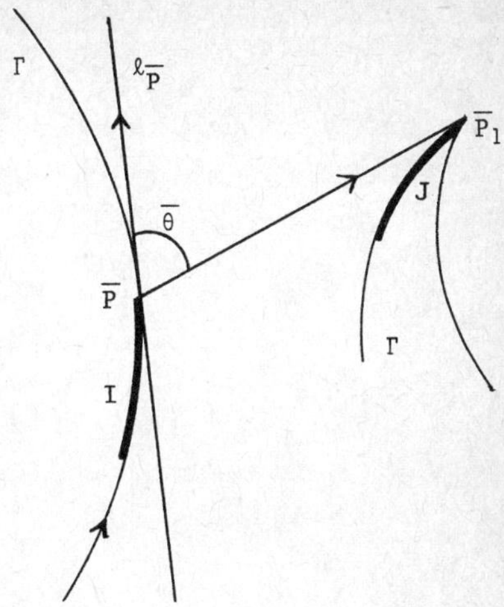

Fig. 6d

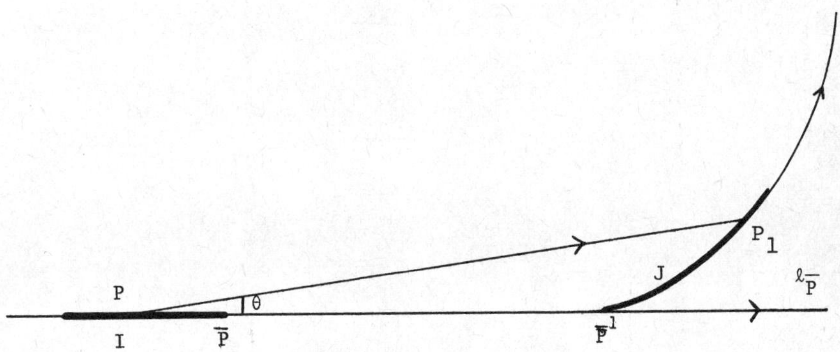

Fig. 6e

Figs. 6a - 6e show five typical situations described in Corollary 4.2. Figs. 6a, 6b and 6e correspond to the case $\theta > \overline{\theta}$ and Figs. 6c - 6d to the case $\theta < \overline{\theta}$ respectively. On Figs. 6a and 6c Γ contains an interval beginning at $\overline{P}^1$ and belonging to the straight line going through $\overline{P}$ and $\overline{P}^1$. On Fig. 6d the billiard trajectory is one-sided tangent to Γ in $\overline{P}^1$. On Fig. 6e the interval going from $\overline{P}$ to $\overline{P}^1$ belongs to Γ.

THEOREM 4.3. *Let* γ *be a strictly convex* C^2 *boundary piece of* Γ. *Let* $\overline{P}$ *be one of the ends of* γ. *Suppose that* γ *is oriented in such a way that for sufficiently small* $\theta > 0$ *one has* $\overline{P}_1 = \overline{P}_1(\overline{P}, \theta) \in \gamma$.

Then $\lim\limits_{\substack{(P,\theta)\to(\overline{P},0)\\ P\in\gamma,\ \theta>0}} \|d\Phi(P,\theta)\| = +\infty$ *holds iff* γ *is of the curvature zero at* $\overline{P}$.

Proof. To simplify the notation we will use lim instead of $\lim\limits_{\substack{(P,\theta)\to(\overline{P},0)\\ P\in\gamma,\ \theta>0}}$

Consider the orthogonal system of co-ordinates (x,y) where the x and the y-axes are defined as follows: the x-axis is the oriented (one-sided) tangent to γ at $\overline{P}$ and the y-axis is an oriented straight line going through point $\overline{P}$ and perpendicular to the x-axis. For an adequate orientation of the y-axis in a sufficiently small neighborhood U of point $\overline{P} = (0,0)$, the curve γ is a graph of strictly convex function g of class C^2 defined for $0 \le x \le x_{\overline{P}}$ where $x_{\overline{P}} > 0$ and such that $g(0) = g'(0) = 0$. The curve γ is of curvature zero at P iff $g''(0) = 0$.

Let now $P \in \gamma \cap U$, $P \neq \overline{P}$ and let P be different from the ends of γ. Point P divides γ into two parts, γ_+ and γ_-, and we will suppose that $\overline{P} \in \gamma_+$. Exactly the same construction as that we have just described with respect to $\overline{P}$ and γ can be also made with respect to P and $\gamma_+ \cap U$. Then, in the corresponding orthogonal system of co-ordinates (x,y), the curve $\gamma_+ \cap U$ is a graph of a strictly convex function f_P of class C^2, defined for $0 \le x \le x_P$ and such that $f_P(0) = f_P'(0) = 0$. (See Fig. 7).

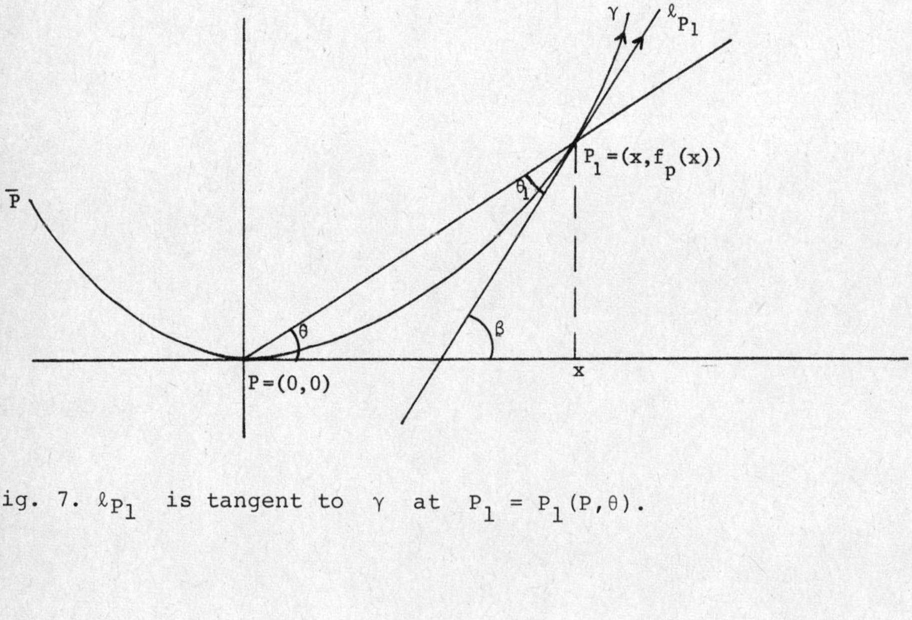

Fig. 7. ℓ_{P_1} is tangent to γ at $P_1 = P_1(P,\theta)$.

It is easy to see that $\lim_{P\to\overline{P}} f_P^{(i)}(x) = g^{(i)}(x)$ for all sufficiently small $x > 0$ and that this convergence is uniform on the interval $\left[0,\frac{x_{\overline{P}}}{2}\right]$, $0 \le i \le 2$. Here $h^{(i)}$ denotes the i-th derivative of h, $h^{(0)} \underset{=}{\overset{def}{=}} h$.

Let us suppose that the curvature of γ at $\overline{P}$ is zero, i.e. let $g''(0) = 0$.

Let us estimate from below for sufficiently small $\theta > 0$ the quantity $\frac{\ell(P,\theta)}{\sin\theta_1}$ where, as we recall, $\ell(P,\theta)$ is the distance between P and $P_1 = P_1(P,\theta)$. Clearly, $\tan\theta_1 = \tan(\beta-\theta) = \frac{\tan\beta - \tan\theta}{1+\tan\beta\tan\theta} \le$

$\tan\beta - \tan\theta = f_P'(x) - \frac{f_P(x)}{x}$. This implies that

$$\frac{\ell(P,\theta)}{\sin\theta_1} = \frac{\tan\theta_1}{\sin\theta_1}\cdot\frac{\ell(P,\theta)}{\tan\theta_1} \ge \frac{1}{2}\cdot\frac{\tan\theta_1}{\sin\theta_1}\cdot\frac{x+f_P(x)}{f_P'(x)-\frac{f_P(x)}{x}} =$$

$$=\frac{1}{2}\cdot\frac{\tan\theta_1}{\sin\theta_1}\cdot\frac{x(x+f_P(x))}{xf_P'(x)-f_P(x)} > \frac{1}{2}\cdot\frac{\tan\theta_1}{\sin\theta_1}\cdot\frac{x^2}{xf_P'(x)-f_P(x)}.$$

Applying to $\frac{x^2}{xf_P'(x)-f_P(x)}$ the Cauchy Mean Value Theorem one obtains that for some $0 < \tilde{x} < x$ one has $\frac{x^2}{xf_P'(x)-f_P(x)} = \frac{2}{f''_P(\tilde{x})}$.

This implies that $\frac{\ell(P,\theta)}{\sin\theta_1} > \frac{\tan\theta_1}{\sin\theta_1}\cdot\frac{1}{f_P''(\tilde{x})}$. As $\lim_{\substack{\theta\to 0\\ \theta>0}}\frac{\tan\theta_1}{\sin\theta_1} = 1$ and $\lim_{\substack{P\to\overline{P}\\ P\in\gamma_+}} f_P''(\tilde{x}) = g''(0) = 0$, in virtue of formula (4.10.) for $d\Phi$, one deduces immediately that if γ is of curvature zero at $\overline{P}$, then

$$\lim \|d\Phi(P,\theta)\| \ge \frac{\ell(P,\theta)}{\sin\theta_1} \ge \lim\frac{1}{f_P''(\tilde{x})} = +\infty.$$

Let now γ be of non zero curvature at $\overline{P}$, i.e. let $g''(0) \ne 0$. Then, as $\tan\theta_1 \ge \frac{1}{2}(\tan\beta - \tan\theta)$ one obtains

$$0 \le \lim\frac{\ell(\overline{P},\theta)}{\sin\theta_1} \le \frac{2}{g''(0)} < +\infty. \tag{4.17.}$$

Now we will prove that

$$\lim \frac{\sin\theta}{\sin\theta_1(\theta,P)} = 1. \tag{4.18.}$$

As $\lim\limits_{\substack{\theta\to 0\\ \theta>0}} \frac{\tan\theta}{\sin\theta} = \lim \frac{\tan\theta_1(P_1,\theta)}{\sin\theta_1(P,\theta)} = 1$, to demonstrate (4.18.) it is sufficient to prove that

$$\lim \frac{\tan\theta}{\tan\theta_1(P,\theta)} = 1. \tag{4.19.}$$

But $\operatorname{tg}\theta = \frac{f_P(x)}{x}$ and $\tan\theta_1 = \tan(\beta-\theta) = \frac{\tan\beta - \tan\theta}{1+\tan\beta\tan\theta} =$

$= \dfrac{f_P'x - \frac{f_P(x)}{x}}{1+f_P'(x)\frac{f_P(x)}{x}}$. From this, just as before, one has $\frac{\tan\theta}{\tan\theta_1(P,\theta)} =$

$= (1+f_P'(x)\cdot\frac{f_P(x)}{x})\cdot\frac{f_P(x)}{xf_P''(x) - f_P(x)} = (1+f_P'(x)\cdot\frac{f_P(x)}{x})\cdot\frac{f_P'(\tilde{x})}{\tilde{x}f_P''(\tilde{x})}$. But

$\lim f_P'(x) = 0$ and $\lim \frac{f_P'(x)}{x} = g''(0)$. From this, one has

$\lim \frac{\tan\theta}{\tan\theta_1(P,\theta)} = \lim \frac{f_P'(x)}{x}\cdot\lim\frac{1}{f_P''x} = 1$. Then (4.19.), and consequently (4.18.), are proved. Now our assertion immediately follows from (4.17.), (4.18.) and from formula (4.10.) for $d\Phi$. ∎

4.4. Now, let us formulate another consequence of Theorem 4.2.

COROLLARY 4.3. If in the assumption of Theorem 4.2. we consider a sufficiently small neighborhood U of the point (P^0,θ^0), then Φ restricted to U preserves the measure $d\mu_0 = |\sin\theta|d\theta d\phi$, i.e. for every measurable subset $S\subset U$, $\mu_0(S) = \mu_0(\Phi(S))$.

Proof. The formula (4.10.) for $d\Phi$ immediately implies that $|\det d\Phi(\phi,\theta)| = \left|\frac{\sin\theta}{\sin\theta_1}\right|$. Now, the Corollary 4.3. follows from the formula of change of variable in double integral. ∎

From the definition of measure μ one obtains immediately that always $\mu_0(M\setminus M_1) = 0$, where M_1 is the subset of M defined at the end of Sec. 3.3. We recall that $|\Gamma|$ denotes the length of Γ. Clearly $\mu_0(M) < +\infty$ iff $|\Gamma| < +\infty$.

COROLLARY 4.4. If $|\Gamma| < +\infty$, then the probability measure

$d\mu = \frac{1}{\mu_0(M)} |\sin\theta|\, d\theta d\phi$ *is* Φ *invariant, i.e. for every measurable subset* $S \subset M$ *one has* $\mu(S) = \mu(\Phi(S))$.

Corollaries 4.3. and 4.4. go back to G.D. Birkhoff (see Chap. VI, Sec. 8 of [Bir]). G.D. Birkhoff proved Corollary 4.3. without explicit computation of elements of the matrix $d\Phi(\phi,\theta)$, but he computes $\det d\Phi(\phi,\theta)$ directly, using the formulas obtained by differentiation of formulas (4.1.) - (4.3.).

Finally, note that Corollary 4.4. and the Birkhoff Ergodic Theorem imply easily that $\nu(Z) = 0$, where ν denotes the Lebesgue measure in $X = \overline{\Omega} \times S^1$.

COROLLARY 4.5. If $|\Gamma| < +\infty$, *then* $\nu(Z) = 0$.

Proof. $Z = Z_A \cup Z_B \cup Z_C$, where Z_A, Z_B and Z_C are sets of such $v \in T_1\Omega$ for which the cases A, B and C described in Sec. 3.1. occure respectively. Clearly, $\nu(Z_A) = \nu(Z_B) = 0$. It remains only to prove that $\nu(Z_C) = 0$.

For this purpose it is enough to prove that $\mu(Z_C \cap M_1) = 0$, as $\mu(M_1) = 1$.

For $w = (z,\theta) \in M$ we will note $\ell(z,\theta) = \ell(w)$. By definition of case C it follows that for $v \in M_1$ one has $v \in Z_C$ iff $\sum_{k=1}^{\infty} \ell(\Phi^k v) < +\infty$. In particular, for $v \in Z_C \cap M_1$, one has $\lim_{n\to\infty} \frac{1}{n} \sum_{k=1}^{n} \ell(\Phi^k v) = 0$. Clearly, $Z_C \cap M_1$ is a Φ invariant set. Suppose that $\mu(Z_C \cap M_1) = c > 0$. Then, from the Birkhoff Ergodic Theorem applied to the space $Z_C \cap M_1$, to the transformation $\Phi : Z_C \cap M_1 \to Z_C \cap M_1$, to the Φ invariant probability measure $\mu_c = \frac{1}{c}\mu$ on $Z_C \cap M_1$, and to the function ℓ, one gets $\lim_{n\to\infty} \frac{1}{n} \sum_{k=1}^{n} \ell(\Phi^k v) = \ell^*(v)$, μ_c-almost everywhere, and $\int_{Z_C \cap M_1} \ell\, d\mu_c = \int_{Z_C \cap M_1} \ell^* d\mu_c$. As ℓ is strictly positive on M_1, $\int_{Z_C \cap M_1} \ell\, d\mu_c > 0$. This is in contradiction with the fact that $\ell^*(v) = 0$ on $Z_C \cap M_1$. So, $\mu(Z_C \cap M_1) = 0$. ∎

Thus, if $|\Gamma| < +\infty$, the mapping Φ is defined μ-almost everywhere on M and Φ preserves the absolutely continuous probability measure μ. In the future, when speaking about ergodic properties of

Φ, we will consider them exclusively with respect to the measure μ.

4.5. Let us make now a few remarks about the orientation of the curves $\{\Gamma_i\}_{i\geq 1}$ and $\{L_j\}_{j\geq 1}$, and of the numbers ε and ε_1 (see formula (4.10.) associated with them.

If Γ is the union of mutually disjoined closed curves H_1, H_2, H_3, ... and if all curves $\{H_i\}_{i\geq 2}$ are inside the curve H_1, then orienting all the curves $\{H_i\}_{i\geq 2}$ (and therefore also their smooth boundary pieces) in the direction opposite to that of H_1 (see Fig. 8), one obtains $\varepsilon = \varepsilon_1 = 1$.

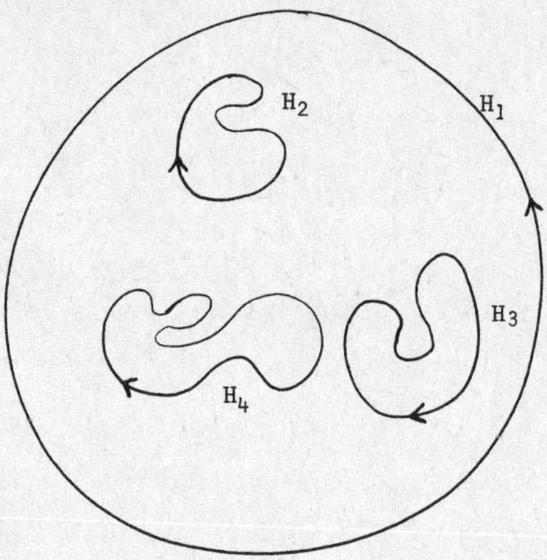

Fig. 8

Nevertheless, generally speaking, in other cases one could always find a $(z,\theta) \in M$ such that in some neighborhoods of z and z_1 respectively, ε and ε_1 are of opposite signs. (See Fig. 9 which shows precisely such a situation).

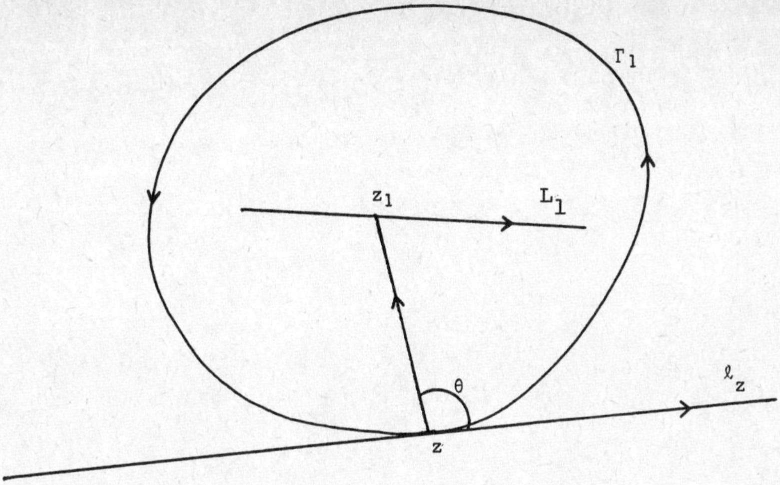

Fig. 9

Notice also that if one changes the orientation of Γ, then the elements of the matrix $d\Phi(\phi,\theta)$, i.e. the matrix (4.9.) change their signs appropriately. It is easy to see that this change of signs is in complete agreement with the rule of transformation of matrix (4.10.) when one changes the orientation of Γ. Indeed, the quantities εk and $\varepsilon_1 k_1$ do not depend the orientation of Γ but intrinsicly depend on Γ only.

4.6. Finally, let us note that very large classes of plane regions defined in this section do not yet exhaust the class of regions for which the results described in this section as well as those of Sec. 5.1. remain true (see Fig. 10).

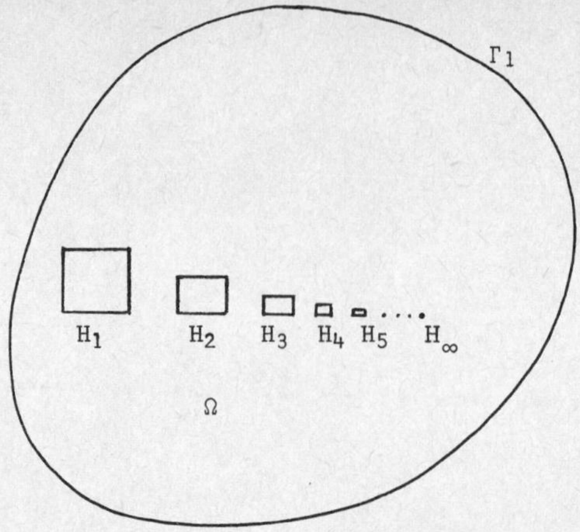

Fig. 10. For all k, $1 \le k < \infty$, H_k are mutually disjoint rectangles. H_∞ is the one point compactification of $\bigcup_{k \ge 1} \Gamma_k$

5. APPLICABILITY OF THE OSELEDEC MULTIPLICATIVE ERGODIC THEOREM

5.1. From now on we will assume, without repeating it, that all considered billiards have boundaries of finite length.

Since we would like to apply the Pesin theory to the billiards, we have first to discuss the applicability of the Oseledec theorem (see Appendix 2) to the map ϕ. Since in the Oseledec theorem one uses $d\Phi$, then from now on we will consider almost always only billiards with piecewise C^2 boundary. However, such a class is too large because in Secs. 5.2. - 5.4. we will describe an example of a simply connected billiard whose boundary is of class C^1 and except for one point also of class C^∞ which does not satisfy the assumption of the Oseledec theorem.

THEOREM 5.1. *Let us consider a billiard with a piecewise C^2 boundary of finite length. If the absolute value of the curvature of its boundary is uniformly bounded, then the Oseledec theorem is applicable to the mapping Φ.*

Proof. We are to prove that

$$\int_M \log^+ \| d\Phi(\phi,\theta) \| |\sin\theta| d\phi d\theta < +\infty \tag{5.1.}$$

and that

$$\int_M \log^+ \| [d\Phi(\phi,\theta)]^{-1} \| \, |\sin\theta| d\phi d\theta < +\infty \qquad (5.2.)$$

where $\log^+ a = \max(0, \log a)$.

Denote by h the mapping of M onto itself given by the formula

$$h(\phi,\theta) = \begin{cases} (\phi, \pi-\theta) & \text{for } 0 \le \theta \le \pi \\ (\phi, 3\pi-\theta) & \text{for } \pi \le \theta \le 2\pi \end{cases}$$

Clearly, $h = h^{-1}$ and $h \circ \Phi \circ h = \Phi^{-1}$ (the so called "time reversal property"). Moreover, as $dh(\phi,\theta) = \begin{pmatrix} 1 & , & 0 \\ 0 & , & -1 \end{pmatrix}$ then $\|dh(\phi,\theta)\| = 2$ for all $(\phi,\theta) \in M$. From the time reversal property and from the equality $\|dh(\phi,\theta)\| = 2$, it follows immediately that (5.1.) is equivalent to (5.2.). Therefore, it is sufficient to prove (5.1.).

From the formula (4.10.) one obtains that for all $(\phi,\theta) \in M_1$ one has $\|d\Phi(\phi,\theta)\| \le \frac{1}{|\sin\theta_1|} \{|k\ell - \varepsilon d| + 1 + |k_1(k\ell - \varepsilon d)| + |d_1 k| + {}$ $+ |k_1| \ell + \ell\}$. As the absolute value of the curvature of Γ is uniformly bounded, one deduces that for some number C, $C < +\infty$ and for all $(\phi,\theta) \in M_1$ one has $\|d\Phi(\phi,\theta)\| \le \frac{C}{|\sin\theta_1|}$. So, we obtain that

$$\iint_M \log^+ \|d\Phi(\phi,\theta)\| \, |\sin\theta| d\phi d\theta \le \iint_M \log^+ \left(\frac{C}{|\sin\theta_1|}\right) |\sin\theta| d\phi d\theta \le$$

$$\iint_M \left|\log \frac{C}{|\sin\theta_1|}\right| |\sin\theta| d\phi d\theta \le |\log c| \iint_M |\sin\theta| d\phi d\theta +$$

$$-\iint_M (\log|\sin\theta_1|) |\sin\theta| d\phi d\theta.$$

As $\iint_M |\sin\theta| d\phi d\theta = \mu_0(M) < +\infty$, hence to prove (5.1.) it is enough to prove that

$$\iint_M (\log|\sin\theta_1(\phi,\theta)|) |\sin\theta| d\phi d\theta > -\infty$$

However, as $d\mu_0 = |\sin\theta| d\phi d\theta$ is a Φ invariant measure, one has

$$\iint_M (\log|\sin\theta_1(\phi,\theta)|) |\sin\theta| d\phi d\theta = \iint_M (\log|\sin\theta|) |\sin\theta| d\phi d\theta.$$

As $\int_0^\pi \log\sin\theta d\theta > -\infty$ one has also $\iint_M (\log|\sin\theta|) |\sin\theta| d\theta d\phi > -\infty$.

■

COROLLARY 5.1. For any billiard whose boundary is the union of a finite number of closed arcs of class C^2 and of finite numbers of closed curves of class C^2, the Oseledec theorem is applicable to the mapping ϕ.

If instead of arcs of class C^2 one considers arcs of class C^k, $1 \le k \le \infty$, a class of billiards as that described in Corollary 5.1. will be called class P_k.

5.2. In the following three subsections we will construct an example of a simply connected billiard Ω whose boundary Γ is of class C^1 and, except for one point, is also of class C^∞ such that

$$\int_M \log^+ \| d\Phi(\phi,\theta) \| \sin\theta d\phi d\theta = +\infty \tag{5.3.}$$

In this example all assumptions of Theorem 5.1. are valid except one: the absolute value of the curvature of the boundary of this billiard is not uniformly bounded, i.e. $\sup_{x\in\Gamma} |k(z)| = +\infty$.

To prove the equality (5.3.), it is sufficient to prove the equality

$$\iint_M (\log^+ |\frac{\partial\theta_1}{\partial\phi}|) \sin\theta d\phi d\theta = +\infty \tag{5.4.}$$

It is precisely this equality that will occur in our example.

We will construct a region Ω such that for some subset $H \subset M$ of positive measure the following conditions will be satisfied:

I. If $(\phi,\theta) \in H$, then ϕ_1 belongs to the interior of some interval of Γ. So, by virtue of formula (5.1.) one has $\frac{\partial\theta_1}{\partial\phi} = -k(\phi)$ for all $(\phi,\theta) \in H$.

II. If $(\phi,\theta) \in H$, then $k(\phi) \ge 1$

III. $$\iint_H (\log k(\phi)) \sin\theta d\theta d\phi = +\infty \tag{5.5.}$$

A possible shape of such a billiard is given for example by the region Ω bounded by the curve Γ obtained by smoothing the angles of a rectangle and by replacing the interval AB belonging to the boundary by some curve L as shown on Fig. 11. We will define this curve exactly in the next subsection. H will be chosen as a suitable subset of set $\{(\phi,\theta); \phi\in L\}$.

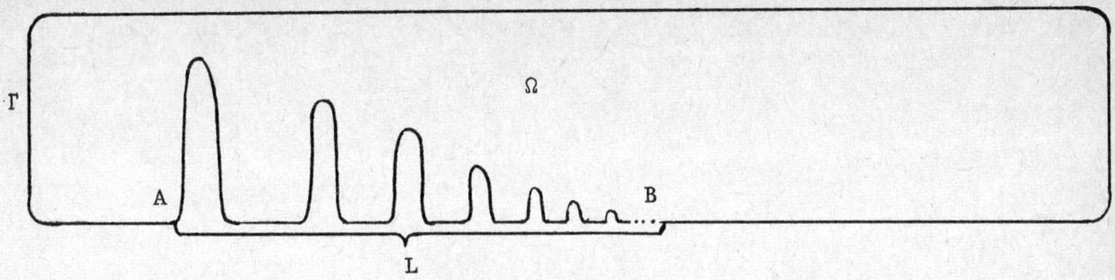

Fig. 11

5.3. The whole construction is based on the following well known and easy to prove facts.

A. The series $\sum_{n=1}^{\infty} \frac{1}{n \log n}$ is divergent.

B. The sequence $\sum_{n=1}^{\infty} \frac{1}{n(\log n)^2}$ is convergent.

C. If $r_n = \frac{1}{n(\log n)^2}$, $n \geq 2$, then $\sum_{n=2}^{\infty} r_n \log r_n = -\infty$.

Consider the following function $f \in C^\infty[-1,1]$ (see Fig. 12).

$$f(x) = \begin{cases} \sqrt{1-x^2} & \text{for } |x| \leq \frac{3}{4} \\ 0 & \text{for } |x| \geq \frac{4}{5} \\ \text{on the interval } [-\frac{4}{5},-\frac{3}{4}] \text{ (respectively } [\frac{3}{4},\frac{4}{5}]) & \\ f \text{ is strictly increasing (respectively decreasing).} & \end{cases}$$

Fig. 12

Let us define for $0 < a < 1$ $f_a(x) = a^{3/2} f(\frac{x}{a})$. Clearly, $f_a \in C^\infty([-a,a])$ and in the interval $[a-,a]$ the function f_a has the same structure as the function f on the interval $[-1,1]$. The only difference is that on the interval $[-\frac{3}{4}a,\frac{3}{4}a]$, the graph of function is an arc of an ellipse instead of that of a circle.

From the usual formula for the curvature of curve $y = h(x)$; $k(x) = \frac{h''(x)}{(1+[h'(x)]^2)^{3/2}}$, it follows easily that for all $x \in (-\frac{3}{4}a,\frac{3}{4}a)$, the absolute values of the curvature of the graph of function f_a is not less that $\frac{1}{\sqrt{a}}$.

The boundary piece L laying between the points A and B (see Fig. 11) will be constructed in the following manner. Let us take an interval $I = [0,4d]$ where $d = \sum_{n=2}^{\infty} r_n$ and where $r_n = \frac{1}{n(\log n)^2}$. For $n \geq 2$, denote by I_n and J_n the closed intervals of length $2r_n$. The interval I can be represented as a union of intervals of $I_2 \cup J_2 \cup I_3 \cup J_3 \cup \ldots \cup I_n \cup J_n \cup \ldots$, as shown on Fig. 13.

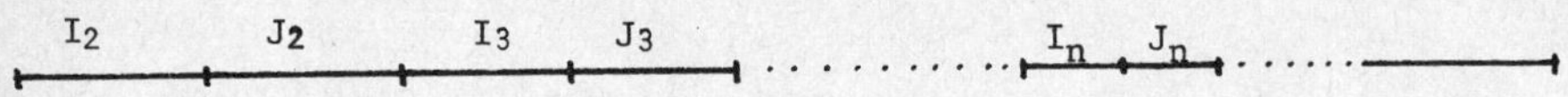

Fig. 13

Define on the interval I the function g as follows:

$$g(x) = \begin{cases} r_n^{3/2} f_{r_n}(x-4\sum_{k=1}^{n-1} r_k - r_n) & \text{for } x \in I_n,\ n \geq 2 \\ 0 & \text{for } x \in J_n,\ n \geq 2 \end{cases}$$

where, by definition, $r_1 = 0$. Define the curve L as a graph of function g.

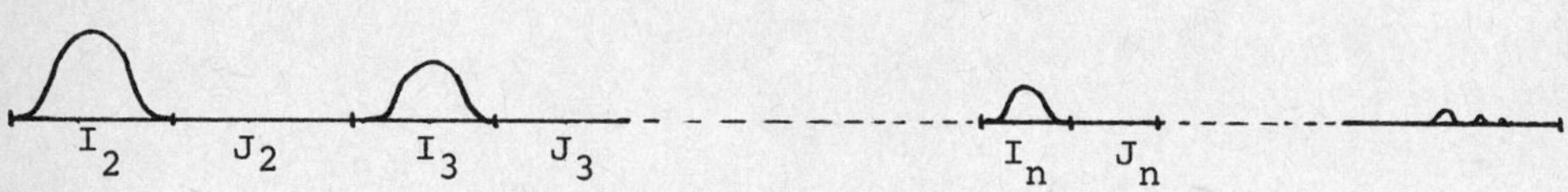

Fig. 14. The shape of the graph of function g

It is easy to check that $g \in C^1(I)$ but $g \notin C^2(I)$, as g does not have the second left derivative at the right end of interval I. Indeed, this follows immediately from the fact that the curvature of the curve L in the centers of intervals I_n tends to infinity when

n tends to the infinity.

Note that, since $g \in C^1(I)$ the curve L is of finite length.

5.4. To finish the construction of Γ it remains to define the size of our rectangle, the size of smoothing of the corners, the set H, and to verify that

$$\iint_H (\log k(\phi) \sin\theta d\phi d\theta = +\infty \tag{5.5.}$$

Let us make the following remark. Since L is a curve of the class C^1 of finite length, instead of an arc length parameter $\{\phi\}$ on L one can consider an equivalent parametrization of class C^1 on Γ which associates a number x to the point $(x,g(x))$. It is easy to see that (5.5.) is equivalent to $\iint_H (\log k(x)) \sin\theta dx\, d\theta = +\infty$ where by $k(x)$ we denote the curvature of L at point $(x,g(x))$.

Let us denote by K_n an interval of length $\frac{3}{4}r_n$ such that $K_n \subset I_n$ and that the left end of K_n coincides with the center of the interval I_n. Consider $x \in K_n$. Denote by L the unique unit vector such that the ray $\{(x,g(x))+tL, t\geq 0\}$ is tangent to the curve L at a point $(y,g(y))$ where $y \in I_{n+1}$ (see Fig. 15). We denote by $\delta(x)$ the angle between vector $A = (0,1)$ and vector L. It is easy to see that there exists $\gamma > 0$ such that for all $n \geq 2$ and for all $x \in K_n$, one has $\delta(x) \geq \gamma$.

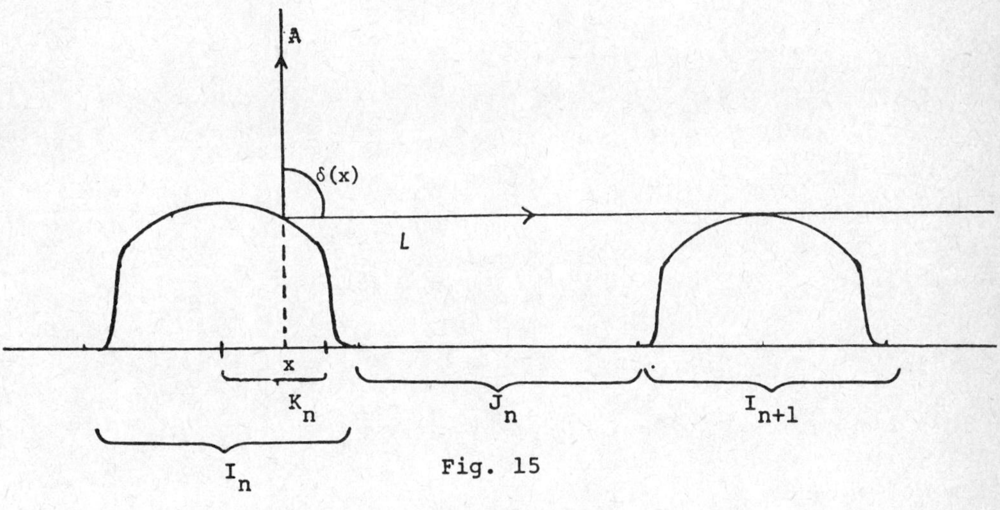

Fig. 15

Let us now define exactly the curve Γ. Consider any rectangle with sides longer than 10d where $d = \sum_{n=2}^{\infty} r_n = \sum_{n=2}^{\infty} \frac{1}{n(\log n)^2}$. Let AB be an interval of length 4d belonging to the lower side of our rectangle and such that its right end coincides with the center of this side (see Fig. 16). Let us substitute the interval AB by the curve L (see Fig. 11). Moreover, let us make such a smoothing of the corners that from the interval Λ dividing the height of the rectangle through the middle, the smoothed arcs are seen at an angle smaller than $\frac{\gamma}{100}$. Without any loss of generality one can assume that the length of any of the four smoothed arcs is smaller than $\frac{1}{4} \cdot \frac{1}{2(\log 2)^2}$. From the last condition we assert that this smoothing takes place outside curve L (see Fig. 16). It is clear that the region bounded by the curve Γ just obtained, satisfies all the assumptions of Theorem 5.1. except one which concerns the curvature.

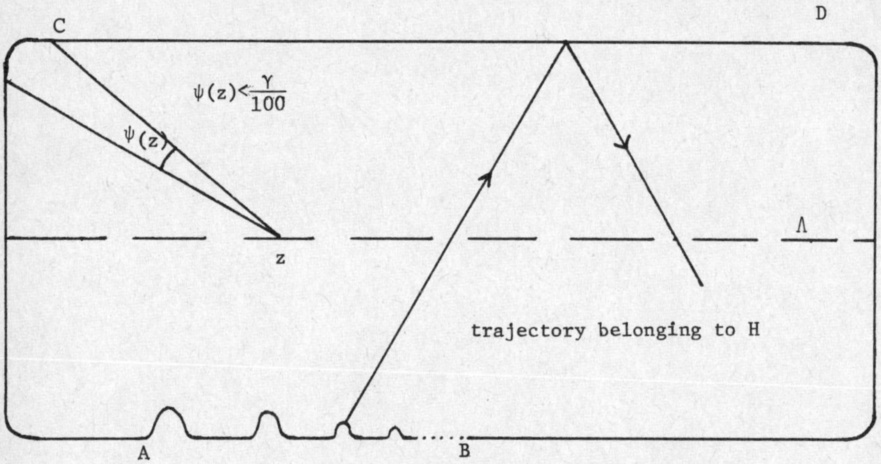

Figure 16

Denote by [C,D] this part of the upper side of our rectangle which has not been smoothed (see Fig. 16).

Define H as the subset of all $(\phi,\theta) \in \Gamma \times (0,\pi)$ such that $\phi \in \{(x,y(x)) \in L; x \in K_n$ for some $n \geq 2\}$ and that $\phi_1 \in [C,D]$ where K_n is the interval defined after (5.6.). Define $K = \bigcup_{n=2}^{\infty} K_n$. Clearly, for any fixed $x \in K_n$, the Lebesque measure of $\{\theta; (x,\theta) \in H\}$ is bigger than $\frac{\gamma}{2}$. Thus, one obtains that $\iint_H (\log k(x)) \sin \theta dx d\theta \geq \iint_k \log k(x)dx) \cdot$ $\cdot (\int_0^{\gamma/2} \sin \theta d\theta) \geq \frac{\gamma^2}{16} \cdot \int_K (\log k(x)dx$, because for $0 < \gamma < \frac{\pi}{2}$ one has

$\sin\gamma \geq \frac{1}{2}\gamma$. However, $\int_K \log k(x)dx = \frac{3}{4}\sum_{n=2}^{\infty} \frac{1}{n(\log n)^2}\cdot\log[n \log n]^{1/2} = +\infty$

because $\sum_{n=2}^{\infty} r_n \log r_n = -\infty$. Therefore, $\iint_H (\log k)\sin\theta d\phi d\theta = +\infty$, i.e. the Oseledec theorem is not applicable to the just constructed billiard.

Let us note that in the example described above the boundary Γ is largely arbitrary, except for the curve L and for the interval $[C,D]$.

5.5. It would be very interesting to know whether the metric entropy $h_\nu(\{T^t\})$ of the billiard flow in the region Ω described above is finite or infinite and if it is finite, is it strictly positive or is it zero?

The curve L was chosen in such a manner as not to contain a strictly convex boundary piece. Clearly, it is possible to modify the example just described in such a way that instead of the curve L we will have a sinusoidal curve of the "damped oscillation" type (see Fig. 17).

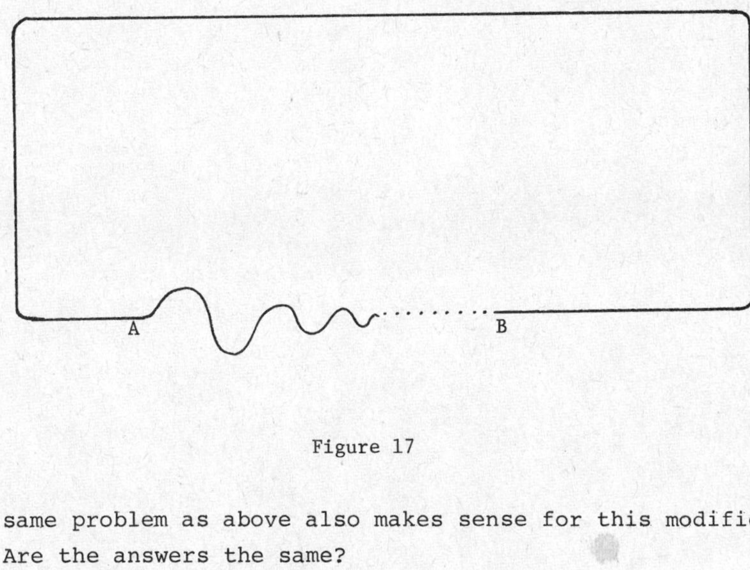

Figure 17

The same problem as above also makes sense for this modified billiard. Are the answers the same?

6. THE SINGULAR SET. THE BILLIARDS OF CLASS P

6.1. After the general considerations of the preceeding two sections we now pass to the central point of this part.

To apply the results of Parts I-IV to the mapping Φ, even the class P_2 of billiards described in Corollary 5.1. is too wide. We will thus define a more restrictive class of billiards, namely a class P. For class P we can already prove the first specific result which is necessary for the application of the results of Parts I-IV.

6.2. Let Ω be a billiard of class P_2. We will now define a subset A_Ω of M which we will call the singular set of the billiard Ω.

Let $\Gamma = (\bigcup_{i=1}^{p} L_i) \cup (\bigcup_{j=1}^{r} \Gamma_j)$ be a representation of Γ as the sum of finite number of smooth closed arcs and closed curves, as described in Sec. 3.1. Beginning from the present section we always suppose that any considered boundary piece γ is contained in some L_i, $1 \le i \le p$, or in some Γ_j, $1 \le j \le r$.

As in Sec. 3.1. let us denote by S the subset of all point which are the ends of arcs $L_1, \ldots, L_r$. Denote by θ_Ω the set of all straight lines in the plane either tangent (perhaps one-sided) to Γ in at least two different points simultaneously or tangent to Γ either at the points of inflection or at the points of S. The cases when a straight line contains some other points of Γ, and in particular an interval of Γ, is not excluded.

The singular set A_Ω is defined as the union of following five subsets, $A_\Omega^1, \ldots, A_\Omega^5$ of M, of which only A_Ω^2 is always nonempty:

$$A_\Omega^1 = \{(z,\theta)\in M:\ z\in S\}$$

$$A_\Omega^2 = \{(z,\theta)\in M:\ \theta=0,\pi\}$$

$$A_\Omega^3 = \{(z,\theta)\in M:\ z_1\in S\}$$

$$A_\Omega^4 = \{(z,\theta)\in M:\ L_{(z,\theta)} \text{ is (one sided) tangent to } \Gamma \text{ at } z_1\}$$

$$A_\Omega^5 = \{(z,\theta)\in M:\ L_{(z,\theta)} \text{ is parallel to a straight line from } \theta_\Omega\}$$

(cf. the beginning of Sec. 3.3).

Let us stress that the singular set A depends, just like the space M, not only on the region Ω but also on the representation $\Gamma = (\bigcup_{i=1}^{p} L_i) \cup (\bigcup_{j=1}^{r} \Gamma_j)$ fixed once and for all.

The set $\tilde{A}_\Omega = \bigcup_{i=1}^{4} A_\Omega^i$ is formed by all points of M in which Φ is not of class C^{k-1} if Γ is piecewise C^k. It is much more $\tilde{A}_\Omega$ than A_Ω which deserves the name of singular set of Φ. The reasons to consider A_Ω^5 as a part of the singular set of Φ are of purely technical nature and will appear only in Sec. 9.

To apply the results of Parts I-IV it is necessary to know that the one dimensional set A_Ω is of finite length. However, for the billiard of class P_2 it is generally not the case. Although A_Ω^1, A_Ω^2 and A_Ω^3 are always of finite length, it is very easy to see that A_Ω^4 and A_Ω^5 can be of infinite length. Nevertheless, in the billiards of class P_2, as we will prove, the length of A_Ω is always finite.

The billiard Ω of class P_k belongs to the class P_k, $1 \le k \le \infty$, if any curve L_i, $1 \le i \le p$, and Γ_j, $1 \le j \le r$, is of class C^k and if it is a union of a finite number of intervals and of a finite number of strictly convex or strictly concave curves.

Clearly, for the billiards of class P_k, the set θ_Ω is finite and it is easy to see that the singular set A_Ω is always a compact subset of M. The following theorem already characterises the set A_Ω.

<u>THEOREM 6.1</u>. <u>If</u> Ω <u>is a billiard of class</u> P_k, $2 \le k \le \infty$, <u>then the singular set</u> A_Ω <u>is a union of a finite number of arcs and of closed curves, all of class</u> C^{k-1} <u>and of finite length, and of finite numbers of isolated points</u>.

<u>Proof</u>. It is clear that the sets A_Ω^1, A_Ω^2 and A_Ω^5 are unions of finite numbers of closed arcs and closed curves, all of class C^{k-1} and of finite length.

Let us examine the set A_Ω^3. First note that A_Ω^3 can also contain isolated points. An example of an isolated point (z,θ) of A_Ω^3 is presented on Fig. 18.

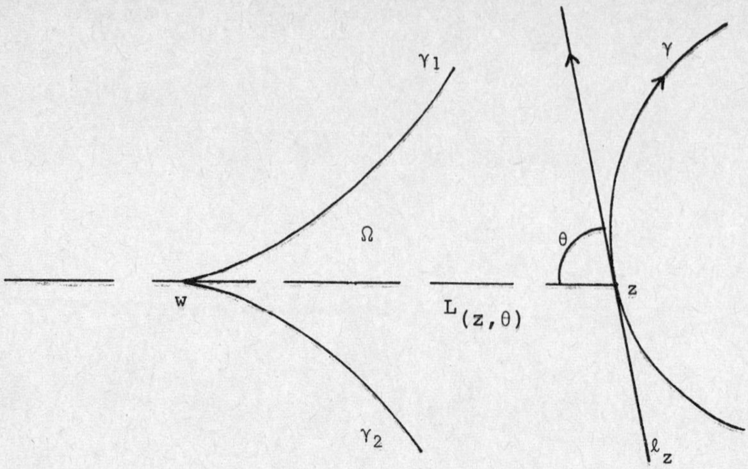

Fig. 18. The boundary pieces γ_1 and γ_2 and the half straight line $L_{(z,\theta)}$ are all mutually tangent in w.

Let us also note that in general A^3_Ω is not compact. Indeed, this is so if Γ contains a closed strictly convex boundary piece γ of which one end w belongs to S, because $A_3(w,\gamma) = \{(z,\theta); z\in\gamma, z_1=w\}$ is not compact in this case. Without a loss of generality we may suppose that γ is oriented in such a way that for $z \in \gamma$ near w one has $\theta(z)$ near π, where $\theta(z)$ is the unique angle such that $z_1(z,\theta(z)) = w$. Clearly, $A_3(w,\gamma) \cup (w,\pi) \stackrel{\text{def}}{=} B_3(w,\gamma)$ is compact and $B_3(w,\gamma) \in A_\Omega$.

Fix $w \in S$. Except for the case of isolated points A^3_Ω, there always exists at w some "sector of visibility" of Γ. As Γ is the union of a finite number of smooth closed arcs and of closed curves, it is sufficient to prove that if γ is a sufficiently small closed boundary piece belonging to some L_i or Γ_j, $1 \le i \le p$, $1 \le j \le r$, and if it is visible from w, then $B_3(w,\gamma)$ is a closed arc of class C^{k-1}. Here "γ is visible from w" means that for all points $z \in \gamma$ the interval going from w to z has no common points with $\Gamma\setminus(w \cup z)$. By subdividing the boundary piece γ into a finite number of smaller parts, one can reduce our problem to considering only the three cases of mutual position of w and of γ as presented on Figs. 19 a-c respectively. In these three figures one uses the following orthogonal system of co-ordinates (x,y); $w = (0,0)$, γ is the graph of a positive convex function f of class C^k one of the ends of γ is on the positive x-axis when $w \notin \gamma$ while γ is tangent to the x-axis in 0

if $w \in \gamma$, i.e. $f'(0) = 0$.

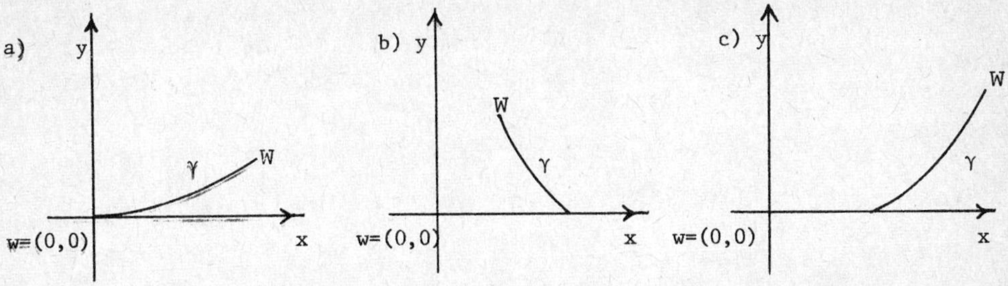

Fig. 19. The case (a) corresponds to $w \in \gamma$. The cases (b) and (c) correspond to $w \notin \gamma$. One does not exclude the case of γ being an interval.

We will consider in detail the case (a) only; the proof of cases (b) and (c) is similar. To dispatch with a trivial situation we will suppose that f is a strictly convex function defined on the interval $[0,A]$ for some $A > 0$. The orientation of γ is from w to W (Fig. 19).

Let ϕ be the arc length parameter on γ calculated with respect to w. Recall that we identify the point $z \in \gamma$ and the corresponding arc length parameter ϕ. For any $\phi \in \gamma$, $\phi \neq 0$, denote the unique oriented angle such that $z_1 = z_1(\phi,\theta) = w$ by $\theta(\phi)$. On the other hand if $z \in \gamma$ then $z = (t,f(t))$ for some t, $0 \leq t \leq A$. By definition $\overline{\theta}(t) = \pi - \theta(\phi)$. Clearly, to prove that $B_3(w,\gamma) = \{(\phi,\theta(\phi)), 0 \leqslant \phi \leq |\gamma|\} \cup (0,\pi)$ is an arc of class C^{k-1} of finite length, it is equivalent to prove the same for $C_3(w,\gamma) \stackrel{\text{def}}{=} \{(t,\overline{\theta}(t)), 0<t\leq A\} \cup \cup (0,0)$. To prove it, we will express $\overline{\theta}(t)$ using the function f (see Fig. 20).

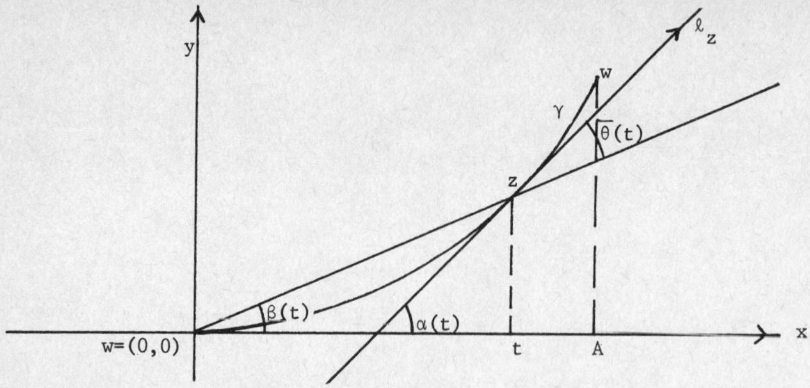

Fig. 20. ℓ_z denotes the oriented tangent to γ at point z.

Clearly, $\overline{\theta}(t) = \alpha(t) - \beta(t)$; $\tan\alpha(t) = f'(t)$ and $\tan\beta(t) = \frac{f(t)}{t}$. Consequently, for $0 < t \le A$ one has

$$\overline{\theta}(t) = \text{Arctan}\, f'(t) - \text{Arctan}\,\frac{f(t)}{t}\,.$$

Moreover let us define, $\theta(0) = 0$. As $f \in C^{k-1}([0,A])$, then $\overline{\theta} \in C^{k-1}([0,A])$, i.e. $C_3(w,\gamma)$ is a closed arc of class C^{k-1}.

Let us consider now the set A^4_Ω. As in the case of set A^3_Ω the set A^4_Ω is in general not compact and may also contain a finite number of isolated points. Like in the case of the set A^3_Ω to prove our assertion it is enough to consider only two boundary pieces, γ and γ_1, fulfilling the following conditions; to every $z \in \gamma$ corresponds the unique angle $\theta = \theta(z)$ such that the half straight line $L_{(z,\theta)}$ is tangent to γ_1 at the point $z_1 = z_1(z,\theta)$ (perhaps one sided).

Let $(F(\phi),G(\phi))$ such that $a < \phi < b$ and $(F_1(\phi_1),G_1(\phi_1))$ such that $c < \phi_1 < d$ are the parimetrisations by arc length of the boundary pieces γ and γ_1 respectively. If $z = (F(\phi^0),G(\phi^0))$ and $z_1 = (F_1(\phi'),G_1(\phi'))$, then

$$\det\begin{pmatrix} F_1(\phi^1)-F(\phi^0), & G_1(\phi^1)-G(\phi^0) \\ F_1'(\phi^1) & G_1'(\phi^1) \end{pmatrix} \overset{\text{def}}{=\!=} R(\phi^0,\phi^1) = 0.$$

By the virtue of the Implicit Function Theorem, for the existence in some neighborhood of ϕ^0 of the function $\phi_1 = \phi_1(\phi)$ of class C^r

such that $\phi_1(\phi^0) = \phi^1$ and $R(\phi,\phi_1(\phi)) = 0$ for all ϕ from this neighborhood, it will suffice that $\frac{\partial R}{\partial \phi_1}(\phi^0,\phi^1) \neq 0$. This is so because

$$\frac{\partial R}{\partial \phi_1}(\phi^0,\phi^1) = \det \begin{pmatrix} F_1(\phi^1)-F(\phi^0), & G_1(\phi^1)-G(\phi^0) \\ F_1''(\phi^1) & , & G_1''(\phi^1) \end{pmatrix}$$

Indeed, the vector $(F(\phi^1)-F(\phi^0),G(\phi^1)-G(\phi^0))$ is parallel to the vector $(F_1'(\phi^1),G_1'(\phi^1))$ and, as ϕ_1 is the arc length parameter, the vector $(F_1''(\phi^1),G_1''(\phi^1))$ is orthogonal to the vector $(F_1'(\phi^1)),G_1'(\phi^1))$. This immediately implies that if γ and γ_1 are two closed disjoined boundary pieces, then the set $A_4(\gamma,\gamma_1) = \{(z,\theta),z\in\gamma,L_{(z,\theta)}$ is (one sided) tangent to $\gamma_1\}$ is a closed arc of class C^{k-1}.

To finish the proof, it remains only to consider the case of two sufficiently small closed convex or concave boundary pieces γ and γ_1 having a common end w as a unique common point. To avoid a trivial situation, we will consider only the case when γ_1 is strictly curvilinear.

If γ and γ_1 are sufficiently small we are in the following situation. Let (x,y) be the orthogonal system of co-ordinates in the plane such that $w = (0,0)$ and γ is tangent to the x-axis at w. Then γ is the graph of a convex function $f \in C^k([0,A])$, $A > 0$, such that $f(0) = f'(0) = 0$ and γ_1 is the graph of some strictly convex or concave function $g \in C^k([B,0])$, $B < 0$.

We will consider the mutual disposition of γ and γ_1 as presented on Fig. 21 but we note that in all the other cases the proof follows along the same line.

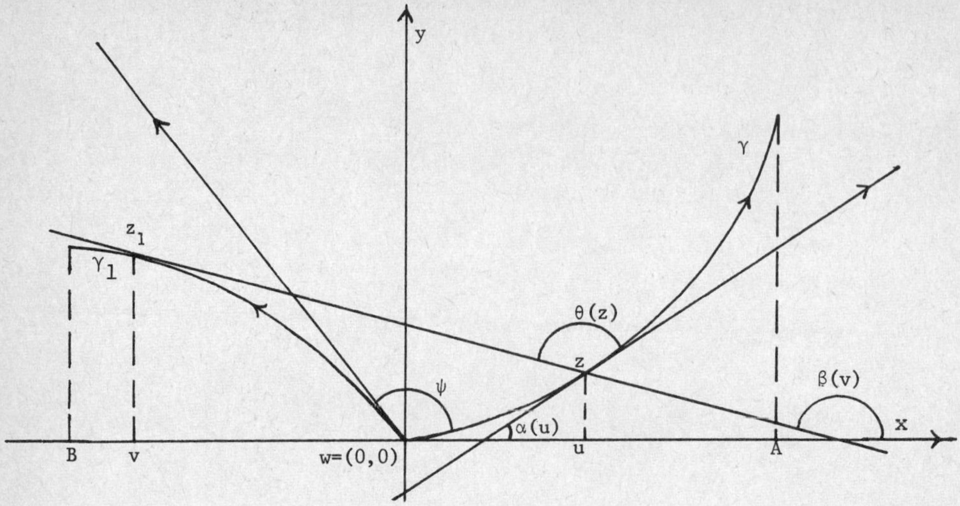

Fig. 21. One does not exclude the case of γ being an interval belonging to the x-axis. $0 \le \Psi < \pi$.

From the preceding considerations it follows $A_4(\gamma,\gamma_1)\setminus(v,\Psi)$ is a half open arc of class C^{k-1}. Moreover, $A_4(\gamma,\gamma_1)$ is a closed arc. To finish the proof one has only to prove that $A_4(\gamma,\gamma_1)$ is of finite length. This is equivalent to showing that the graph $\{(u,\bar{\bar{\theta}}(u));0<u\le A$ is of finite length, where $\bar{\bar{\theta}}(u) \stackrel{\text{def}}{=\!=} \theta(z)$ (see Fig. 21). Clearly, $\bar{\bar{\theta}}(u) = \beta(v) - \alpha(u)$ where $\alpha(u) = \text{Arctan } f'(u)$ and $\beta(v) = \text{Arctan } g'(v)$. Moreover, from above it follows that $v = v(u)$ is a function of class C^{k-1}. Let us note $\beta(v(u)) = \delta(u)$. δ and α are bounded, continuous, monotonically decreasing functions of u defined on interval $[0,A]$. Thus the graphs of δ and of α are of finite length and consequently the same is true for the graph of $\bar{\bar{\theta}} = \delta - \alpha$. ∎

Notice that using the convex functions described in Sec. 7.8. it is possible to find a strictly convex function g of class C^∞, such that for $f = 0$, $\lim_{u\to 0} \delta'(u)$ does not exist. This means that in general $A_4(\gamma,\gamma_1)$ is not a closed arc of class C^1.

6.3. Let X be a metric space with metric ρ. For $E \subset X$ and $\varepsilon > 0$ we denote by $U_\varepsilon(E)$ the open ε-neighborhood of E, i.e. $U_\varepsilon(E) = \{z \in X;\ \rho(z,e) < \varepsilon$ for some $e \in E\}$.

It is well known, and easy to prove, that there exists an absolute constant $C > 0$ such that for every rectifiable arc $E \subset \mathbb{R}^2$, there exists $\varepsilon_0 = \varepsilon_0(E) > 0$ such that for every ε, $0 < \varepsilon < \varepsilon_0$, one has

$$\mathrm{Vol}(U_\varepsilon(E)) \le C\varepsilon|E| \tag{6.1.}$$

Theorem 6.1. and (6.1.) imply directly the following theorem.

THEOREM 6.2. If Ω is a billiard of class P_2, then the set A_Ω is a closed subset of M and there exist $C_1 > 0$ and $\varepsilon_0 > 0$ such that for every $0 < \varepsilon < \varepsilon_0$, one has

$$\mu(U_\varepsilon(A_\Omega)) \le C_1\varepsilon.$$

Let us recall that μ is the Φ invariant probability measure defined in the Corollary 4.4.

7. THE BILLIARDS OF CLASS Π. THE RATE OF GROWTH OF $\|d\Phi\|$ and $\|d^2\Phi\|$

7.1. It is essential for the theory described in Part I-IV to use the first and second derivatives of the mapping. Thus, if one wants to apply this theory to billiards, it is necessary to consider the billiards with curves $\{L_i\}_{1\le i\le p}$ and $\{\Gamma_j\}_{1\le j\le r}$ of class C^3. In Secs. 7-8 this will always be assumed. Let us recall that $N_\Omega = M\backslash A_\Omega$, where A_Ω is the singular set of Ω. From Corollary 4.1. we obtain for the billiards of the class P_3 that $\Phi : N_\Omega \to M$ is a diffeomorphic imbedding of class C^2.

The only case when the study of the mapping $\Phi : N_\Omega \to M$ reduces in a natural way to the study of diffeomorphisms of compact manifold, is the case of strictly convex curve Γ of class C^k, $k \ge 2$ with everywhere non-zero curvature. In this case Φ admits an extension to a diffeomorphism of class C^{k-1} of the two-dimensional torus onto itself. This torus is obtained by identifying in M the points $(\phi,0)$ and (ϕ,π). This property of the billiards in the region bounded by the strictly convex smooth curves with everywhere non-zero curvature is systematically used in the study of some ergodic properties of such billiards. See for example [Laz], [Dou] and [Dvo].

Except for the case described above any attempt to apply the Pesin theory to the mapping $\Phi : N_\Omega \to M$ inevitably meets an obstacle that Φ can in no way be considered as a diffeomorphism of compact manifold. Moreover, generally speaking, $\|d\Phi\|$ is not bounded near the boundary of N_Ω, i.e. near the singular set A_Ω. As it follows from Corollary 4.1. and Theorem 4.3., this unbounded growth of $\|d\Phi\|$ is equivalent to the following geometrical situation: either Γ contains a strictly concave boundary piece of the first kind (see Sec. 2), or a strictly convex boundary piece of the first kind with a point of zero curvature, or a strictly convex boundry piece of the second kind.

Nevertheless, as it is proved in Secs. 7-9 for the billiards of class Π which are a particular case of the billiards of class P and which are defined in Sec. 7.8. $\|d\Phi\|$ and $\|d^2\Phi\|$ cannot grow too fast near the singular set. From this it follows immediately that all results from Parts I-IV are applicable to the billiards of class Π. This enables us to deduce that the principal results of Pesin from $[\text{Pes}]_{1-3}$ remain true for the billiards of class Π as well.

7.2. The definition of the billiards of class Π uses a particular class of convex curves which we are going to define now. This definition is based on what we will call the condition F.

Let us denote by $V(a)$, $a > 0$, the set of all strictly convex functions $f \in C^1([-a,a])$ such that $f(0) = f'(0) = 0$. For such functions $f(x) > 0$ for $x \neq 0$.

It is geometrically clear that if $f \in V(a)$ and if $s,t \in [-a,a]$, $s \neq t$, then $f'(s) - \frac{f(s) - f(t)}{s - t}$ and $\frac{f(s) - f(t)}{s - t} - f'(t)$ are both non-zero and of the same sign; thus $\frac{(s-t)f'(s) - f(s) + f(t)}{f(s) - f(t) - (s-t)f'(t)} > 0$.

Let $f \in V(a)$. We will say that f satisfies the condition F on the interval $[-a,a]$ if there exists a number $C > 0$ such that for every $s, t \in [-a,a]$, $s \neq t$ one has

$$R_f(s,t) \stackrel{\text{def}}{=} \frac{(s-t)\, f'(s) - f(s) + f(t)}{f(s) - f(t) - (s-t)\, f'(t)} \geq C \qquad (7.1.)$$

Any such positive number C will be called an exponent of f (on interval $[-a,a]$). Let us remark that, as s and t play a completely symmetrical role in the definition of conditon F, the inequality (7.1.) implies that for every $s, t \in [-a,a]$, $s \neq t$, one has

$$0 < C \leq R_f(s,t) \leq \frac{1}{C} < +\infty \qquad (7.2.)$$

Thus, necessarily, $0 < C \leq 1$.

If instead of the interval $[-a,a]$ one considers the intervals $[0,a]$ and $[-a,0]$ and the strictly convex functions $f \in C^1([0,a])$ and $f \in C^1([-a,0])$, such that $f(0) = f'(0) = 0$, one obtains the condition F on intervals $[0,a]$ or $[-a,0]$ respectively.

To fix a framework made to avoid repetitions, we will consider in Sec. 7.2 - 7.4 only the condition F on the interval $[-a,a]$. Nevertheless, in the Proposition 7.1., Theorem 7.1. and Corollary 7.1. one can replace the interval $[-a,a]$ by the interval $[0,a]$ or $[-a,0]$ with only minor modifications. We leave this to the reader.

Let us note that if $f \in V(a) \cap C^k[-a,a]$, if $f^{(i)}(0)=0$ for $2 \le i \le k-1$, and if $f^{(k)}(0) \neq 0$, then necessarily k is even and $f^{(k)}(0) > 0$.

An example of a function from $V(a)$ which does not satisfy the condition F is given by a well known function:

$$f(x) = \begin{cases} e^{-1/x^2} & \text{for } 0 < |x| \le a \\ 0 & \text{for } x = 0 \end{cases} \tag{7.3.}$$

For this function (7.2.) is not satisfied when $t = 0$. Other, much more interesting examples of such functions, will be described in Sec. 7.8.

Let us note also that, as remarked by M. Misiurewicz, the function

$$g(x) = \begin{cases} x^2 & \text{for } x \le 0 \\ x^4 & \text{for } x > 0 \end{cases}$$

does not satisfy condition F on any interval $[-a,a]$. Indeed $\lim_{t \to +0} R_g(-t^2,t) = +\infty$.

7.3. In the present paper we do not deal with the search for the possibly largest class of functions satisfying the condition F. We will only prove the following Proposition and Theorem.

PROPOSITION 7.1. *Let* $f \in V(a) \cap C^2([-a,a])$, $a > 0$, *and let* $f''(0) \neq 0$. *Then, there exists* a_0, $0 < a_0 \le a$ *such that* f *satisfies the condition* F *on the interval* $[-a_0,a_0]$.

THEOREM 7.1. *Let* $f \in V(a) \cap C^{k+2}([-a,a])$, $a > 0$, $k \ge 2$, *let* $f^{(i)}(0) = 0$ *for* $2 \le i \le k-1$ *and let* $f^{(k)}(0) \neq 0$. *Then, there exists* a_0, $0 < a_0 \le a$ *such that* f *satisfies the condition* F *on the interval* $[-a_0,a_0]$.

We do not know whether in Theorem 7.1. the assumptions $f \in C^{k+2}([-a,a])$ can be replaced by the assumption $f \in C^k([-a,a])$.

By the simple compactness argument one obtains from Theorem 7.1. the following Corollary.

COROLLARY 7.1. *If* $f \in V(a)$, $a > 0$ *and if* f *is real-analytic on the interval* $[-a,a]$, *then* f *satisfies the condition* F *on the whole*

<u>interval</u> $[-a,a]$.

<u>Proof of Proposition 7.1</u>. As $f \in C^2([-a,a])$ and $f''(0) > 0$, there exists a_0, $0 < a_0 \le a$, such that for all $s, t \in [-a_0,a_0]$, $s \neq t$ one has $\frac{f''(s)}{f''(t)} \ge \frac{1}{2}$.

Let us fix $s \in [-a_0,a_0]$ and let us note $G(t) = (s-t)\,f'(s) + -f(s) + f(t)$ and $H(t) = f(s) - f(t) - (s-t)\,f'(t)$. Then, from Cauchy Mean Value Theorem one has $R_f(s,t) = \frac{(s-t)f'(s) - f(s) + f(t)}{f(s) - f(t) - (s-t)f'(t)} = \frac{G(t)}{H(t)} = \frac{G(t) - G(s)}{H(t) - H(s)} = \frac{G'(t_1)}{H'(t_1)} = \frac{f'(t_1) - f'(s)}{t_1 - s} \cdot \frac{1}{f''(t_1)}$ for some t_1 belonging to the open interval linking t and s. But $\frac{f'(t_1) - f'(s)}{t_1 - s} = f''(t_2)$ for some t_2 belonging to the open interval linking t_1 and s. Thus, finally,

$$R_f(s,t) = \frac{f''(t_2)}{f''(t_1)} \ge \frac{1}{2}, \text{ if only } s, t \in [-a_0,a_0]. \qquad \blacksquare$$

The proof of Theorem 7.1. is based on the Lemmas 7.1. and 7.2. formulated below.

<u>LEMMA 7.1</u>. <u>If</u> $k \ge 2$ <u>is even, then the function</u> $f(x) = x^k$ <u>satisfies the condition</u> F <u>on the interval</u> $[-1,1]$.

<u>Proof</u>. Let $s, t \in [-1,1]$, $s \neq t$, $t \neq 0$ and $r = \frac{s}{t}$. For $f(x) = x^k$ one has $R_f(s,t) = \frac{(k-1)s^k - kts^{k-1} + t^k}{s^k - kst^{k-1} + (k-1)t^k} = \frac{(k-1)r^k - kr^{k-1} + 1}{r^k - kr + (k-1)} \overset{\text{def}}{=\!=} \Phi(r)$.
As k is even, for $s \neq t$, $R_f(s,t) > 0$. Consequently, $\Phi(r) > 0$ for $r \neq 1$. The use of the de l'Hospital rule gives $\lim_{r\to 1} \Phi(r) = 1$. On the other hand $\lim_{r\to\pm\infty} \Phi(r) = k - 1$. Thus, Φ can be considered as a strictly positive continuous function on $\overline{\mathbb{R}} = \mathbb{R} \cup \{-\infty\} \cup \{+\infty\}$ and consequently $\inf_{r\in\mathbb{R}} \Phi(r) > 0$. $\blacksquare$

The following lemma is a simple consequence of the Taylor Formula and we omit its proof.

<u>LEMMA 7.2</u>. <u>Let</u> $k \ge 2$ <u>be even and let</u> $f \in V(a) \cap C^{k+2}([-a,a])$, $a > 0$, <u>be such that</u> $f^{(i)}(0) = 0$ <u>for</u> $2 \le i \le k - 1$ <u>and</u> $f^{(k)}(0) \neq 0$. <u>Then, the function</u> g <u>given by the formula</u>

$$g(x) = \begin{cases} \sqrt[k]{f(x)} & \text{for } 0 \le x \le a \\ -\sqrt[k]{f(x)} & \text{for } -a \le x < 0 \end{cases}$$

<u>is of class</u> $C^2([-a,a])$ <u>and</u> $g'(0) = \sqrt[k]{\frac{f^{(k)}(0)}{k!}} > 0.$

The function $f(x) = x^k + x^{k+1}|x|$ shows that in Lemma 7.2. the class C^{k+2} cannot be replaced by the class C^{k+1}.

Lemma 7.2. allows to consider the function f as the function $F(z) = z^k$ in the new variable $z = g(x)$. This idea is the basis of the proof of Theorem 7.1.

7.4. <u>Proof of Theorem 7.1.</u> In virtue of Lemma 7.2. we can write $f(x) = [g(x)]^k$ where $g \in C^2([-a,a])$ and where $g'(0) = \sqrt[k]{\frac{f^k{}_0}{k!}} > 0$. As $g'(0) > 0$, the function h inverse to g, $h = g^{-1}$, exists and h is also of class C^2 on some interval $[-q_0,q_0]$, $q_0 > 0$. Moreover, $h'(0) > 0$. We chose q_0, $0 < q_0 \le 1$ so small that for every $u, v \in [-q_0,q_0]$ one has $\frac{1}{2} \le \frac{h'(u)}{h'(v)} \le 2$. Let us note $a = g(s)$ and $b = g(t)$, where $a, b \in [-q_0,q_0]$. We now write:

$$\begin{aligned} R_f(s,t) &= \frac{(s-t)f'(s) - f(s) + f(t)}{f(s) - f(t) - (s-t)f'(t)} = \\ &= \frac{[h(a)-h(b)]\, k[g(s)]^{k-1}g'(s) - [g(s)]^k + [g(t)]^k}{[g(s)]^k - [g(t)]^k - [h(a)-h(b)]\, k[g(t)]^{k-1}g'(t)} = \\ &= \frac{ka^{k-1}[h(a)-h(b)]\,\frac{1}{h'(a)} - a^k + b^k}{a^k - b^k - kb^{k-1}[h(a)-h(b)]\,\frac{1}{h'(b)}} = \\ &= \frac{h'(b)}{h'(a)} \cdot \frac{ka^{k-1}[h(a)-h(b)] - (a^k-b^k)h'(a)}{(a^k-b^k)h'(b) - kb^{k-1}[h(a)-h(b)]} \overset{\text{def}}{=} \\ &= \frac{h'(b)}{h'(a)} \cdot T_f(a,b). \end{aligned}$$

As we know from Lemma 7.1. the function $F(x) = x^k$ satisfies the condition F on the interval $[-1,1]$. This means that there exists $C_1 > 0$ such that for all $a, b \in [-1,1]$, $a \ne b$, one has

$$R_F(a,b) = \frac{(k-1)a^k - kba^{k-1} + b^k}{(k-1)b^k - kab^{k-1} + a^k} \ge C_1.$$

As $\frac{1}{2} \le \frac{h'(a)}{h'(b)} \le 2$, to prove the condition F for f in some neighborhood of zero is equivalent to prove that there exist $C_2 > 0$ and q_1, $0 < q_1 \le q_0$, such that for all $a, b \in [-q_1 q_1]$, $a \ne b$ one has

$$\frac{T_f(a,b)}{R_F(a,b)} \ge C_2. \tag{7.4.}$$

$\frac{T_f(a,b)}{R_F(a,b)} = \frac{M(a,b)}{N(a,b)}$ where $M(a,b) = \frac{ka^{k-1}[h(a)-h(b)] - (a^k-b^k)h'(a)}{(k-1)a^k - kba^{k-1} + b^k}$ and

$$N(a,b) = \frac{(a^k-b^k)h'(b) - kb^{k-1}[h(a)-h(b)]}{(k-1)b^k - kab^{k-1} + a^k} = M(b,a).$$

Instead of proving (7.4.) directly we will prove that there exists C_3, $0 < C_3 < \infty$ and q_2, $0 < q_2 \le q_0$ such that for all $a, b \in [-q_2, q_2]$, $a \ne b$ one has

$$C_3 \le M(a,b) \le \frac{1}{C_3} \tag{7.5.}$$

Clearly, (7.5.) implies (7.4.).

$$M(a,b) = \frac{ka^{k-1}\,\frac{h(a)-h(b)}{a-b} - \frac{a^k-b^k}{a-b}\,h'(a)}{ka^{k-1} - \frac{a^k-b^k}{a-b}} = \frac{ka^{k-1}h'(z) - \frac{a^k-b^k}{a-b}\,h'(a)}{ka^{k-1} - \frac{a^k-b^k}{a-b}}$$

for some z belonging to the open interval linking a and b. Thus,

$$M(a,b) = h'(z)\,\frac{ka^{k-1} - \frac{a^k-b^k}{a-b}\,\frac{h'(a)}{h'(z)}}{ka^{k-1} - \frac{a^k-b^k}{a-b}} =$$

$$= h'(z)\left\{1 + \left(1 - \frac{h'(a)}{h'(z)}\right)\frac{a^k-b^k}{(k-1)a^k - kba^{k-1} + b^k}\right\} =$$

$$= h'(z)\left\{1 + \frac{1}{h'(z)}\cdot\frac{h'(z)-h'(a)}{z-a}\cdot\frac{(a^k-b^k)(z-a)}{(k-1)a^k - kba^{k-1} + b^k}\right\} =$$

$$= h'(z)\left\{1 + \frac{h''(z_1)}{h'(z)}\cdot\frac{(a^k-b^k)(z-a)}{(k-1)a^k - kba^{k-1} + b^k}\right\}$$

where z_1 belongs to the open interval linking a and b.

As $h \in C^2([-q_0,q_0])$, h'' is uniformly bounded on the interval $[-q_0,q_0]$.

Moreover, as we know, $\lim_{\substack{a\to 0\\ b\to 0}} h'(z) = h'(0) > 0$. Thus, for sufficiently small $\delta > 0$, $\frac{h''(z_1)}{h'(z)}$ is uniformly bounded when $|a| < \delta$, $|b| < \delta$ and to finish the proof of (7.5.), it is then sufficient to prove that

$$\lim_{\substack{a\to 0\\ b\to 0\\ a\neq b}} L(a,b) = 0 \tag{7.6.}$$

where $L(a,b) = \frac{(a^k-b^k)(a-b)}{(k-1)a^k - kba^{k-1} + b^k}$.

Let us suppose that $a \neq 0$ and let us note $r = \frac{b}{a}$. Then, $L(a,b) = a\frac{(1-r^k)(1-r)}{(k-1) - kr + r^k} = a\Psi(r)$. The unique real root of the denominator of $L(a,b)$ is $r = 1$. $r = 1$ is also a root of the numerator of $L(a,b)$. The use of the de l'Hospital rule gives $\lim_{r\to 1} \Psi(r) = \frac{2}{k-1}$. Thus, the function Ψ, as a continuous function, is bounded on every interval $[-A,A]$ by some positive constant C_A, where $0 < A < +\infty$. Consequently

$$\sup_{\substack{a\neq 0, a\neq b\\ \left|\frac{b}{a}\right| \leq A}} |L(a,b)| \leq aC_A \tag{7.7.}$$

On the other hand, there exists $A_0 > 0$ such that if $|r| \geq A_0$, then $|\Psi(r)| \leq 2|r|$. Thus if $\left|\frac{b}{a}\right| \geq A_0$, then $|L(a,b)| \leq 2|b|$. This last inequality together with the inequality (7.7.) proves (7.6.). ∎

7.5. Let γ denote a strictly convex oriented smooth plane curve. Let $z \in \gamma$ and let $\ell_1(z)$ denote the oriented tangent to γ at point z (one-sided tangent if z is one of the ends of γ). Let $\ell_2(z)$ denote the straight line orthogonal to $\ell_1(z)$ passing through point z. Orienting in a suitable way the straight line $\ell_2(z)$ one can always assume that in the orthogonal system of co-ordinates (x,y) corresponding to the oriented straight lines $\ell_1(z)$ and $\ell_2(z)$, in a neighborhood of point z the curve γ is the graph of a smooth strictly convex function f_z. f_z is defined only for $x \geq 0$ or $x \leq 0$ if z is one of the ends of γ. In particular, $f_z(x) > 0$ for all $x \neq 0$.

We will say that γ satisfies the condition F with exponent C if at any point $z \in \gamma$ the function f_z satisfies this contition in some neighborhood of zero (at the ends of γ one considers the respective one-sided neighborhoods of zero).

A billiard of class P_k, $1 \le k \le \infty$ belongs to the class Π_k if there exists a representation $\Gamma = (\bigcup_{i=1}^{p} L_i) \cup (\bigcup_{j=1}^{r} \Gamma_j)$ and a constant $C > 0$ such that for all strictly convex boundary pieces γ of Γ (see Sec. 2) belonging to some $L_i (1 \le i \le p)$ or to some $\Gamma_j (1 \le j \le r)$, the condition F with exponent C is satisfied.

The class Π_3 will be denoted simply by Π. Natural examples of billiards belonging to the class Π are the billiards for which all the curves $\{\Gamma_i\}_{1 \le i \le p}$ and $\{L_j\}_{1 \le j \le r}$ are real analytic; this is a immediate consequence of Corollary 7.1. Another natural example is given by the billiards of class P_3 such that the curvature of their boundaries is always non-vanishing. This follows directly from Proposition 7.1.

The class Π contains in particular all the plane billiards considered in $[\text{Sin}]_2$ and $[\text{Bun}]_{1-4}$.

7.6. Let us recall that for $w = (\phi,\theta) \in M$ and $\tilde{w} = (\tilde{\phi},\tilde{\theta}) \in M$, the distance $\rho(w,\tilde{w})$ is defined at the end of Sec. 3.2.

Recall also that $\left\|\begin{matrix} a,b \\ c,d \end{matrix}\right\| = |a| + |b| + |c| + |d|$. Analogously, we fix the norm $\|d^2\Phi(\phi,\theta)\| = \left|\frac{\partial^2\phi_1}{\partial\phi^2}\right| + \left|\frac{\partial^2\phi_1}{\partial\theta^2}\right| + 2\left|\frac{\partial^2\phi_1}{\partial\phi\partial\theta}\right| + \left|\frac{\partial^2\theta_1}{\partial\theta^2}\right| + \left|\frac{\partial^2\theta_1}{\partial\theta^2}\right| + 2\left|\frac{\partial^2\theta_1}{\partial\phi\partial\theta}\right|$, where $\Phi(\phi,\theta) = (\phi_1(\phi,\theta), \theta_1(\phi,\theta))$.

The aim of the present and the next two sections is to prove the following theorem which is the main results of the present paper.

THEOREM 7.2. *If Ω is a billiard of class Π then, for all the points $w \in N_\Omega$ the following estimates hold:*

$$\|d\Phi(w)\| \le \frac{C_1}{\rho(w,A_\Omega)} \tag{7.8.}$$

$$\|d^2\Phi(w)\| \le \frac{C_2}{[\rho(w,A_\Omega)]^3} \tag{7.9.}$$

where C_1 and C_2 are constants depending only on Ω and where $\rho(v,A_\Omega) = \inf_{a \in A_\Omega} \rho(v,a)$. *In fact, for the validity of estimate* (7.8.) *it suffices that Ω belongs to the class Π_2.*

Remark that the choice of norms $\|\ \|$ has an influence only on the size of C_1 and C_2 but not on the existence or non-existence of such constants.

Let us note also that in Sec. 7.9. a class of billiards for which estimate (7.9.) fails, is described. Such billiards can be found even among strictly convex billiards with C^∞ boundary.

The proof of Theorem 7.2. is an immediate consequence of the following two lemmas where $r_1 = r_1(w) = |d_1(w)| = |\sin \theta_1(w)|$.

LEMMA 7.3. Let Ω be a billiard of class P_3. Then, for all the points $w \in N_\Omega$ the following estimation holds:

$$\|d\Phi(w)\| \le \frac{a_1}{r_1} \tag{7.10.}$$

$$\|d^2\Phi(w)\| \le \frac{a_2}{r_1^3} \tag{7.11.}$$

where a_1 and a_2 are constants depending only on the billiard Ω.

In fact, for the validity of (7.10.) it suffices that Ω belongs to the class P_2.

LEMMA 7.4. Let Ω be a billiard of class Π_1. Then, for all the points $w \in N_\Omega$ the following inequality holds:

$$r_1 \ge b\rho(w, A_\Omega)$$

where $b > 0$ is a constant depending only on the billiard Ω.

The proof of Lemma 7.3. easily follows from the formula (4.10.) for $d\Phi$ and has a purely formal character unlike Lemma 7.4. whose proof is a really geometrical one. The proof of Lemma 7.4. is given in Secs. 8 and 9. It should be emphasized that this proof uses only C^1 differentiability of all the curves $\{L_i\}_{1\le i\le p}$ and $\{\Gamma_j\}_{1\le j\le r}$ instead of C^3 differentiability of these curves which is necessary for Lemma 7.3.

7.7. Proof of Lemma 7.3.

Let us recall that

$$d\Phi(\phi,\theta) = \begin{pmatrix} \varepsilon_1 \dfrac{k\ell - \varepsilon d}{d_1} , & \varepsilon_1 \dfrac{\ell}{d_1} \\[2ex] \varepsilon_1 k_1 \dfrac{k\ell - \varepsilon d}{d_1} - k, & \varepsilon_1 \dfrac{k_1 \ell}{d_1} - 1 \end{pmatrix} \tag{4.10.}$$

Since Ω belongs to the class P_3, then from formula (4.10.) the estimation (7.10.) obviously follows. From the rule $\left(\frac{f}{g}\right)' = \frac{f'g - fg'}{g^2}$ it follows immediately that to prove the estimation (7.11.) it suffices to show that the absolute values of all the quantities from the following table are less that $\frac{a}{r_1}$, where $a > 0$ is a constant depending only on Ω.

$\frac{\partial k}{\partial \phi}(z)$	$\frac{\partial k_1}{\partial \phi}(z)$	$\frac{\partial d}{\partial \phi}(z)$	$\frac{\partial d_1}{\partial \phi}(z)$	$\frac{\partial \ell}{\partial \phi}(z)$
$\frac{\partial k}{\partial \theta}(z)$	$\frac{\partial k_1}{\partial \theta}(z)$	$\frac{\partial d}{\partial \theta}(z)$	$\frac{\partial d_1}{\partial \theta}(z)$	$\frac{\partial \ell}{\partial \theta}(z)$

This is evident for $\frac{\partial k}{\partial \theta}$ and $\frac{\partial d}{\partial \phi}$ because they are always equal to zero. For $\frac{\partial k}{\partial \phi}, \frac{\partial d}{\partial \theta}, \frac{\partial k_1}{\partial \phi} = \frac{\partial k_1}{\partial \phi_1} \cdot \frac{\partial \phi_1}{\partial \phi}$ $\frac{\partial k_1}{\partial \theta} = \frac{\partial k_1}{\partial \phi_1} \cdot \frac{\partial \phi_1}{\partial \theta}$, $\frac{\partial d_1}{\partial \phi} = \frac{\partial d_1}{\partial \theta_1} \cdot \frac{\partial \theta_1}{\partial \phi}$ and for $\frac{\partial d_1}{\partial \theta} = \frac{\partial d_1}{\partial \theta_1} \cdot \frac{\partial \theta_1}{\partial \theta}$, this follows trivially from the formula (4.10.) for $d\Phi$ and from the condition P_3, indeed, all $\{L_i\}_{1 \le i \le p}$ and $\{\Gamma_j\}_{1 \le j \le r}$ are curves of class C^3. It remains to examine $\frac{\partial \ell}{\partial \phi}$ and $\frac{\partial \ell}{\partial \theta}$.

In the notation of Sec. 4.1. one has $\ell = \sqrt{[F(\phi_1)-F(\phi)]^2 + [G(\phi_1)-G(\phi)]^2}$. Therefore, $\frac{\partial \ell}{\partial \theta}(\phi,\theta) = \frac{\partial \ell}{\partial \phi_1}(\phi,\phi_1(\phi,\Theta)) \cdot \frac{\partial \phi_1}{\partial \theta} =$

$$\frac{[F(\phi_1)-F(\phi)]\, F'(\phi_1) + [G(\phi_1)-G(\phi)]\, G'(\phi_1)}{\ell} \cdot \frac{\partial \phi_1}{\partial \theta} =$$

$$\frac{\varepsilon_1}{d_1} \{[F(\phi_1)-F(\phi)]F'(\phi_1)+[G(\phi_1)-G(\phi)]G'(\phi_1)\} \quad \text{because} \quad \frac{\partial \phi_1}{\partial \theta} = \varepsilon_1 \frac{\ell}{d_1}.$$

Thus, from condition P_3 the required estimation for $\frac{\partial \ell}{\partial \phi}$ immediately follows. Analogously, one has $\frac{\partial \ell}{\partial \phi}(\phi,\theta) =$

$$\frac{[F(\phi_1)-F(\phi)] \cdot [F'(\phi_1)\frac{\partial \phi_1}{\partial \phi} - F'(\phi)]}{\ell} + \frac{[G(\phi_1)-G(\phi)] \cdot [G'(\phi_1)\frac{\partial \phi_1}{\partial \phi} - G'(\phi)]}{\ell}.$$

As Ω belongs to class P_3, from the formula (4.10.) for $d\Phi$ it follows that there exists an $\alpha > 0$ such that for all $(\phi,\theta) \in N_\Omega$ one has: $|F'(\phi_1)\frac{\partial\phi_1}{\partial\phi} - F'(\phi)| \le \frac{\alpha}{r_1}$ and $|G'(\phi_1)\frac{\partial\phi_1}{\partial\phi} - G'(\phi)| \le \frac{\alpha}{r_1}$. From this we obtain the required estimation

$$\left|\frac{\partial \ell}{\partial\phi}(\phi,\theta)\right| \le \frac{\alpha}{r_1} \cdot \frac{|F(\phi_1) - F(\phi)| + |G(\phi_1) - G(\phi)|}{\ell} \le \frac{2\alpha}{r_1} = \frac{a}{r_1} \qquad \blacksquare$$

7.8. We will now describe the Misiurewicz example of strictly convex function $f \in C^\infty([-1,1])$ such that $f(0) = f'(0) = 0$ and that the condition F is not satisfied for f. In fact, the construction described below provides a whole class of such functions. Using some of these functions we will construct in Sec. 7.9. exmaples of strictly convex billiards with C^∞ boundary for which the estimate (7.8.) fails.

Before we pass to this construction remark that if for $f \in C^1([0,a])$, $a > 0$,

$$\liminf_{x\to 0} \frac{xf'(x)}{f(x)} = 1 \tag{7.12.}$$

then f can not satisfy the condition F. This follows immediately from (7.1.) when $t = 0$. Precisely, for the examples described here, (7.12.) is satisfied. Notice that this is not true for function (7.3.) which also fails to satisfy the condition F. On the other hand, for any strictly increasing convex function $f \in C^1([0,a])$ such that $f(0) = 0$ and for every x, $0 < x \le a$, one has

$$\frac{xf'(x)}{f(x)} > 1 \tag{7.13.}$$

Let us fix two arbitrary sequences, $\{a_n\}_{n\ge1}$ and $\{b_n\}_{n\ge1}$, of positive numbers, such that $\lim_{n\to\infty} a_n = \lim_{n\to\infty} b_n = 0$, $a_n < b_n < a_{n-1}$ for every $n \ge 2$ and such that $\lim_{n\to\infty} \frac{a_n}{b_n} = 0$, or equivalently that

$$\lim_{n\to\infty} \frac{b_n}{b_n - a_n} = 1 \tag{7.14.}$$

Let us choose an arbitrary function $h \in C^\infty([0,1])$ such that $h(x) = 0$ for $x \in \{0\} \cup \bigcup_{n=1}^{\infty} [a_n,b_n]$ and $h(x) > 0$ for $x \notin \{0\} \cup \bigcup_{n=1}^{\infty} [a_n,b_n]$. If such a function h is fixed, then one can find a function $g \in C^\infty([0,1])$ such that $g(0) = 0$ and $g(x) > 0$ if $x \neq 0$ except perhaps an at most countable set of numbers, and that

$$\lim_{n\to\infty} \frac{\int_0^{b_n} g(t)\,dt}{\int_0^{b_n} h(t)\,dt} = 0.$$ Let us fix such a function g.

Let us define now the second derivative f'' of f by $f'' = h + g$. Let us define $f'(x) = \int_0^x f''(t)\,dt$ and $f(x) = \int_0^x f'(t)\,dt$. It is clear that $f \in C^\infty([0,1])$, that $f(0) = 0$, and that f is strictly convex.

We will prove now that

$$\lim_{n\to\infty} \frac{b_n f'(b_n)}{f(b_n)} = 1. \tag{7.15.}$$

This implies that the function f so defined does not satisfy the condition F (see 7.12.).

Let us define $H'(x) = \int_0^x h(t)\,dt$, $H(x) = \int_0^x H'(t)\,dt$, $G'(x) = \int_0^x g(t)\,dt$ and $G(x) = \int_0^x G'(t)\,dt$. Remark that H and G are convex on the interval $[0,1]$ and

$$\lim_{n\to\infty} \frac{b_n H'(b_n)}{H(b_n)} = 1 \tag{7.16.}$$

Indeed, by virtue of the definition of function h, one deduces immediately that for every $n \geq 1$ one has

$$H(b_n) > \int_{a_n}^{b_n} H'(t)\,dt = (b_n - a_n)\, H'(b_n) \tag{7.17.}$$

because the function H' is constant on the interval $[a_n, b_n]$. Consequently, $\frac{b_n H'(b_n)}{H(b_n)} \leq \frac{b_n}{b_n - a_n}$ and thus (7.16.) follows from (7.13.) and (7.14.). Unfortunately, although the function H is convex, it is not strictly convex in any neighborhood of zero.

Pass now to the proof of (7.15.). $1 \leq \frac{b_n f'(b_n)}{f(b_n)} \leq \frac{b_n H'(b_n) + b_n G'(b_n)}{H(b_n)}$ $= \frac{b_n H'(b_n)}{H(b_n)} + \frac{b_n G'(b_n)}{H(b_n)}$. By virtue of (7.16.), in order to prove (7.15.) it is suffice to prove that

$$\lim_{n\to\infty} \frac{b_n G'(b_n)}{H(b_n)} = 0. \tag{7.18.}$$

Indeed, by virtue of (7.17.) one obtains that $0 \leq \frac{b_n G'(b_n)}{H(b_n)} \leq$

$\frac{b_n}{b_n - a_n} \cdot \frac{G'(b_n)}{H'(b_n)}$ for all $n \geq 1$. Thus, (7.18.) follows from (7.14). and from the very definition of the function g.

7.9. Suppose that in addition to the conditions described in Sec. 7.4. one has $\lim_{n\to\infty} \frac{a_n}{b_n^2} = 0$ and $g(b_n) = 0$ for all n. Let us note by γ the graph of the corresponding function f. Let $P = (0,0) \in \gamma$ and $R_n = (b_n, f(b_n)) \in \gamma$, $n \geq 1$.

Consider now an arbitrary strictly convex billiard Ω with C^∞ boundary $\Gamma = \Gamma_1$ such that $\gamma \in \Gamma$. Let us note by θ_n^1 such angles that $\Phi(P,\theta_n) = (R_n,\theta_n^1)$, $n \geq 1$. It is very easy to verify that $\lim_{n\to\infty} |\sin \theta_n| \frac{\ell(P,\theta_n)}{|\sin \theta_n^1|} = +\infty$. In virtue of the formula (4.10.) this immediately implies that $\lim_{n\to\infty} |\sin \theta_n| \cdot \|d\Phi(P,\theta_n)\| = +\infty$. Consequently, for such a billiard the estimate (7.8.) fails.

8. PROOF OF LEMMA 7.4. PART ONE: ELEMENTARY CONFIGURATIONS

8.1. In the present and in the following sections the elements of A_Ω will be called singular elements.

Recall that $N_\Omega = M \backslash A_\Omega$, that A_Ω is a compact subset of M and that $\Phi : N_\Omega \to M$ is a homeomorphic imbedding. This immediately implies that if Q is a compact subset of N_Ω, then there exists a number $b_Q > 0$ such that for all $w \in Q$ one has $r_1 \geq b_Q \rho(w, A_\Omega)$. As A_Ω is a compact subset of M, to prove Lemma 7.4. it suffices to prove that for any singular element P there exists its neighborhood $U(P)$ in M and a number $b_P > 0$ such that for all points $w \in U(P) \cap N_\Omega$ one has

$$r_1 \geq b_P \rho(w, A_\Omega) \tag{8.1.}$$

The inequality (8.1.) is not evident only in the case when for every neighborhood $U(P)$

$$\inf_{w \in U(P) \cap N_\Omega} \theta_1(w) = 0 \tag{8.2.}$$

The proof of Lemma 7.4. consists of three steps. The first step consists of a description of all "elementary configurations" of boundary pieces for which (8.2.) holds; this is given in Secs. 8.2. and 8.3. The "elementary configurations" are like bricks of which one can compose any configuration of boundary pieces. The next step consists of the reduction of Lemma 7.4. to the demonstration of Lemma 8.1. which asserts the validity of the main inequality (8.1.) for the elementary configuration. This reduction is described in Sec. 8.4. The last step consists of proving Lemma 8.1. and this is the content of Sec. 9. Notice that the condition F which appears in the definition of the billiards of class Π is used only in the proof of Lemma 8.1.

8.2. Denote by π the canonical projection of M onto Γ. Indeed, if $w \in M$ then either $w = (v,\theta) \in M_{L_i}$ or $w = (v,\theta) \in M_{\Gamma_j}$ for some i or j and by definition $\pi(w) = v$. Recall that $\ell(w)$ denotes the plane distance between v and v_1 if v_1 is defined (see Sec. 3.3.). In this section we will always denote $\pi(P) = z$.

To find all geometrically distinct configurations of boundary pieces and of singular elements $P = (z,\theta)$ for which (8.2.) holds, we will consider in detail the following three cases which exhaust all cases in which (8.2.) takes place.

(a) z belongs to a strictly convex closed boundary piece γ and $P = (z,0)$ or $P = (z,\pi)$.

(b) z is the beginning of two closed boundary pieces γ and γ_1 such that z is their only common point and such that the billiard trajectories starting from all points of $\gamma\setminus\{z\}$ to $\gamma_1\setminus\{z\}$ do exist.

(c) z is the beginning of a closed boundary piece γ and $\bar{z}$ is the beginning of the closed boundary piece γ_1 which is disjoint from γ. Moreover, the straight line L passing through z and $\bar{z}$ is unilaterally tangent to γ_1 at $\bar{z}$ and the billiard trajectories going from all points of $\gamma\setminus\{z\}$ to $\gamma_1\setminus\{\bar{z}\}$ do exist.

Let us pass now to the detailed description of all cases occurring in (a), (b) and (c) respectively,

(a) This case, noted A_1, is presented on Fig. 22.

(b) If z is a point of differentiability of the arc $\gamma \cup \gamma_1$, then in the sufficiently small neighborhood of z, except for the case when $\gamma \cup \gamma_1$ is a strictly convex arc, we have the cases $A_2 - A_5$ (Fig. 22). If z is the point of non-differentiability of arc $\gamma \cup \gamma_1$, then in the sufficiently small neighborhood of z, we have cases $B_1 - B_9$ (Fig. 23) if γ and γ_1 are transversal at z, and the cases $C_1 - C_5$ (Fig. 24) if γ and γ_1 are tangent at z.

These various cases differ by their respective type of convexity of γ and γ_1, and by their relative position.

(c) We can classify all the configurations of boundary pieces γ and γ_1 for which (8.2.) holds in a similar way as in (b) but also taking into account the mutual disposition of L, of γ and γ_1, as well as the transversality or tangency of γ and L at z. Then, if we restrict ourselves to sufficiently small γ and γ_1, we obtain the cases $D_1 - D_4$, $E_1 - E_4$ and $F_1 - F_3$ represented on Figs. 25 - 27 respectively.

To this list of configurations we will add the following one, denoted by G: γ and γ_1 are two disjoint boundary pieces such that for every $v \in \gamma$ and θ such that $v_1 = v_1(v,\theta) \in \gamma_1$, the half line $L_{(v,\theta)}$ is transversal to γ_1.

We will call the 36 cases listed above the "elementary configurations". Before we pass to the list of the elementary configurations, let us make a few remarks.

Firstly, let us point out that z can be a point of ramification of Γ in which various boundary pieces begin.

In cases A_1, C_1 and C_2 it turns out that in addition to the singular element $P = (z,0)$ there exists another singular element $p_1 = (z,\pi)$ connected with these configurations of boundary pieces.

We point out that in great many cases in any neighborhood $U(P)$ one has a continuous family of singular elements $P_v = (v,\theta_v)$, where $v \in \gamma$, $v \neq z = \pi(P)$ and v is sufficiently close to z, all belonging to A^4_Ω, such that (8.2.) is satisfied. In Case C_5 even two of such families, $P_v = (v,\theta_v)$ and $\overline{P}_v = (v,\overline{\theta}_v)$, exist. Finally, notice that $U(P)$ also contain two continuous families of singular elements: $q_v = (v,0)$ and $\overline{q}_v = (v,\pi)$ where $v \in \gamma$, $v \neq z$ and v is sufficiently close to z. Nevertheless, except the configuration A_1 and C_1 these singular elements will not play any role in our considerations and we will simply ignore them.

8.3. Let us turn now to Figs. 22 - 27 which give the relative position of γ and γ_1. To exhaust all possible trajectories going from γ to γ_1, we will suppose that γ and γ_1 are the boundary pieces of the second kind. To avoid an exception, in case A_1 one defines $\gamma_1 = \gamma$. Even if z or $\overline{z}$ is a point of ramification of the boundary, on Figs. 22 - 27 we show only boundary pieces γ and γ_1 which are essential for our considerations. As we restrict ourselves to sufficiently small neighborhoods of z and of $\overline{z}$, all boundary pieces γ and γ_1 laying in these neighborhoods are either strictly convex or intervals. The corresponding type of convexity of these boundary pieces will always clearly appear on the figures and will in principle not be described separately.

The arrows on Figs. 22 - 27 show the orientation of the boundary pieces γ and γ_1 fixed by us. If we consider a configuration which is not shown on a separate figure but is obtained from a given one by changing γ into γ_1 and γ_1 into γ, we will always assume that in such a case the orientation of the boundary pieces remain the same as on the figure. It is clear that the truthfulness of the inequality (8.1.) is completely independent from the chosen orientation of γ and γ_1.

We will denote by ℓ and ℓ_1 the oriented tangents at points z and $\bar{z}$ to the curves γ and γ_1 respectively. In the figure captions we will denote by P, P_1, P_v and $\bar{P}_v$ the corresponding singular elements without any additional comment. In these captions only the singular elements P and P_1 are explicitly described.

With any elementary configuration one can associate the co-ordinate systems (x,y) related to the boundary piece γ and to the straight lines $\ell_1(z)$ and $\ell_2(z)$ (see Sec. 7.5.). In all the elementary configurations the boundary pieces γ and γ_1 are so small that the following assumption is always verified. In the described co-ordinate system, γ is the graph of some monotonic function of class C^1 defined on a closed interval; moreover, except for the case when γ_1 is an interval belonging to the y-axis, the same is also true for $\bar{\gamma}_1$.

Moreover, if z is a point of nondifferentiability of Γ then in some neighborhood of z, γ belongs to some L_i or Γ_j. We will note $(z,\theta)_\gamma = (z,\theta)_{L_i}$, $L_{(z,\theta)_\gamma} = L_{(z,\theta)_{L_i}}$ if $\gamma \subset L_i$ and $(z,\theta)_\gamma = (z,\theta)_{\Gamma_j}$, $L_{(z,\theta)_\gamma} = L_{(z,\theta)_{\Gamma_j}}$ if $\gamma \subset \Gamma_j$.

Configurations A_1 - A_5

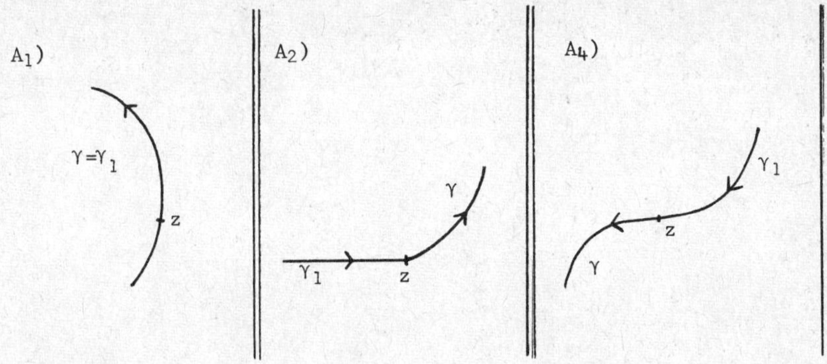

Fig. 22. A_1 : $\gamma = \gamma_1$ is a strictly convex boundary piece. The case when z is at the end of γ is not excluded. $P = (z,0)_\gamma$, $P_1 = (z,\pi)_\gamma$.

A_2 : γ is a strictly convex boundary piece which begins at z, γ_1 is an interval which begins at z and z is the point of differentiability of arc $\gamma \cup \gamma_1$. $P = (z,0)_\gamma$.

A_3 : Like A_2, but the roles of γ and γ_1 are reversed. $P = (z,\pi)_\gamma$.

A_4 : γ and γ_1 are two boundary pieces beginning at z and z is is the point of differentiability of arc $\gamma \cup \gamma_1$. z is the point of inflection. $P = (z,\pi)_\gamma$ and the family P_v exists.

A_5 : Like A_4, but the arc $\gamma \cup \gamma_1$ is strictly convex, $P = (z,\pi)_\gamma$.

Configurations B_1 - B_9

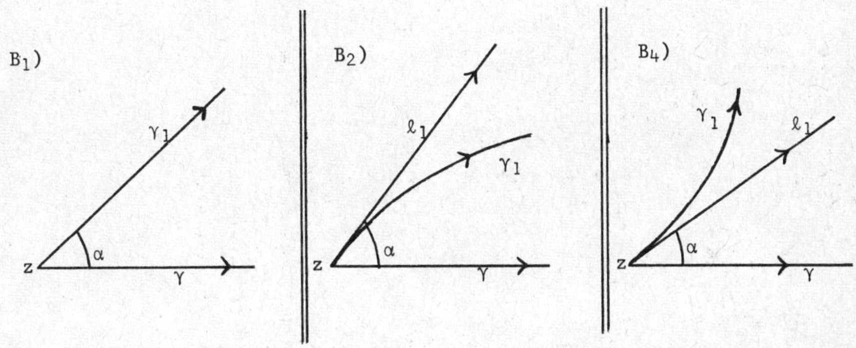

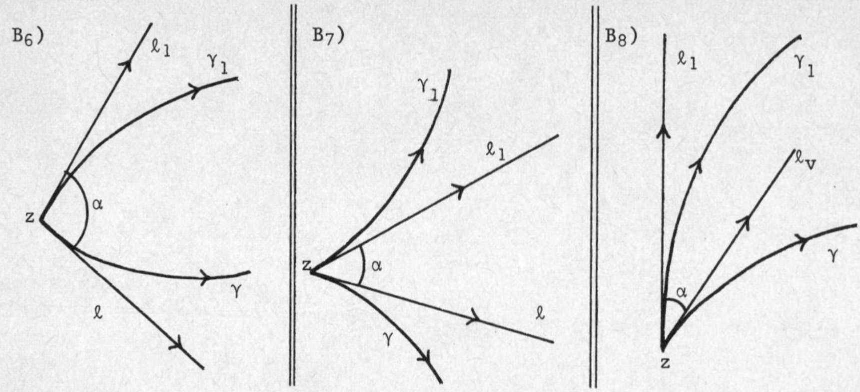

Fig. 23. At point z the boundary pieces γ and γ_1 are transversal. ℓ and ℓ_1 denote the half straight lines tangent at z to γ and γ_1 respectively. α denotes the angle between ℓ and ℓ_1 : $0 < \alpha < \pi$.

B_1 : $P = (z,\alpha)_\gamma$, $P_1 = (z,\pi)_\gamma$.

B_2 : $P = (z,2\pi-\alpha)_\gamma = (z,\pi)_{\gamma_1}$.

B_3 : Like B_2, role of γ and γ_1 are reversed. $P = (z,0)_{\gamma_1}$.

B_4 : $P = (z,\alpha)_\gamma$ and the family P_v exists.

B_5 : Like B_4, role of γ and γ_1 are reversed. $P = (z,0)_{\gamma_1}$.

B_6 : $P = (z,\alpha)_\gamma$.

B_7 : $P = (z,\alpha)_\gamma$ and the family P_v exists.

B_8 : $P = (z,\alpha)_\gamma$.

B_9 : Like B_8, role of γ and γ_1 are reversed. $P = (z,\alpha)_{\gamma_1}$ and the family P_v exists.

Configurations C_1 - C_5

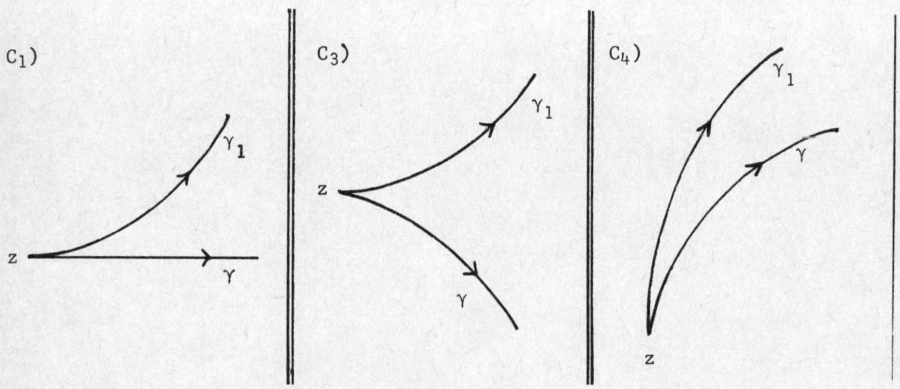

Fig. 24. At point z the boundary pieces γ and γ_1 are tangent and the angle between them is zero.

C_1 : $P = (z,0)_\gamma$, $P_1 = (z,\pi)_\gamma$ and the family P_v exists.

C_2 : Like C_1, role of γ and γ_1 are reversed. $P = (z,0)_{\gamma_1}$, $P_1 = (z,\pi)_{\gamma_1}$.

C_3 : $P = (z,0)_\gamma$ and the family P_v exists.

C_4 : $P = (z,0)_\gamma$.

C_5 : Like C_4, role of γ and γ_1 are reversed. $P = (z,0)_\gamma$, $P_1 = (z,\pi)_\gamma$ and two families, P_v and $\overline{P}_v$, exist.

Configurations D_1 - D_4

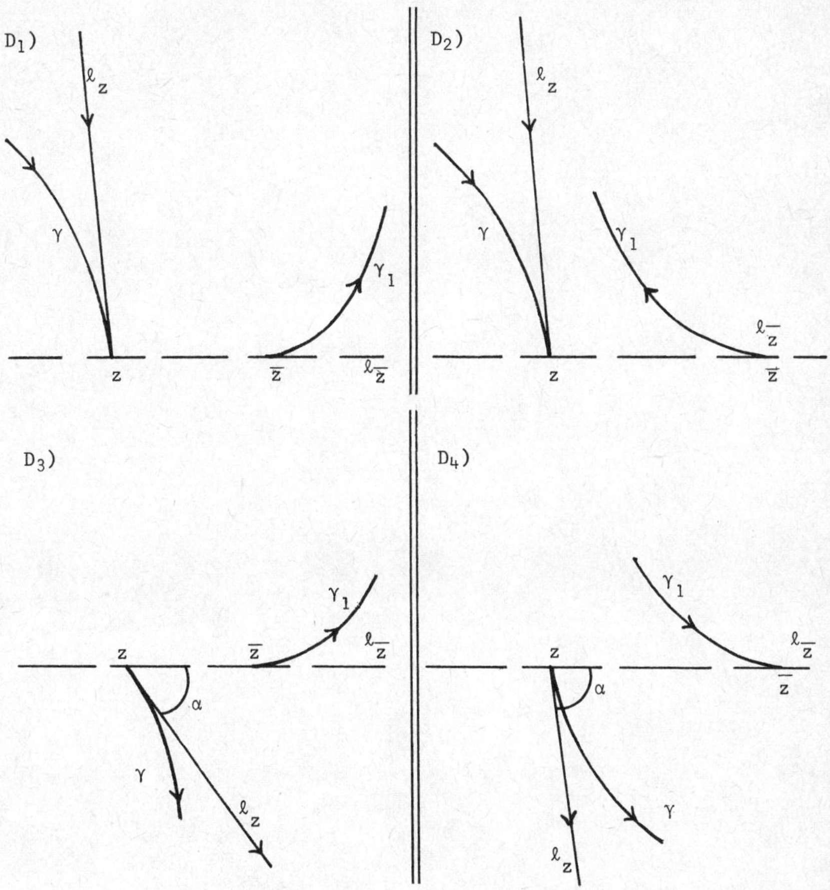

Fig. 25. γ is an arbitrary boundary piece transversal to the straight line $\ell_{\overline{z}}$ passing through the points z and $\overline{z}$. α is the oriented angle between the straight lines ℓ_z and $\ell_{\overline{z}}$. $0 < \alpha < \pi$.

D_1 : $P = (z,\pi-\alpha)_\gamma$.
D_2 : $P = (z,\pi-\alpha)_\gamma$ and the family P_v exists.
D_3 : $P = (z,\alpha)_\gamma$ and the family P_v exists.
D_4 : $P = (z,\alpha)_\gamma$.

Configurations $E_1 - E_4$

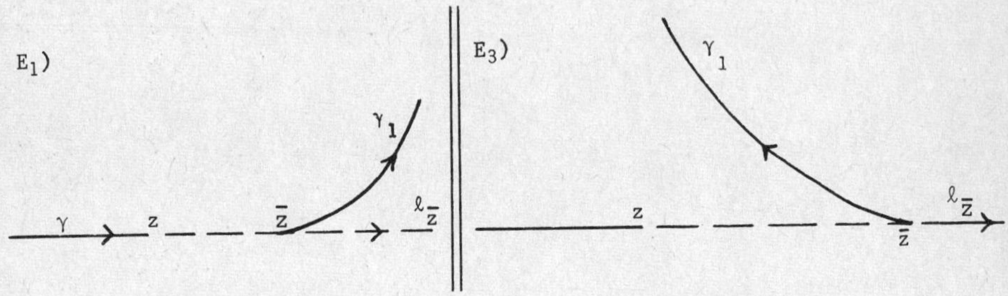

Fig. 26. E_1 : γ is an interval and, seen from γ, γ_1 is a strictly convex curve tangent to $\ell_{\bar{z}}$ in $\bar{z}$.

E_2 : Like E_1, role of γ and γ_1 are reversed.
E_3 : Like E_1, but γ_1 is a strictly concave curve.
E_4 : Like E_3, role of γ and γ_1 are reversed.

It is clear that all these cases might be considered as a particular case of A_2, A_3, B_4 and B_5 respectively. Therefore, we will not consider them separately.

Configurations $F_1 - F_8$

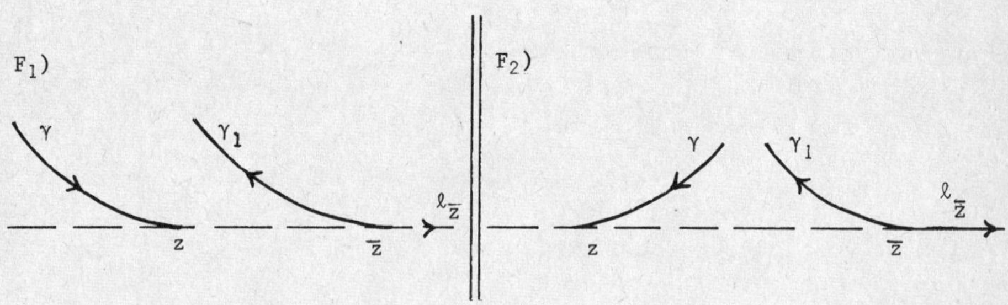

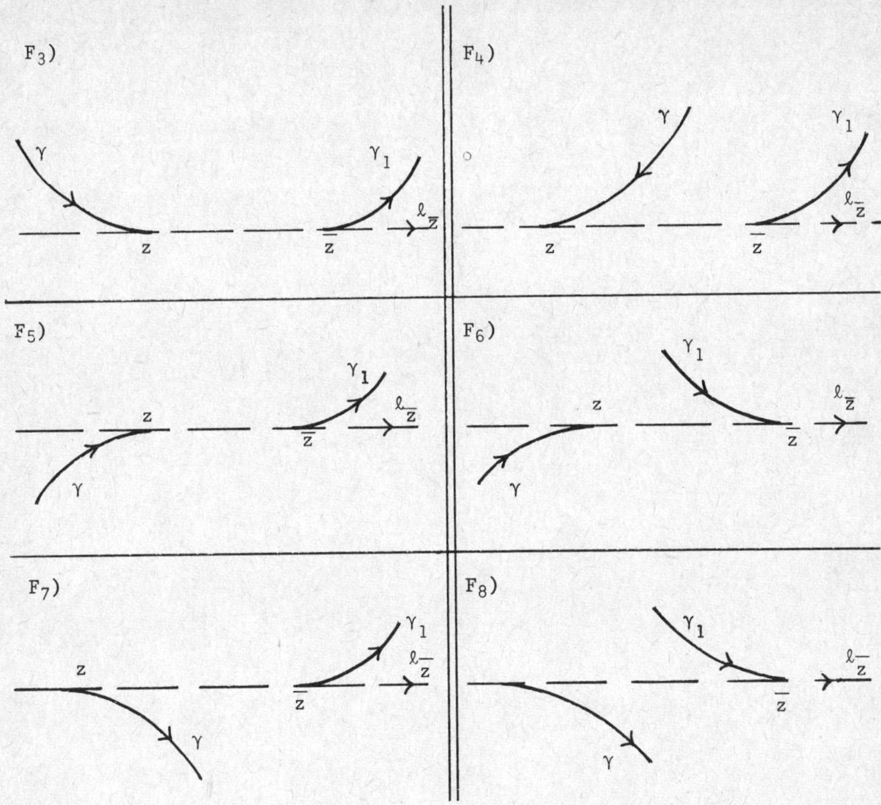

Fig. 27. Except for the fact that γ is tangent to $\ell_{\bar z}$ at z, these cases are exactly the same as $D_1 - D_4$. Nevertheless, we make a distinction with respect to the type of convexity of γ and γ_1.

In cases F_2, F_3, F_4 and F_8 there is only one singular element $P = (z,0)_\gamma$. In cases F_1, F_5, F_6 and F_7 one has $P = (z,0)_\gamma$ and the family P_v always exists.

8.4. Fix now a billiard Ω of class Π_1. To any two boundary pieces γ and γ_1 of Γ we associate the subset $L_{\gamma,\gamma_1} = \{(v,\theta)\in N_\Omega; v\in\gamma, v_1\in\gamma_1\}$.

<u>LEMMA 8.1</u>. <u>For any elementary configuration of boundary pieces</u> γ <u>and</u> γ_1 <u>of</u> Γ <u>there exists a number</u> $b_{\gamma,\gamma_1} > 0$ <u>such that for all</u> $w \in L_{\gamma,\gamma_1}$ <u>one has</u>

$$r_1 = r_1(w) \geq b_{\gamma,\gamma_1}\,\rho(w,A_\Omega) \qquad (8.3.)$$

The Lemma 8.1. is proved in Sec. 9. It is easy to see that it implies (8.1.). We leave details for the reader.

9. PROOF OF LEMMA 7.4. PART TWO: PROOF OF THE MAIN INEQUALITY

9.1. From the compactness arguments it follows immediately that to prove Lemma 8.1. it is sufficient to prove the following Lemma.

LEMMA 9.1. *For any elementary configuration of boundary pieces* γ *and* γ_1 *of* Γ *and for every singular element* s *related to it there exists a neighborhood* $U(s)$ *of* s *and a number* $b_s > 0$ *such that for all* $w \in U(s) \cap L_{\gamma,\gamma_1}$ *one has*

$$r_1 = r_1(w) \geq b_s \rho(w, A_\Omega). \qquad (9.1.)$$

Thus, to prove the Theorem 7.1. it remains to prove only Lemma 9.1.

The proof of Lemma 9.1. consists of the verification that the inequality (9.1.) is true for any elementary configuration and for any singular element related to it. We have nothing to prove for configuration G and since the configurations $E_1 - E_4$ are particular cases of the configurations A_2, A_3, B_4 and B_5 respectively, it remains only to study the remaining 31 configurations. In this section these configurations will be simply denoted by their symbol introduced in Sec. 8.3.

In fact, we will prove separately the inequality (9.1.) only for a few elementary configurations because for the other ones either the proofs are analogous or they follow from the Proposition 9.1. formulated below.

All the proofs are of very elementary nature and, except for the condition F, only the simplest geometrical considerations are used in them. Notice that we will never define explicitly the neighborhoods in M of singular elements for which the estimate (9.1.) will be proved. For a singular element s all these neighborhoods will be denoted by the same symbol U(s). Nonetheless, this will never lead to ambiguity.

9.2. First of all let us prove the following proposition.

PROPOSITION 9.1. *Let the boundary pieces* γ *and* γ_1 *form an elementary configuration. If* γ_1 *seen from* γ *is strictly concave, then for any singular element* $P_v = (v, \theta_v) \in L_{\gamma,\gamma_1} \cap A^4_\Omega$ *which is not an*

<u>isolated point of</u> $L_{\gamma,\gamma_1} \cap A^4_\Omega$, <u>the inequality</u> (9.1.) <u>holds with</u> $b_{P_v} = 1/2$.

<u>Proof</u>. It is geometrically evident and it was rigorously demonstrated in the proof of Theorem 6.1 that there exists a neighborhood U_v of v on γ such that for every $\tilde{v} \in U_v$ there exists an angle $\theta_{\tilde{v}}$ such that $P_{\tilde{v}} = (\tilde{v}, \theta_v) \in L_{\gamma,\gamma_1} \cap A^4_\Omega$ and such that $\theta_{\tilde{v}}$ is a continuous function of $\tilde{v}$ (see Fig. 28 where γ is strictly convex).

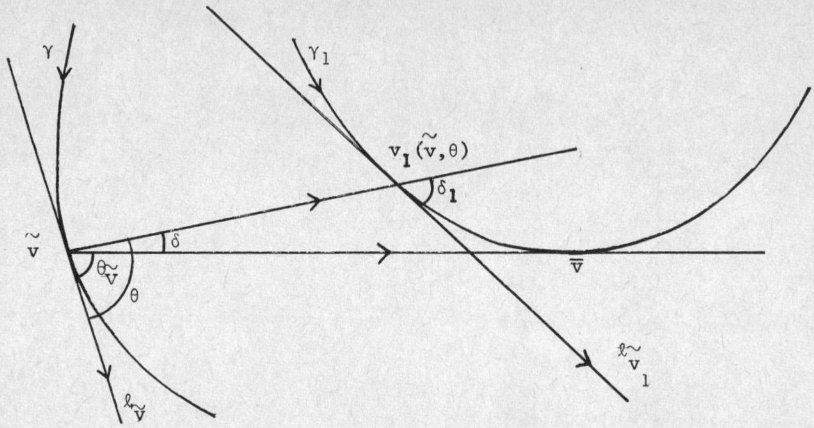

Fig. 28. $\delta = |\theta_v - \theta|$, $0 < \delta_1 < \pi/2$, $\sin \delta_1 = |\sin \theta_1(\tilde{w})|$.

Let us denote $\tilde{w} = (\tilde{v}, \theta)$. Thus, from the concavity of γ_1 one has that $\rho(\tilde{w}, A_\Omega) \le \rho(\tilde{w}, P_{\tilde{w}}) = \delta \le \delta_1 \le 2 \sin \delta_1 = 2|\sin \theta_1(w)|$, if only δ is sufficiently small, i.e. if $0 < \delta \le \delta_0$ where δ_0 does not depend on $\tilde{v} \in U_v$. ∎

Notice that in this proof the case when v is a common point of γ and γ_1 is not excluded.

This proposition immediately implies the validity of Lemma 7.4. for B_4, B_7, B_9, C_1, C_3, C_5, D_2, F_1 and F_2. The remaining twenty two cases will be considered separately.

9.3. A_1 - In this case for any point v belonging to the interior of γ one has two singular elements, $q_v = (v,0)_\gamma$ and $\bar{q}_v = (v,\pi)_\gamma$. At the ends of γ only one such element exists. From the compactness of γ it easily follows that to prove the inequality (9.1.) for A_1 it is sufficient to prove that for every point $v \in \gamma$ there exist neighborhoods $U(q_v)$ and $U(\bar{q}_v)$ in M and a number $b_v > 0$ such that for every $w = (u,\theta) \in U(q_v) \cap L_{\gamma,\gamma}$ and $\bar{w} = (\bar{u},\bar{\theta}) \in U(\bar{q}_v) \cap L_{\gamma,\gamma_1}$ one has

$$r_1 = r_1(w) \ge b_v \rho(w, A_\Omega) \qquad (9.2.)$$

and

$$\bar{r}_1 = r_1(\bar{w}) \ge b_v \rho(\bar{w}, A_\Omega)$$

As q_v and $q_{\bar{v}}$ play a completely symmetrical role in the above inequalities, it is sufficient to prove only (9.2.)

To prove (9.2.) it is enough to show that there exists a number $C_v > 0$ and a sufficiently small neighborhood $U(q_v)$ such that for all $w = (u,\theta) \in U(q_v) \cap L_{\gamma,\gamma}$ one has

$$\tan \theta_1(u,\theta) \geq C_v \tan \theta \qquad (9.3.)$$

From (9.3.) one deduces immediately that (9.2.) is true with $b_v = \frac{C_v}{2}$ in $U(q_v) \cap L_{\gamma,\gamma}$.

Without any restriction of generality we can assume that the boundary piece γ is so small that the following condition is verified. In the co-ordinate system (x,y) related to the straight lines $\ell_1(v)$ and $\ell_2(v)$, γ is the graph of some convex function of class C^1.

Now, we will prove (9.3.). Consider the points $w = (v,\theta)$ where $v \in \gamma$ and θ is sufficiently small. Let us introduce the co-ordinate system (x,y) related to the straight line $\ell_1(v)$ and $\ell_2(v)$, introduced in Sec. 7.5. while defining the condition F. In this co-ordinate system in some neighborhood of the point $v = (0,0)$, γ is the graph of a strictly convex smooth function $g = f_v$ satisfying the condition F. When v is an end of γ, then one considers the corresponding one-sided neighborhoods of $v = (0,0)$. Here we will consider the case when v is not an end of γ.

Let $w = (u,\theta) \in L_{\gamma,\gamma}$. Let $u = (s,f_v(s))$ and let $u_1 = u_1(u,\theta) = (t,f_v(t))$. We will distinguish four cases: (a) $s \leq 0 < t$ and $f(s) < f(t)$, (b) $s < 0 \leq t$ and $f(s) \geq f(t)$, (c) $s, t \geq 0$, (d) $s, t \leq 0$.

Consider first the case (a) which is presented on Fig. 29.

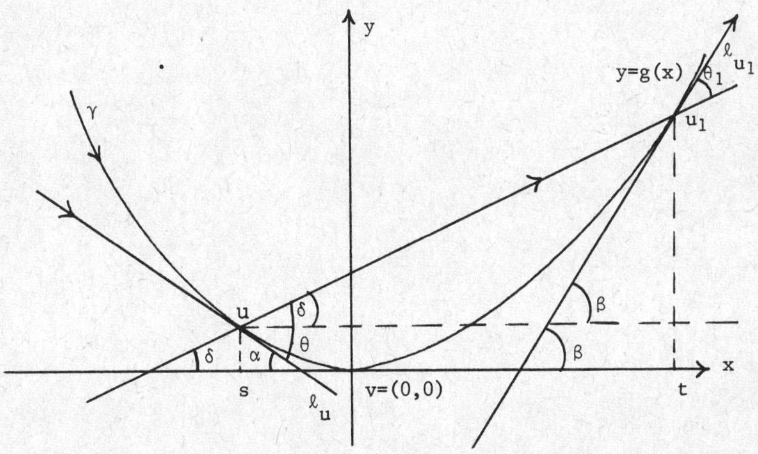

Fig. 29

Clearly, $\theta = \alpha + \delta$ and $\theta_1 = \beta - \delta$ where $\tan\delta = \dfrac{f(t) - f(s)}{t - s}$.
Thus $\tan\theta = \tan(\alpha+\delta) = \dfrac{\tan\alpha + \tan\delta}{1-\tan\alpha\tan\delta} \geq \tan\alpha + \tan\delta = -g'(s)$ $+\dfrac{g(t) - g(s)}{t - s}$ and $\tan\theta_1 = \tan(\beta-\delta) = \dfrac{\tan\beta - \tan\delta}{1 + \tan\beta\tan\delta} \geq \dfrac{1}{2}(g'(t) - \dfrac{g(t)-g(s)}{t - s})$, if only one considers s and t sufficiently small.

To prove (9.3.) in the case (a) under consideration, it is sufficient to prove that for some $C_v > 0$ and for all s, t, $s \leq 0 < t$, $|s|$ and t sufficiently small, one has

$$\frac{g'(t) - \dfrac{g(t) - g(s)}{t - s}}{\dfrac{g(t) - g(s)}{t - s} - g'(s)} \geq C_v \tag{9.4.}$$

However (9.4.) states nothing but the validity of the condition F for the function $g = f_v$. Thus, (9.4.) is true and one can take as C_v the exponent of γ.

In the cases (b), (c) and (d), the proofs are completely analogous. In all these cases one arrives to (9.4.). We omit the details.

It follows from this proof that the inequality (8.3.) holds for A_1 iff γ satisfies the condition F.

A_2 - Let $w = (v,\theta) \in L_{\gamma,\gamma_1}$. We will use the co-ordinate system (x,y) related to the straight line $\ell_1(z)$ and $\ell_2(z)$. For $x \geq 0$, γ is a graph of strictly convex smooth function g defined on the interval $[0,A]$.

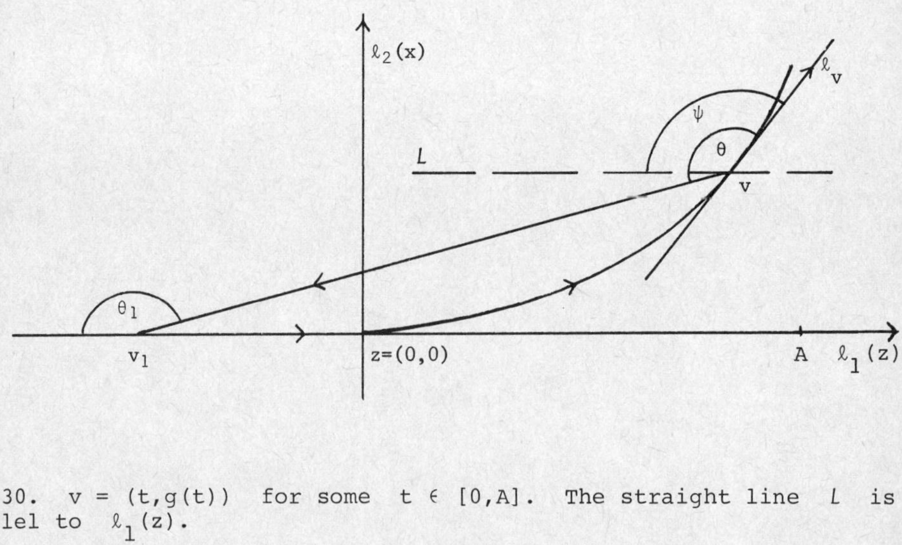

Fig. 30. $v = (t,g(t))$ for some $t \in [0,A]$. The straight line L is parallel to $\ell_1(z)$.

By virtue of the definition of A_Ω^5, $(v,\Psi) \in A_\Omega^5 \subset A_\Omega$ where Ψ is defined on Fig. 30. This implies that $\rho((v,\theta),(v,\Psi)) \geq \rho((v,\theta),A_\Omega)$. On the other hand $\rho((v,\theta),(v,\Psi)) = \pi - \theta_1$ where $0 < \theta_1 < \pi/2$. Thus, $\pi - \theta_1 \geq \rho((v,\theta),A_\Omega)$ and the inequality (8.3.) immediately follows.

The proof for A_2 is a very typical one as an example of the use of singular elements of A_Ω^5. With very minor modifications the same reasoning will appear in most of the proofs to follow.

A_3 - One has here singular elements $\bar{q}_v = (v,\pi)$ where $v \in \gamma$. Let $\bar{w} = (v,\theta) \in L_{\gamma,\gamma_1}$. As γ_1 is a strictly convex curve, then $\pi/2 > \theta_1(w) > \theta_1(\tilde{w}) > 0$ where $\tilde{w} = (z,\theta)$. However, from (9.3.) one has that $\tan\theta_1(\tilde{w}) \geq C_z \tan\theta$ and in consequence $d_1(w) = \sin\theta_1(w) \geq b\theta$ where $b > 0$ depends on γ_1 only. Thus, the inequality (8.3.) is proved.

A_4 - Let $w = (v,\theta) \in L_{\gamma,\gamma_1}$. We will consider two cases separately: the first when $0 < \theta < \pi$ and the second when $\pi < \theta < 2\pi$.

Let us consider the first case (see Fig. 31). In this case there exists a unique θ_v such that $P_v = (v,\theta_v) \in A_\Omega^4$ and the inequality (9.1.) then follows from Proposition 9.1. (see Fig. 31).

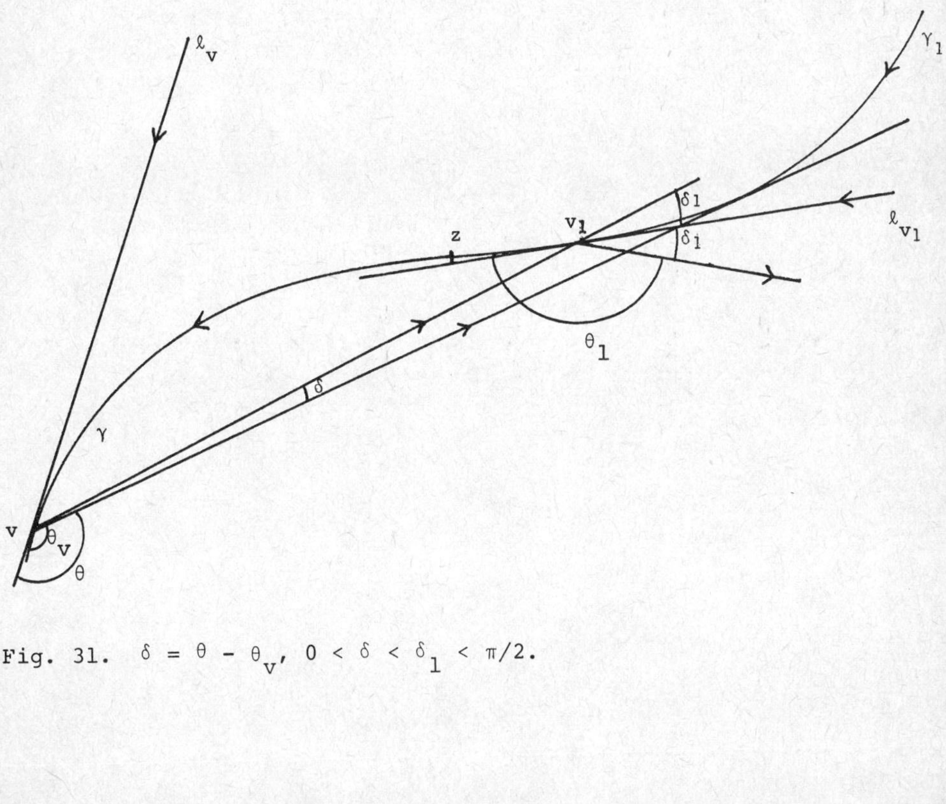

Fig. 31. $\delta = \theta - \theta_v$, $0 < \delta < \delta_1 < \pi/2$.

In the second case (see Fig. 32) to every $w = (v,\theta) \in L_{\gamma,\gamma_1}$ we associate $\overline{w} = (\overline{v},\pi-\beta)_{\ell_z}$, (i.e. $\overline{v} \in \ell_z$), $\overline{v}$ belongs to the interval $[v,v_1]$, and moreover w and $\overline{w}$ are parallel.

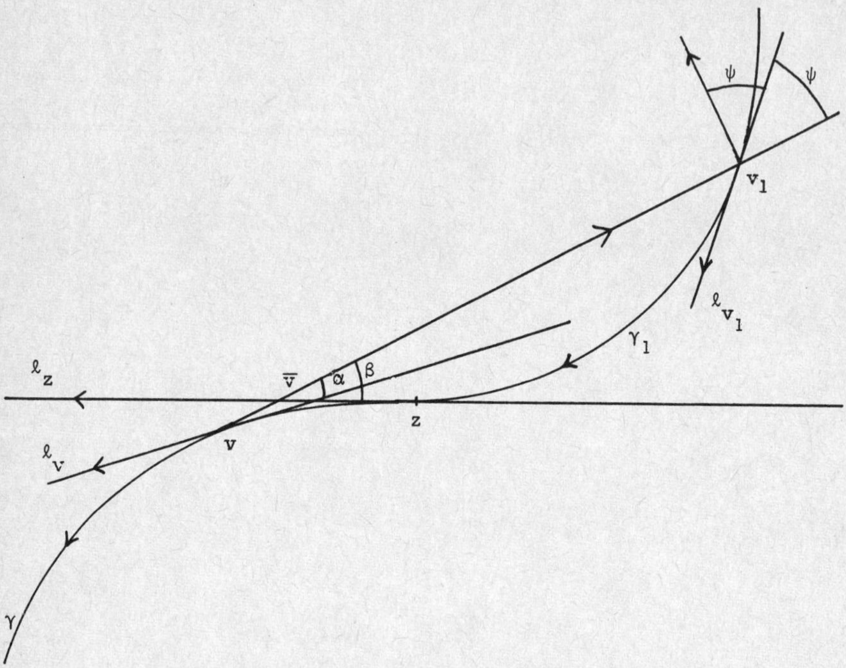

Fig. 32. $\alpha = \theta - \pi$, $\Psi = \pi - \theta_1$.

Then, from A_3 we know that $\sin \Psi \geq b\beta$ where $b > 0$ is a constant depending on γ_1 only. As $\beta > \alpha$ and $\sin \alpha = \sin \theta$, the inequality (9.1.) follows.

A_5 - With the minor modifications the proof goes along the same lines as in A_2.

B_1 - Here there is only one singular element, $P = (z,\alpha)_\gamma$.

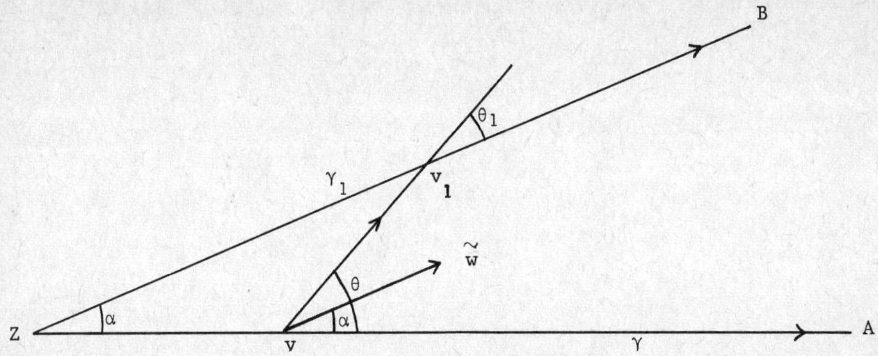

Fig. 33. $0 < \alpha < \pi$, $\gamma = [z,A]$, $\gamma_1 = [z,B]$, $\tilde{w} = (v,\alpha) \in A^5_\Omega$.

Let $w = (v,\theta) \in L_{\gamma,\gamma_1}$ where v is sufficiently close to z. Then, if $\theta - \alpha < \pi/2$, one has $\rho(w,A_\Omega) \le \rho(w,\tilde{w}) = \theta_1 \le 2 \sin \theta_1$ (see fig. 33). The inequality (9.1.) is then proved.

B_2 - In this case there is only one singular element $P = (z,\alpha)_\gamma$. Let $w = (v,\theta) \in L_{\gamma,\gamma_1}$ where v is sufficiently close to z. We will distinguish two cases: the first one when $0 < \theta \le \alpha$, and the second one when $\alpha < \theta < \pi$.

In the first case the proof easily follows from A_1. From the strict convexity of γ_1 one has $0 < \tilde{\delta}_1 < \delta < \pi/2$ (see Fig. 34). On the other hand, from A_1 we know that there exists $b > 0$ such that for all θ sufficiently close to α one has $\sin \tilde{\delta}_1 \ge b\rho(\tilde{w},p) = b(\alpha-\theta)$ where $\tilde{w} = (z,\theta)_\gamma$. Thus, finally, for $w = (v,\theta) \in L_{\gamma,\gamma_1}$ sufficiently close to $P = (z,\alpha)_\gamma$ one has $|\sin \theta_1| = \sin \delta_1 \ge b(\alpha-\theta) = b\rho(w,w_v) \ge \rho(w,A_\Omega)$.

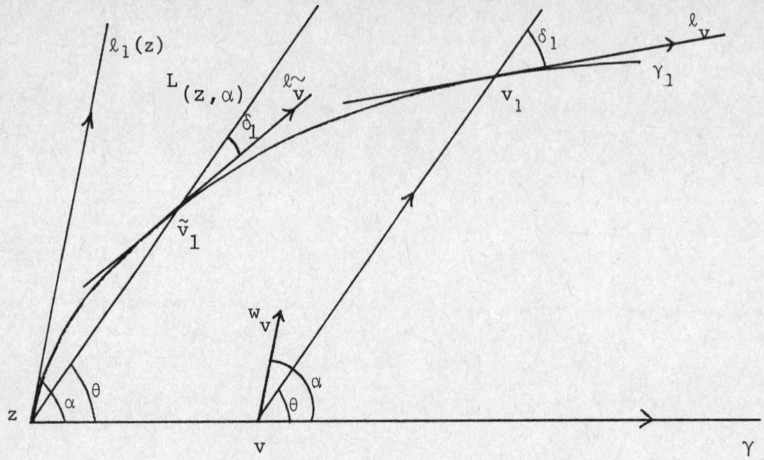

Fig. 34. $0 < \alpha < \pi$, $\sin \delta_1 = |\sin \theta_1|$, the intervals $[z,\tilde{v}_1]$ and $[v,v_1]$ are parallels, $w_v = (v,\alpha) \in A^5_\Omega$.

In the second case, for $w = (v,\theta)$ sufficiently close to $P = (z,\alpha)_\gamma$ one has $\pi/2 > \delta_1 > \theta - \alpha$ in virtue of the convexity of γ_1 (see Fig. 35).

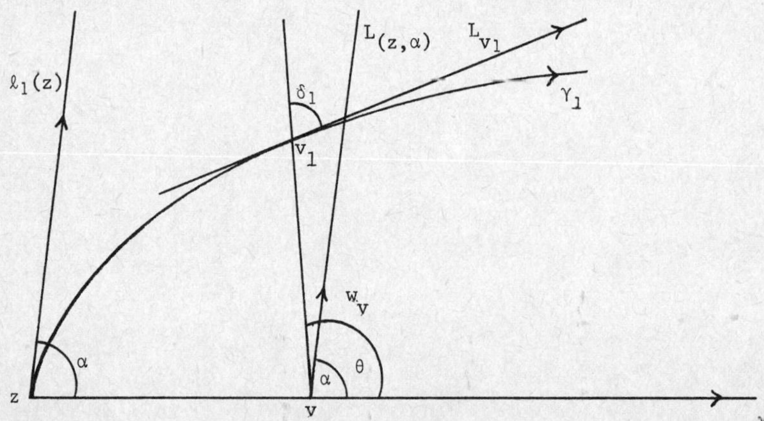

Fig. 35. $0 < \alpha < \pi$, $\delta_1 \geq \theta - \alpha > 0$.

Moreover, if $\delta_1 < \pi/2$, then $\sin \delta_1 = |\sin \theta_1|$. Thus, $\rho(w,A) \leq \rho(w,w_v) = \theta - \alpha \leq \delta_1 \leq 2|\sin \theta_1|$. So, the inequality (9.1.) is proved.

B_3 - Here there is only one singular element $P = (z,0)_\gamma$. Let

$w = (v,\theta) \in L_{\gamma,\gamma_1}$ and let $w_v \in A_\Omega^5$ be as defined on Fig 36.

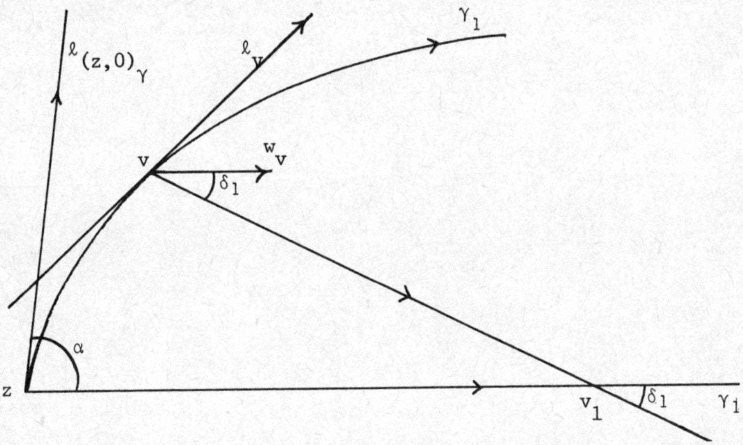

Fig. 36. $0 < \alpha < \pi$, $\rho(w,w_v) = \delta_1$.

Then, if w is sufficiently close to P, one has $0 < \delta_1 < \pi/2$ and $\sin \delta_1 = |\sin \theta_1|$, and consequently $\rho(w,A_\Omega) \leq \rho(w,w_v) = \delta_1 \leq 2 \sin \delta_1 = 2|\sin \theta_1|$. Thus, the inequality (9.1.) is proved.

B_5 - The proof is exactly the same as in B_2.

B_6 and B_8 - The proofs are exactly the same as in B_2 if one defines w_v to be a vector parallel to $\ell_1(z)$ directed toward γ_1.

C_2 - Here there are two singular elements, $P = (z,0)\gamma_1$ and $P_1 = (z,\pi)\gamma_1$. As far as P is concerned, the proof is exactly the same as in B_2. As for P_1, one has that if $w = (v,\theta) \in L_{\gamma,\gamma_1}$ is sufficiently close to P_1, then $\rho(w,A_\Omega) \leq \rho(w,\tilde{w}) = \delta \leq 2 \sin \delta \leq 2 \sin \delta_1 = 2|\sin \theta_1|$, where $\tilde{w} = (v,\pi) \in A_\Omega$ (see Fig. 37). Thus the inequality (9.1.) is proved.

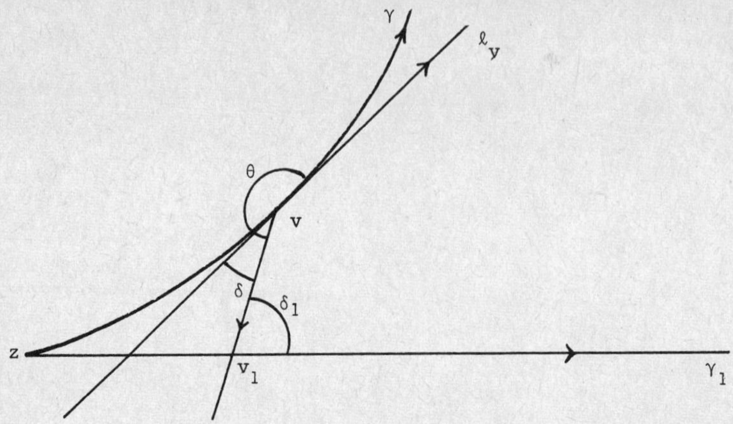

Fig. 37. $0 < \delta_1 < \pi/2$, $\sin \delta = |\sin \theta|$, $\sin \delta_1 = |\sin \theta_1|$.

C_4 - Here we have one singular element $P = (z,0)_\gamma$ and the proof is the same as in C_2 (see Fig. 38).

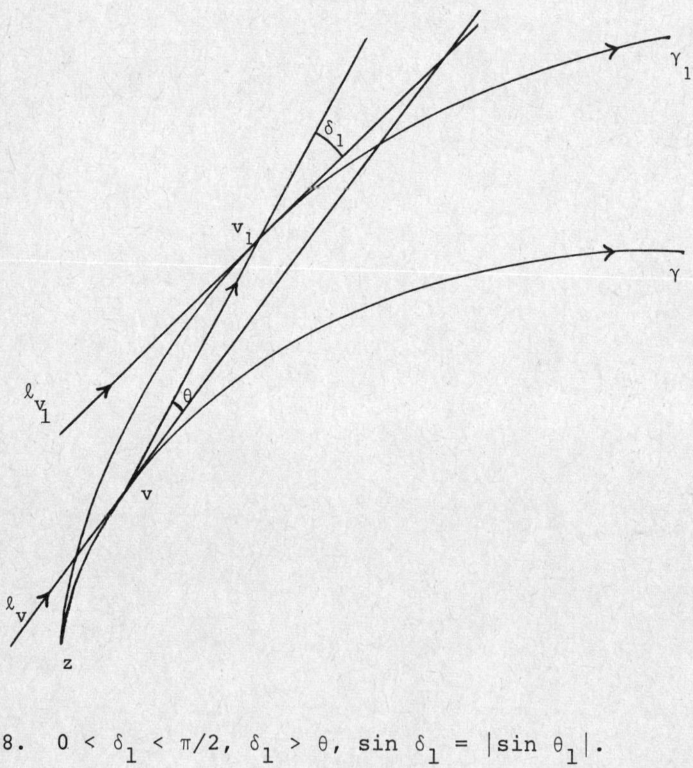

Fig. 38. $0 < \delta_1 < \pi/2$, $\delta_1 > \theta$, $\sin \delta_1 = |\sin \theta_1|$.

Before proceeding to the proof of D_1, D_3 and D_4, let us notice that in D_1 and D_3 the straight line $\ell_{\bar{z}} \in \theta_\Omega$ because $\bar{z}$ is here always either a point of inflection or a point of nondifferentiability of the boundary. In D_4 two possibilities occur: either $\ell_{\bar{z}} \in \theta_\Omega$ or $\bar{z}$ is a point of differentiability of Γ which is not a point of inflection. In the latter case the proof follows directly from Proposition 9.1. Thus, in D_4 we can also suppose, without any restriction of generality, that $\ell_{\bar{z}} \in \theta_\Omega$.

D_1 - In this case there is one singular element $P = (z,\pi-\alpha)_\gamma$. The proof is substantially the same as in B_2. Like in B_2 we distinguish two cases represented on Fig. 39 and Fig. 40 respectively. We omit the details.

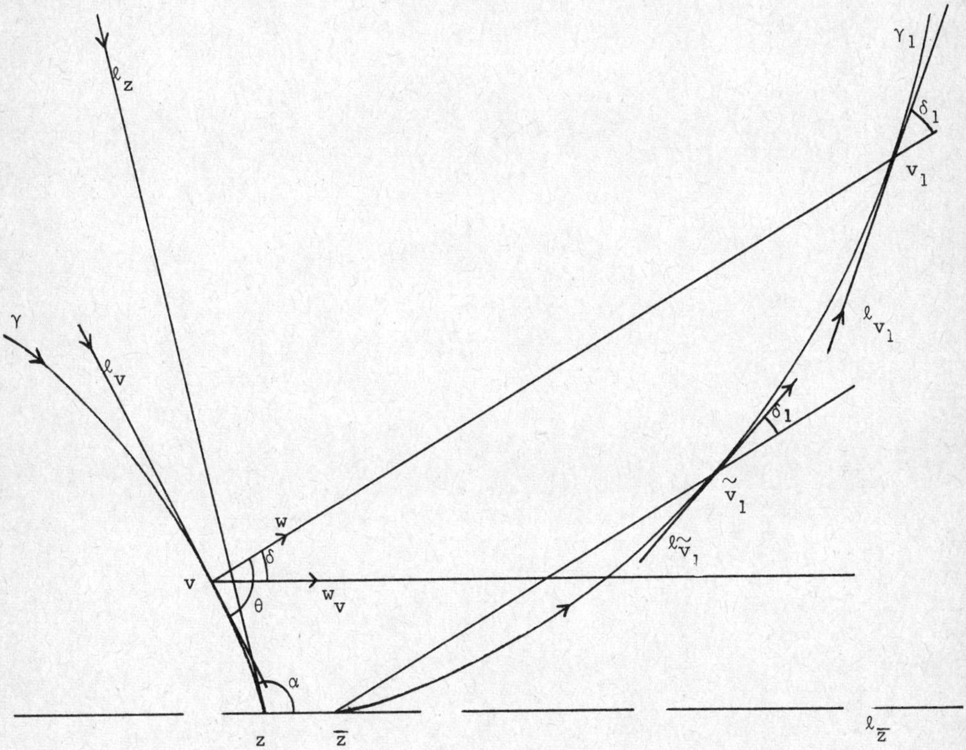

Fig. 39. $0 < \alpha < \pi$, the intervals $[v,v_1]$ and $[\bar{z},\tilde{v}_1]$ are parallels, $0 < \tilde{\delta}_1 < \delta_1 < \pi/2$, $\sin \delta_1 = |\sin \theta_1|$, $w_v \in A^5_\Omega$ is parallel to $\ell_{\bar{z}}$.

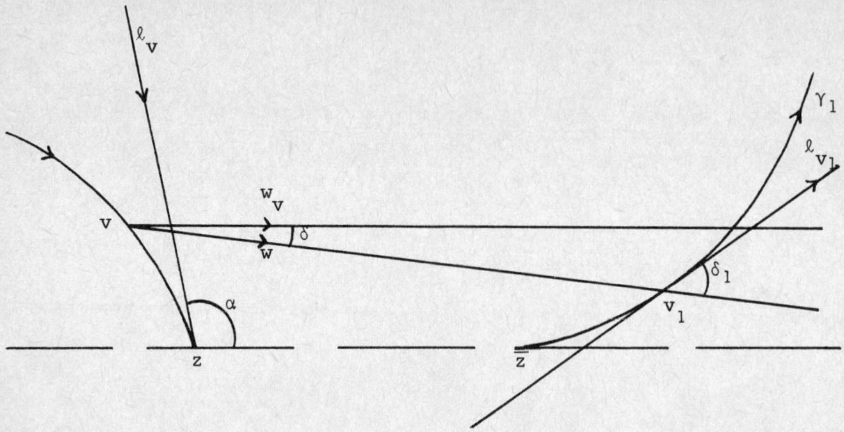

Fig. 40. $0 < \alpha < \pi$, $w_v \in A^5_\Omega$ is parallel to $\ell_{\bar{z}}$, $0 < \delta < \delta_1 < \pi/2$, $\sin \delta_1 = |\sin \theta_1|$.

Notice that in this proof we have never used the assumption that $\alpha \neq 0, \pi$. Thus, the above proof gives the inequality (9.1.) also for F_3 and F_4.

D_3 - In this case there is the singular element $P = (z,\alpha)_\gamma$ and the family $P_v = (v,\theta_v)$ exists (see Fig. 41). As far as the singular element P_v is concerned, the proof follows from Proposition 9.1. Consider now $w = (v,\theta) \in L_{\gamma,\gamma_1}$ sufficiently close to P. Now two cases are possible: either the interval $[v,v_1]$ cuts the straight line $\ell_{\bar{z}}$ or it does not. The first case is presented on Fig. 41.

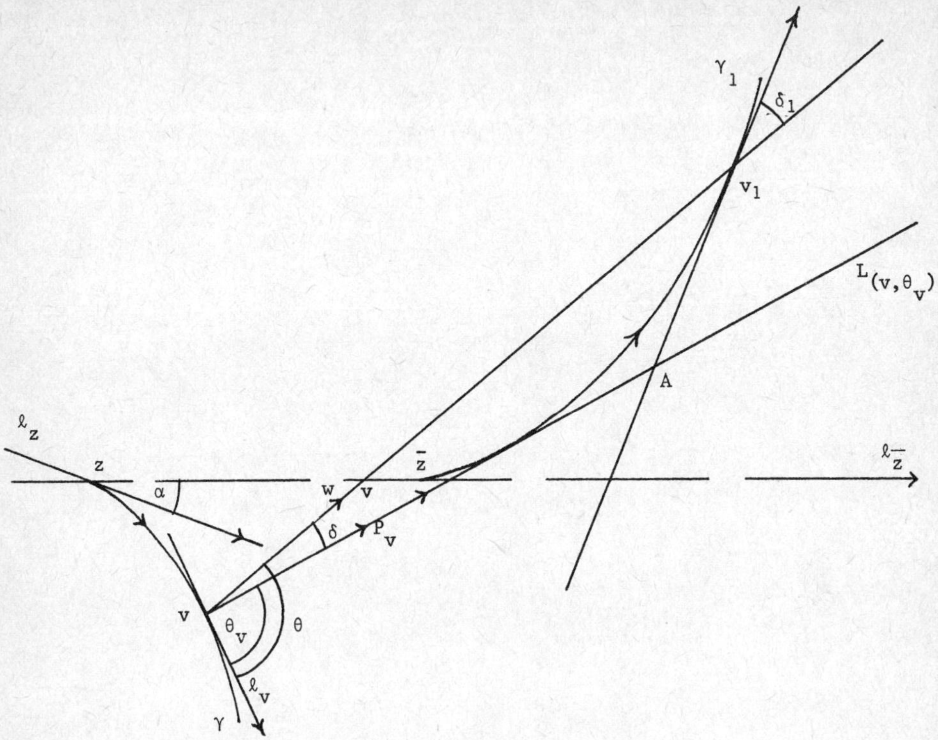

Fig. 41. $0 < \alpha < \pi$, $\rho(w,P_v) = \theta - \theta_v = \delta$, $\sin\delta_1 = |\sin\theta_1|$.

In this case, because $\dfrac{\sin\delta}{\sin\delta_1} = \dfrac{|[v_1,A]|}{|[v,A]|}$ then for all $v \in \gamma$, $v \neq z$, v sufficiently close to z, and for all $\theta > \theta_v$

$$\frac{\rho(w,A_\Omega)}{|\sin\theta_1|} \le \frac{\rho(w,P_v)}{|\sin\theta_1|} \le \frac{\delta}{\sin\theta_1(w)} \le 2\,\frac{\sin\delta}{\sin\theta_1(w)} = 2\,\frac{|[v_1,A]|}{|[v,A]|} \le$$

$$\frac{4}{|[z,\bar z]|}\cdot|[v_1,A]| \le \frac{4}{|[z,\bar z]|} \overset{\text{def}}{=} C < +\infty.$$ If the interval $[v,v_1]$ does not cut $\ell_{\bar z}$, then from Proposition 9.1. one has $\rho(w,A_\Omega) \le 2\sin\theta_1(w)$. Thus, the inequality (9.1.) is proved.

Notice that in this proof we never used the assumption that $\alpha \neq 0$. Thus, the above proof gives the inequality (9.1.) also for F_7.

D_4 - In this case there is only one singular element $P = (z,\alpha)_\gamma$. This case is presented on Fig. 42.

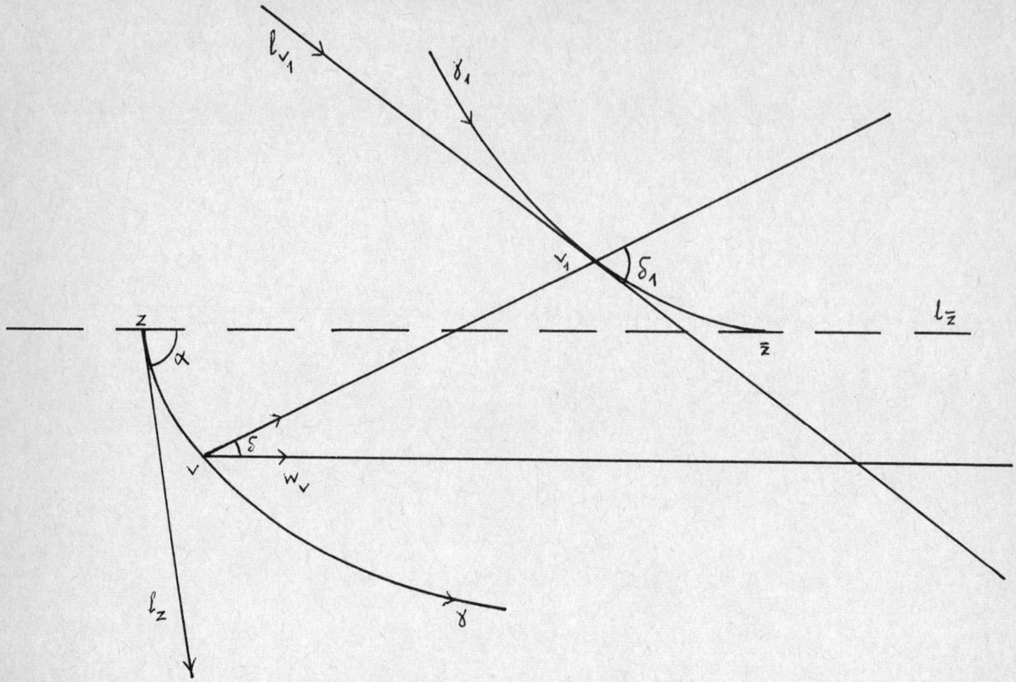

Fig. 42. $0 < \alpha < \pi$, $0 < \delta < \delta_1 < \pi/2$, $w_v \in A^5_\Omega$ is parellel to ℓ_z, $\sin \delta_1 = |\sin \theta_1|$.

Notice that if $w = (v,\theta) \in M$ is sufficiently close to P, then $0 < \delta_1 < \pi/2$. As usually, for such w one has $\rho(w,A_\Omega) \le \rho(w,w_v) = \delta < \delta_1 \le 2 \sin \delta_1 = 2|\sin \theta_1|$. Thus the inequality (9.1.) is proved.

Notice that in this proof we never used the assumption that $\alpha \neq 0$. Thus, the above proof gives the inequality (9.1.) also for F_8.

F_5 - In this case there is the singular element $P = (z,0)_\gamma$ and the family P_v always exists. Consider $w = (v,\theta) \in L_{\gamma,\gamma_1}$ where w is sufficiently close to P. Three possibilities may occur here: (a) the straight line $L_{(z,\theta)}$ cuts the straight line $\ell_{\bar{z}}$ at the right of $\bar{z}$, (b) it cuts the interval $(z,\bar{z})$, (c) it cuts $\ell_{\bar{z}}$ at the left of z.

In the cases (a) and (b), the proof is exactly the same as in D_3. As for the case (c), the proof is the same as for the singular element in C_2.

F_6 - Here the singular elements are exactly the same as in F_5.

Among the cases enumerated in F_5 only cases (b) and (c) are represented here. In the case (b) the proof is exactly the same as in D_4. As far as the case (c) is concerned, the proof follows directly from Proposition 9.1. ∎

10. FINAL REMARKS

10.1 Let us consider the flat two dimensional torus $T^2 = \mathbb{R}^2/\mathbb{Z}^2$ which will be identified with the unit square with identified opposite sides. Let us delete from the torus the disc D_r bounded by the circle Γ_r centered at $(1/2,1/2)$ and of radius r, $0 < r < \frac{1}{2}$. Let us denote the obtained manifold with boundary by H. One can consider the billiard in H extensively studied by Ja. C. Sinai in $[\text{Sin}]_2$.

The main difference between the billiard in H and billiards studies in Secs. 2 - 9 is that the time between two consecutive collisions with Γ is not bounded; as one says, here the horizon is not bounded. If we proceed in case of the billiard in H as we proceeded in Sec. 6, we will obtain a singular set of infinite length.

To avoid that, we insert in H the supplementary sections for the billiard flows which are not the scatterers. Indeed, let us denote by $L_1 - L_4$ the four sides of the unit square. Any trajectory of the billiard in H undergoes two consecutive crossing of $L = \bigcup_{i=1}^{4} L_i$ during a time period of length at most 3. Now let us define like in Sec. 2, $M = (\bigcup_{i=1}^{4} M_{L_i}) \cup M_\Gamma$, where $M_{L_i} = L_i \times [0,\pi]$ and $M_\Gamma = \Gamma_r \times [0,\pi]$, where all linear elements of M are directed toward the interior of H with obvious identification of the linear elements which belong simultaneously to two different M_{L_i}.

The mapping Φ is now defined exactly in the same way as in Sec. 2. If $(\phi,\theta) \in M$ then $\Phi(\phi,\theta) = (\phi_1,\theta_1)$ where (ϕ_1,θ_1) correspond to the reflection, when the colision with Γ_r takes place before the crossing of L and (ϕ_1,θ_1) correspond to the crossing of L in the opposite case.

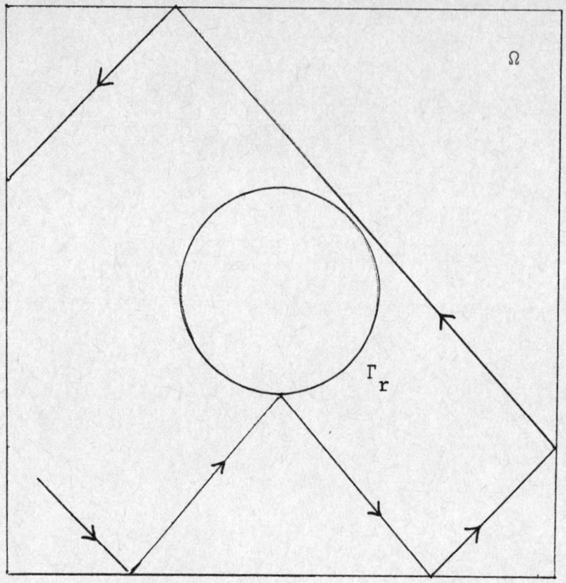

Fig. 43. A trajectory of the plane billiard in the domain Ω bounded by the unit square and the circle Γ_r is designed.

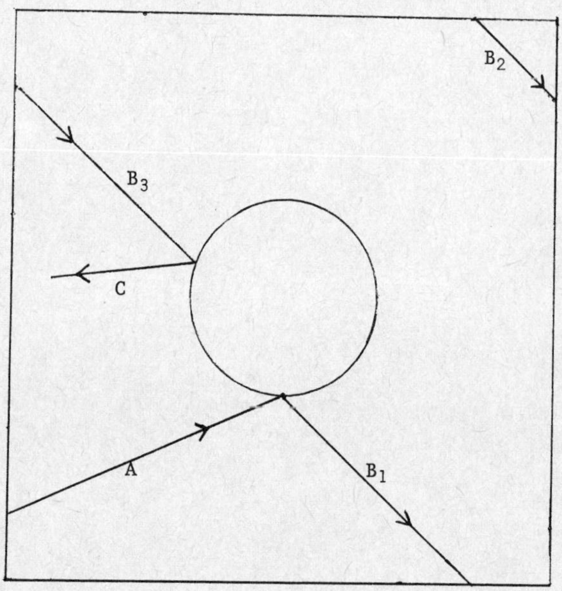

Fig. 44. A trajectory of the toral billiard in H. The intervals of the trajectory between two consecutive collisions are marked by the same letter, the change of the index indicates the crossing of L.

It is easy to see that for such defined mapping Φ, all results from Secs. 2 - 9 remain true word by word.

Let us remark that the above reduction of the case of "unbounded horizon" is not specific to the considered example and can be carried out for more general situations.

Finally let us note that instead of section L one can choose many other sections which also provide the reduction to the situation when our results apply.

APPENDIX 2. OSELEDEC MULTIPLICATIVE ERGODIC THEOREM

A.1. The proof of Oseledec theorem formulated below can be found in the original paper [Ose]. Other proofs can be found in [Rag]; see also $[\text{Led}]_2$ and $[\text{Rue}]_{2,3}$. The infinite dimensional versions of the Oseledec theorem are proved in $[\text{Mañ}]_2$ and in $[\text{Rue}]_3$.

By $M(m,\mathbb{R})$ we denote the set of all $m \times m$ matrices with real entries and by $GL(m,\mathbb{R}) \subset M(m,\mathbb{R})$ the group of all invertible matrices.

THEOREM. *Let μ be a probability measure defined on a σ-field $\mathfrak{M}$ of some subsets of space X. Let $T : X \to X$ be a measurable, μ preserving transformation*

T.1. *Let $A : X \to M(m,\mathbb{R})$ be a measurable mapping such that*

$$\int_X \log^+ \|A(x)\| d\mu(x) < +\infty.$$

where $\log^+ a = \max(\log a, 0)$. Then there exists a T invariant measurable subset $Y \subset X$ (i.e. $T(Y) \subset Y$), $\mu(Y) = 1$ such that:

(T.1.1) *For every point $x \in Y$ there exist a measurable filtration $\{0\} = L_0(x) \subset L_1(x) \subset L_2(x) \subset \ldots \subset L_{s(x)}(x) = \mathbb{R}^m$, where $\{L_i(x)\}_{1 \le i \le s(x)}$ are linear subspaces of $\mathbb{R}^m$ such that $L_i(x) \neq L_{i+1}(x)$ and $A(x)L_i(x) - L_i(Tx)$ and that for all vectors $v \in L_i(x) \setminus L_{i-1}(x)$, $1 \le i \le s(x)$, the following limit exists*

$$\lim_{n \to +\infty} \frac{\log \|A(n,x)v\|}{n} = \chi_i(x) \tag{1}$$

where $A(n,x) = A(T^{n-1}x) \cdot A(T^{n-2}x) \cdot \ldots \cdot A(x)$;

$$-\infty \le \chi_1(x) < \chi_2(x) < \ldots < \chi_{s(x)}(x) < +\infty.$$

Moreover for every linear subspace $F \subset L_i(x)$, $F \cap L_{i-1}(x) = \{0\}$, the convergence in (1) is uniform for vectors $v \in F$, $\|v\| = 1$.

The numbers $\{\chi_i(x)\}_{1 \le i \le s(x)}$ are called the Lyapunov characteristic exponents. The number $k_i(x) = \dim L_i(x) - \dim L_{i-1}(x)$ is called the multiplicity of the exponent $\chi_i(x)$.

(T.1.2) For every point $x \in Y$, let us denote by $\lambda_1(x) \le \lambda_2(x) \le \ldots \le \lambda_m(x)$ all Lyapunov characteristic exponents at x, where any exponent $\chi_i(x)$ is repeated $k_i(x)$ times; i.e. the first $k_1(x)$ of them coincides with $\chi_1(x)$, the following $k_2(x)$ with $\chi_2(x)$, etc. Now, for $x \in Y$, let $f_1(x),\ldots,f_m(x)$ be an arbitrary family of bases of $\mathbb{R}^m$ such that for every $i, 1 \le i \le m$,

$$\lim_{n\to\infty} \frac{1}{n} \log \|d\phi_x^n(f_i(x))\| = \lambda_i(x).$$

Then, if $m \ge 2$, for every two non-empty disjoint subsets $P,Q \subset \{1,2,\ldots,m\}$ we have:

$$\lim_{n\to\infty} \frac{1}{n} \log \gamma(d\phi_x^n(E_P),\ d\phi_x^n(E_Q)) = 0,$$

where E_P and E_Q denote the linear subspaces of $\mathbb{R}^m$ spanned by the vectors $\{f_i(x)\}_{i\varepsilon P}$ and $\{f_i(x)\}_{i\varepsilon Q}$ respectively and where the angle γ is defined in Sec. 2.1_I (see Sec. 3.1 of $[\text{Rue}]_3$).

T.2. If moreover T is invertible, $A(x) \in GL(m,\mathbb{R})$ and

$$\int_M \log^+\|A^{-1}(x)\|\, d\mu(x) < +\infty$$

then there exists a T-invariant measurable subset $Y_1 \subset X$ (i.e. $T(Y_1) = Y_1$), $\mu(Y_1) = 1$, such that

(T.2.1) for every point $x \in Y_1$ there exists a μ measurable decomposition

$$\mathbb{R}^m = \bigoplus_{i=1}^{s(x)} H_i(x),$$

such that $A(x)H_i(x) = H_i(Tx)$, $1 \le i \le s(x)$ and that uniformly for all vectors $v \in H_i(x)$, $\|v\| = 1$ one has the finite limits

$$\lim_{n\to\pm\infty} \frac{\log\|A(n,x)v\|}{n} = \chi_i(x), \quad i = 1,\ldots,s(x)$$

where for $n < 0$ we take $A(n,x) = A(T^{-(n-1)}x)\cdot\ldots\cdot A(T^{-1}x)\cdot A(x)$.

(T.2.2) <u>For every point</u> $x \in Y_1$

$$\lim_{n\to\pm\infty} \frac{1}{n} \log \gamma(d\phi_x^n(E_P), d\phi_x^n(E_Q)) = 0,$$

<u>where</u> E_P <u>and</u> E_Q <u>have the same meaning as in</u> (T.1.2).

Let us note that now $k_i(x) = \dim H_i(x)$.

A.2. Let $B : \mathbb{R}^m \to \mathbb{R}^m$ be a linear mapping. For every k, $1 \le k \le m$, let us denote by $B^{\wedge k}$ the k-th exterior power of B and let us take $B^{\wedge 0} = \mathrm{Id}$. Let us define the operator $B^{\wedge} : \bigoplus_{k=0}^{m} \Lambda^k \mathbb{R}^m \to \bigoplus_{k=0}^{m} \Lambda^k \mathbb{R}^m$ by $B^{\wedge} = \bigoplus_{k=0}^{m} B^{\wedge k}$.

The following proposition of Ruelle ([Rue]$_1$) plays an important role in Sec 12 of Part II and in Parts III and IV.

<u>PROPOSITION</u>. <u>Let</u> X, $\mathfrak{M}$, μ, T, $A(x)$ <u>and</u> Y <u>be as in</u> (T.1). <u>Then for every</u> $x \in Y$

$$\lim_{n\to\infty} \frac{1}{n} \log\| (A(n,x))^{\wedge} \| = \sum_{\chi_i(x)>0} \chi_i(x) k_i(x). \tag{2}$$

<u>Proof</u>. Let us denote by $\lambda_1(x) \le \lambda_2(x) \le \ldots \le \lambda_m(x)$ all Lyapunov characteristic exponents at x, where any exponent $\chi_i(x)$ is repeated $k_i(x)$ times; i.e. the first $k_1(x)$ of them coincide with $\chi_1(x)$ the following $k_2(x)$ with $\chi_2(x)$, etc.

From (T.1.1) and (T.1.2) it easily follows that for every k, $1 \le k \le m$,

$$\lim_{n\to\infty} \frac{1}{n} \log\| (A(n,x))^{\wedge k} \| = \sum_{i=m-k+1}^{m} \lambda_i(x)$$

and this implies (2). We leave the details to the reader. □

REFERENCES

[Abr] L. M. Abramov, V. A. Rohlin, The Entropy of Skew Product of the Measure Preserving Transformation(in Russian), Vestnik Leningradskogo Universiteta, No. 7, 1962, 5-13.

[Adl] R. Adler, A Note on the Entropy of Skew Product Transformations, Proc. Amer. Math. Soc, 14, 1963, 665-669.

[Akh] N. I. Akhiezer, I. M. Glazman, Theory of Linear Operators in Hilbert Space, Ungar, New York, 1961.

$[\text{Ano}]_1$ D. V. Anosov, Geodesic Flows on Closed Riemannian Manifolds of Negative Curvature (in Russian), Trudy Math. Inst. Steklova, 90, 1967. English transl. in Proc. Steklov Inst. Math., 90, 1967,

$[\text{Ano}]_2$ D. V. Anosov, Ya. G. Sinai, Certain Smooth Ergodic Systems (in Russian), Usp. Mat. Nauk, 22, No. 5, 1967, 107-172. English transl. in Russ. Math. Surveys, 22, No. 5, 1967, 103-167.

[Bir] G. D. Birkhoff, Dynamical Systems, Amer. Math. Soc. Colloq. Publ. Vol. IX (Providence, Rhode Island, 1927. Reprinted 1966).

$[\text{Bri}]_1$ M. I. Brin, Ya. B. Pesin, Partially Hyperbolic Dynamical Systems (in Russian), Izv. Akad. Nauk SSSR, Ser. Mat., 38, No. 1, 1974, 170-212. English transl. in Math. of the USSR - Izvestija, 8, No. 1, 1974, 177-218.

$[\text{Bri}]_2$ M. Brin, Z. Nitecki, to be published.

$[\text{Bow}]_1$ R. Bowen, Equilibrium States and the Ergodic Theory of Anosov Diffeomorphisms, Lect. Notes in Math. 470, Springer-Verlag, 1975.

$[\text{Bow}]_2$ R. Bowen, D. Ruelle, The Ergodic Theory of Axiom A Flows. Invent. Math. 29, 1975, 181-202.

$[\text{Bun}]_1$ L. A. Bunimovich, On Billiards Close to Dispersing (in Russian), Mat. Sb., 94, 1974, 49-73. English transl. in Math. of the USSR - Sb. 23, 1974, 45-67.

$[\text{Bun}]_2$ L. A. Bunimovich, On the Ergodic Properties of Some Billiards (in Russian), Funk. Anal. i jego prilozh., 8, 1974, 73-74. English transl. in Funct. Anal. and its Appl., 8, 1974, 73-74.

$[\text{Bun}]_3$ L. A. Bunimovich, On the Ergodic Properties of Nowhere Dispersing Billiards, Comm. Math. Phys., 65, 1979, 295-312.

$[\text{Bun}]_4$ L. A. Bunimovich, Ya. G. Sinai, On a Fundamental Theorem in the Theory of Dispersing Billiards (in Russian), Mat. Sb., 90, 1973, 415-431. English transl. in Math. of the USSR - Sb., 19, 1973, 407-424.

$[\text{Bun}]_5$ L. A. Bunimovich, Ya. G. Sinai, Stochasticity of Attractor in Lorenz Model in Nonlinear Waves, Ed. Nauka, Moscow, 1978, (in Russian), 212-226.

[Byl] B. F. Bylov, R. E. Vinograd, D. M. Grobman, V. V. Nemyckij, The Theory of Lyapunov Exponents (in Russian), Nauka, Moscow, 1966.

[Cam] M. Campanino, E. Presutti, Perturbative Ergodic Theorem in Hilbert Spaces, preprint.

[Che] J. Cheeger, D. G. Ebin, Comparison Theorems in Riemannian Geometry, North Holland Publ. Co. Amsterdam, 1975.

[Car] H. Cartan, Calcul Différentiel, Herman, Paris, 1960.

[Der] Y. Derriennic, Sur le Théorème Ergodique Sous-additif. Compte Rendus Acad. Sc. Paris. 281 A, 1975, 985-988.

[Dou] R. Douady, Applications du théorème des tores invariants (thèse de 3 ème cycle, Université Paris VII, 1982)

[Dvo] M. M. Dvorin, V. F. Lazutkin, The Existence of an Infinite Number of Elliptic and Hyperbolic Periodic Trajectories for a Convex Billiard (in Russian), Funkt. Anal. i jego prilozh., 7, No. 2, 1973, 20-27. English transl. in Funct. Anal. and its Appl., 7, No.2, 1973, 103-109.

[Fat] A. Fathi, M. R. Herman, J. C. Yoccoz, A Proof of Pesin's Stable Manifold Theorem, in Geometric Dynamics, 177-215, Lecture Notes in Math. 1007, Springer, 1983.

[Gal] G. Gallavotti, Lectures on the Billiard, in Dynamical Systems, Theory and Applications, ed. J. Moser, Lecture Notes in Physics 38, Springer, 1975, 236-295.

[Hal] B. Halpern, Strange Billiard Table, Trans. Amer. Math. Soc,. 232, 1977, 297-295.

[Gro] D. Gromoll, W. Klingenberg, W. Meyer, Riemannsche Geometrie im Grossen Lecture Notes in Math. 55, Springer, Berlin, 1968.

$[\text{Kat}]_1$ A. B. Katok, Dynamical Systems with Hyperbolic Structure, in IX Summer Math. School, ed. Naukova Dumka, Kiev, 1976, 125-211, (in Russian). English transl. in Three papers on dynamical systems by A. G. Kušinrenko, A. B. Katok and V. M. Alekseev, Amer. Math. Soc., 1981.

$[\text{Kat}]_2$ A. B. Katok, Lyapunov Exponents, Entropy and Periodic Orbits for Diffeomorphisms, Publications Math. IHES 51, 1980, 137-173.

[Kol] A. N. Kolmogorov, V. M. Tihomirov, ε-entropy and ε-capacity of Subsets of Functional Spaces (in Russian), Uspehi Mat. Nauk, 14, No. 2, 1959, 3-86. English transl. Amer. Math. Soc. Transl. Vol. 17, Second series, 1961.

[Laz] V. P. Lazutkin, The Existence of Caustics for Billiard Problem in Convex Domain (in Russian), Izv. Akad. Nauk SSSR, Ser. Mat., 37, No. 1, 1973, 186-216. English transl. in Math. of the USSR-Izvestija, 7, No. 1, 1973, 185-214.

$[\text{Led}]_1$ F. Ledrappier, J.-M. Strelcyn, A Proof of the Estimation from below in Pesin's entropy formula, Ergod. Th. 8c Dynam. Sys. 2, 1982, 203-219.

$[\text{Led}]_2$ F. Ledrappier, Quelques Proprietés des Exposants Caractéristiques, in Lectures Notes in Math 1097, 305-396, Springer Verlag, 1984.

$[\text{Led}]_3$ F. Ledrappier, Some Properties of Absolutely Continuous Invariant Measures on an Interval. Ergodic Th. 8c Dynam. Sys. 1, 1981, 77-93.

$[\text{Led}]_4$ F. Ledrappier, Propriétés ergodiques des mesures de Sinaï, C. R. Acad. Sci. Paris, 294, 1982, 593-595.

$[\text{Led}]_5$ F. Ledrappier, Propriétés ergodiques des mesures de Sinaï, Publ. Math. IHES, 59, 1984, 163-188.

$[\text{Led}]_6$ F. Ledrappier, L. S. Young, The Metric Entropy of Diffeomorphisms, Bull. Amer. Math. Soc., 11, 1984, 343-346. See also preprint MSRI, Berkeley, 1984.

[Liv] A. N. Livsic, Ya. G. Sinai, On Invariant Measures Compatible with the Smooth Structure for Transitive U-systems. Soviet. Math. Dokl, 13, 1972, 1656-1659.

[Mañ]$_1$ R. Mañe, A Proof of Pesin's Formula, Ergodic Theory & Dynam. Sys. 1, 1981, 95-102.

[Mañ]$_2$ R. Mañe, Lyapunov Exponents and Stable Manifolds for Compact Transformations, in Geometric Dynamics, 522-577, Lect. Notes in Math. 1007, Springer, 1983.

[Mil] V. M. Millionschikov, Two Theorems on Systems in Variations of Smooth Dynamical Systems (in Russian), Matemat. Zametki 5, No. 1, 1969, 49-54. English transl. Math. Notes 5, No. 1, 1969.

[Nit] Z. Nitecki, Differentiable Dynamics. An Introduction to the Orbit Structure of Diffeomorphism, The M.I.T. Press, Cambridge, Mass. 1971.

[Ose] V. I. Oseledec, The Multiplicative Ergodic Theorem. The Lyapunov Characteristic Numbers of Dynamical Systems (in Russian). Trudy Mosk. Mat. Obsch., 19, 1968, 179-210. English transl. in Trans. Mosc. Math. Soc., 19, 1968, 197-231.

[Pes]$_1$ Ya. B. Pesin, Families of Invariant Manifolds Corresponding to Nonzero Characteristic Exponents (in Russian), Izv. Akad. Nauk SSSR, Ser. Mat., 40, No. 6, 1976, 1332-1379. English transl. in Math. of the USSR - Izvestija, 10, No. 6, 1976, 1261-1305.

[Pes]$_2$ Ya. B. Pesin, Lyapunov Characteristic Exponents and Smooth Ergodic Theory in Russian, Usp. Mat. Nauk, 32, No. 4, 1977, 55-112. English transl. in Russ. Math. Surveys, 32, No. 4, 1977, 55-114.

[Pes]$_3$ Ya. B. Pesin, Description of π-Partition of a Diffeomorphism with Invariant Measure, Math. Notes of the USSR Acad. of Sci., 22, 1977, 1, 506-515.

[Pet] K. Petersen, Ergodic Theory, Cambridge University Press, 1983.

[Prz] F. Przytycki, Ergodicity of Toral Linked Twist Mappings, Ann. Scient. Ec. Norm. Sup., 4^e serie, 16, 1983, 345-354.

[Rag] M. S. Raghunathan, A Proof of Oseledec's Multiplicative Ergodic Theorem, Israel J. of Math. 32, 1979, 356-362.

[Rob] C. Robinson, L. S. Young, Nonabsolutely Continuous Foliations for an Anosov Diffeomorphism, Invent. Math. 61, 1980, 159-176.

[Roh]$_1$ V. A. Rohlin, On the Fundamental Ideas of Measure Theory, (in Russian), Mat. Sbornik, 25, 1949, 107-150. English transl. in Amer. Math. Soc. Transl., Series 1, No. 10, 1962, 1-54.

[Roh]$_2$ V. A. Rohlin, Selected Topics from the Metric Theory of Dynamical Systems (in Russian, Usp. Mat. Nauk, 4, 1949, No. 2, 57-128. English transl. in Amer. Math. Soc. Transl., Series 2, 49, 1966, 171-240.

[Roh]$_3$ V. A. Rohlin, Lectures on the Theory of Entropy of Transformations with Invariant Measure (in Russian), Usp. Mat. Nauk 22, No. 5, 1967, 3-56. English transl., Russian Math. Surverys 22, No. 5, 1967, 1-52.

[Rog] C. A. Rogers, Hausdorff Measures, Cambridge University Press, 1970.

[Rue]$_1$ D. Ruelle, An Inequality for the Entropy of Differential Maps, Bol Soc. Bras. Math. 9, No. 1, 1978, 83-87.

[Rue]$_2$ D. Ruelle, Ergodic Theory of Differentiable Dynamical Systems, Publ. Math. IHES, 50, 1979, 275-306.

[Rue]$_3$ D. Ruelle, Characteristic Exponents and Invariant Manifolds in Hilbert Space, Ann. of Math. 115, 1982, 243-290.

[Rue]$_4$ D. Ruelle, A Measure Associated with Axiom A Attractor, Amer. J. Math. 98, 1976, 619-654.

[Rue]$_5$ D. Ruelle, Sensitive Dependence on Initial Conditions and Turbulent Behavior of Dynamical Systems, Ann. N.Y. Acad. Sci. 316, 1978, 408-416.

[Sch] K. Schmidt, Cocycles of Ergodic Transformation Groups, Macmillan (India), 1977.

[Shi] G. E. Shilov, B. L. Gurevich, Integral, Measure and Derivative: A Unified Approach, Dover Publ., N.Y., 1977.

[Sin]$_1$ Ya. G. Sinaï, Classical Dynamical Systems with Countable Multiple Lebesgue Spectrum II (in Russian), Izv. Akad. Nauk SSSR, Ser. Matem, 30, 1966, 15-68. English transl. in Amer. Math. Soc. Transl. Series 2, Vol. 68, 1968, 34-88.

[Sin]$_2$ Ya. G. Sinaï, Dynamical Systems with Elastic Reflections. Ergodic Properties of Dispersing Billiards (in Russian), Usp. Mat. Nauk, 25, No. 2, 1970, 141-192. English transl. in Russ. Math. Surveys, 25, No. 2, 1970, 137-189.

[Sin]$_3$ Ya. G. Sinai, Gibbs Measure in Ergodic Theory, Russian Math. Survey 27, No. 4, 1972, 21-69.

[Sma] S. Smale, Differentiable Dynamical Systems, Bull. Amer. Math. Soc. 73, 1967, 747-817.

[Str] J.-M. Strelcyn, Les applications differentiables avec singularities; les sous-varietes invariantes, l'entropie et les billards in J.-P. Dufour, Séminaire sud-rhodanien de géometrie: Singularités, feuilletages et mécanique hamiltonienne, 93-111, Collection "Travaux en cours," Hermann ed., Paris, 1985.

[Woj]$_1$ M. Wojtkowski, Invariant Families of Cones and Lyapunov Exponents, Ergodic Th. 8c Dynam. Sys. 5, 1985, 145-161.

[Woj]$_2$ M. Wojtkowski, Principles for the design of billiards with nonvanishing Lyapunov Exponents (preprint, 1984)

Note added in proof

Recently A. Vetier in papers

[Vet]$_1$ A. Vetier, Sinai-billiard in potential field (Construction of stable and unstable fibers), in Limit Theorems in Probability and Statistics, Vol. II, 1079-1146, Ed. by P. Révész, Colloq. Mat. Soc. J. Bolyai 38, North Holland Publ. Comp., 1984.

[Vet]$_2$ A. Vetier, Sinai billiard in potential field (Absolute continuity), in Proc. of 3rd Pannonian Sympos. on Math. Stat., 341-351, Publ. House of Hung. Acad. of Sci., Budapest, 1983,

studied directly a class of dynamical systems with singularities, which seems to belong to the class of systems studied in the present book.

See also

[Kub]$_1$ I. Kubo, Perturbed billiard systems I, Nagoya Math. J. 61 (1976), 1-57.

[Kub]$_2$ I. Kubo, H. Murata, Perturbed billiard systems II; Bernoulli properties, Nagoya Math. J. 81 (1981), 1-25.

[Rob] M. Robnik, M. V. Berry, Classical billiards in magnetic fields, J. Phys. A: Math. Gen. 18 (1985), 1361-1378.

Vol. 1062: J. Jost, Harmonic Maps Between Surfaces. X, 133 pages. 1984.

Vol. 1063: Orienting Polymers. Proceedings, 1983. Edited by J.L. Ericksen. VII, 166 pages. 1984.

Vol. 1064: Probability Measures on Groups VII. Proceedings, 1983. Edited by H. Heyer. X, 588 pages. 1984.

Vol. 1065: A. Cuyt, Padé Approximants for Operators: Theory and Applications. IX, 138 pages. 1984.

Vol. 1066: Numerical Analysis. Proceedings, 1983. Edited by D.F. Griffiths. XI, 275 pages. 1984.

Vol. 1067: Yasuo Okuyama, Absolute Summability of Fourier Series and Orthogonal Series. VI, 118 pages. 1984.

Vol. 1068: Number Theory, Noordwijkerhout 1983. Proceedings. Edited by H. Jager. V, 296 pages. 1984.

Vol. 1069: M. Kreck, Bordism of Diffeomorphisms and Related Topics. III, 144 pages. 1984.

Vol. 1070: Interpolation Spaces and Allied Topics in Analysis. Proceedings, 1983. Edited by M. Cwikel and J. Peetre. III, 239 pages. 1984.

Vol. 1071: Padé Approximation and its Applications, Bad Honnef 1983. Prodeedings. Edited by H. Werner and H.J. Bünger. VI, 264 pages. 1984.

Vol. 1072: F. Rothe, Global Solutions of Reaction-Diffusion Systems. V, 216 pages. 1984.

Vol. 1073: Graph Theory, Singapore 1983. Proceedings. Edited by K.M. Koh and H.P. Yap. XIII, 335 pages. 1984.

Vol. 1074: E.W. Stredulinsky, Weighted Inequalities and Degenerate Elliptic Partial Differential Equations. III, 143 pages. 1984.

Vol. 1075: H. Majima, Asymptotic Analysis for Integrable Connections with Irregular Singular Points. IX, 159 pages. 1984.

Vol. 1076: Infinite-Dimensional Systems. Proceedings, 1983. Edited by F. Kappel and W. Schappacher. VII, 278 pages. 1984.

Vol. 1077: Lie Group Representations III. Proceedings, 1982–1983. Edited by R. Herb, R. Johnson, R. Lipsman, J. Rosenberg. XI, 454 pages. 1984.

Vol. 1078: A.J.E.M. Janssen, P. van der Steen, Integration Theory. V, 224 pages. 1984.

Vol. 1079: W. Ruppert. Compact Semitopological Semigroups: An Intrinsic Theory. V, 260 pages. 1984

Vol. 1080: Probability Theory on Vector Spaces III. Proceedings, 1983. Edited by D. Szynal and A. Weron. V, 373 pages. 1984.

Vol. 1081: D. Benson, Modular Representation Theory: New Trends and Methods. XI, 231 pages. 1984.

Vol. 1082: C.-G. Schmidt, Arithmetik Abelscher Varietäten mit komplexer Multiplikation. X, 96 Seiten. 1984.

Vol. 1083: D. Bump, Automorphic Forms on GL (3,IR). XI, 184 pages. 1984.

Vol. 1084: D. Kletzing, Structure and Representations of Q-Groups. VI, 290 pages. 1984.

Vol. 1085: G.K. Immink, Asymptotics of Analytic Difference Equations. V, 134 pages. 1984.

Vol. 1086: Sensitivity of Functionals with Applications to Engineering Sciences. Proceedings, 1983. Edited by V. Komkov. V, 130 pages. 1984

Vol. 1087: W. Narkiewicz, Uniform Distribution of Sequences of Integers in Residue Classes. VIII, 125 pages. 1984.

Vol. 1088: A.V. Kakosyan, L.B. Klebanov, J.A. Melamed, Characterization of Distributions by the Method of Intensively Monotone Operators. X, 175 pages. 1984.

Vol. 1089: Measure Theory, Oberwolfach 1983. Proceedings. Edited by D. Kölzow and D. Maharam-Stone. XIII, 327 pages. 1984.

Vol. 1090: Differential Geometry of Submanifolds. Proceedings, 1984. Edited by K. Kenmotsu. VI, 132 pages. 1984.

Vol. 1091: Multifunctions and Integrands. Proceedings, 1983. Edited by G. Salinetti. V, 234 pages. 1984.

Vol. 1092: Complete Intersections. Seminar, 1983. Edited by S. Greco and R. Strano. VII, 299 pages. 1984.

Vol. 1093: A. Prestel, Lectures on Formally Real Fields. XI, 125 pages. 1984.

Vol. 1094: Analyse Complexe. Proceedings, 1983. Edité par E. Amar, R. Gay et Nguyen Thanh Van. IX, 184 pages. 1984.

Vol. 1095: Stochastic Analysis and Applications. Proceedings, 1983. Edited by A. Truman and D. Williams. V, 199 pages. 1984.

Vol. 1096: Théorie du Potentiel. Proceedings, 1983. Edité par G. Mokobodzki et D. Pinchon. IX, 601 pages. 1984.

Vol. 1097: R.M. Dudley, H. Kunita, F. Ledrappier, École d'Éte de Probabilités de Saint-Flour XII – 1982. Edité par P.L. Hennequin. X, 396 pages. 1984.

Vol. 1098: Groups – Korea 1983. Proceedings. Edited by A.C. Kim and B.H. Neumann. VII, 183 pages. 1984.

Vol. 1099: C.M. Ringel, Tame Algebras and Integral Quadratic Forms. XIII, 376 pages. 1984.

Vol. 1100: V. Ivrii, Precise Spectral Asymptotics for Elliptic Operators Acting in Fiberings over Manifolds with Boundary. V, 237 pages. 1984.

Vol. 1101: V. Cossart, J. Giraud, U. Orbanz, Resolution of Surface Singularities. Seminar. VII, 132 pages. 1984.

Vol. 1102: A. Verona, Stratified Mappings – Structure and Triangulability. IX, 160 pages. 1984.

Vol. 1103: Models and Sets. Proceedings, Logic Colloquium, 1983, Part I. Edited by G.H. Müller and M.M. Richter. VIII, 484 pages. 1984.

Vol. 1104: Computation and Proof Theory. Proceedings, Logic Colloquium, 1983, Part II. Edited by M.M. Richter, E. Börger, W. Oberschelp, B. Schinzel and W. Thomas. VIII, 475 pages. 1984.

Vol. 1105: Rational Approximation and Interpolation. Proceedings, 1983. Edited by P.R. Graves-Morris, E.B. Saff and R.S. Varga. XII, 528 pages. 1984.

Vol. 1106: C.T. Chong, Techniques of Admissible Recursion Theory. IX, 214 pages. 1984.

Vol. 1107: Nonlinear Analysis and Optimization. Proceedings, 1982. Edited by C. Vinti. V, 224 pages. 1984.

Vol. 1108: Global Analysis – Studies and Applications I. Edited by Yu.G. Borisovich and Yu.E. Gliklikh. V, 301 pages. 1984.

Vol. 1109: Stochastic Aspects of Classical and Quantum Systems. Proceedings, 1983. Edited by S. Albeverio, P. Combe and M. Sirugue-Collin. IX, 227 pages. 1985.

Vol. 1110: R. Jajte, Strong Limit Theorems in Non-Commutative Probability. VI, 152 pages. 1985.

Vol. 1111: Arbeitstagung Bonn 1984. Proceedings. Edited by F. Hirzebruch, J. Schwermer and S. Suter. V, 481 pages. 1985.

Vol. 1112: Products of Conjugacy Classes in Groups. Edited by Z. Arad and M. Herzog. V, 244 pages. 1985.

Vol. 1113: P. Antosik, C. Swartz, Matrix Methods in Analysis. IV, 114 pages. 1985.

Vol. 1114: Zahlentheoretische Analysis. Seminar. Herausgegeben von E. Hlawka. V, 157 Seiten. 1985.

Vol. 1115: J. Moulin Ollagnier, Ergodic Theory and Statistical Mechanics. VI, 147 pages. 1985.

Vol. 1116: S. Stolz, Hochzusammenhängende Mannigfaltigkeiten und ihre Ränder. XXIII, 134 Seiten. 1985.

Vol. 1117: D.J. Aldous, J.A. Ibrag
Probabilités de Saint-Flour XIII – 1
IX, 409 pages. 1985.

Vol. 1118: Grossissements de filtrat
Seminaire, 1982/83. Edité par Th. Je

Vol. 1119: Recent Mathematical Methods in Dynamic Programming. Proceedings, 1984. Edited by I. Capuzzo Dolcetta, W.H. Fleming and T. Zolezzi. VI, 202 pages. 1985.

Vol. 1120: K. Jarosz, Perturbations of Banach Algebras. V, 118 pages. 1985.

Vol. 1121: Singularities and Constructive Methods for Their Treatment. Proceedings, 1983. Edited by P. Grisvard, W. Wendland and J.R. Whiteman. IX, 346 pages. 1985.

Vol. 1122: Number Theory. Proceedings, 1984. Edited by K. Alladi. VII, 217 pages. 1985.

Vol. 1123: Séminaire de Probabilités XIX 1983/84. Proceedings. Edité par J. Azéma et M. Yor. IV, 504 pages. 1985.

Vol. 1124: Algebraic Geometry, Sitges (Barcelona) 1983. Proceed-

Vol. 1137: Xiao G., Surfaces fibrées en courbes de genre deux. IX, 103 pages. 1985.

Vol. 1138: A. Ocneanu, Actions of Discrete Amenable Groups on von Neumann Algebras. V, 115 pages. 1985.

Vol. 1139: Differential Geometric Methods in Mathematical Physics. Proceedings, 1983. Edited by H. D. Doebner and J. D. Hennig. VI, 337 pages. 1985.

Vol. 1140: S. Donkin, Rational Representations of Algebraic Groups. VII, 254 pages. 1985.

Vol. 1141: Recursion Theory Week. Proceedings, 1984. Edited by H.-D. Ebbinghaus, G.H. Müller and G.E. Sacks. IX, 418 pages. 1985.

Vol. 1142: Orders and their Applications. Proceedings, 1984. Edited by I. Reiner and K. W. Roggenkamp. X, 306 pages. 1985.

Vol. 1143: A. Krieg, Modular Forms on Half-Spaces of Quaternions. XIII, 203 pages. 1985.

Vol. 1144: Knot Theory and Manifolds. Proceedings, 1983. Edited by D. Rolfsen. V, 163 pages. 1985.

Winkler, Choquet Order and Simplices. VI, 143 pages.

minaire d'Algèbre Paul Dubreil et Marie-Paule Malliavin. 1983–1984. Edité par M.-P. Malliavin. IV, 420 pages. 1985.

Vol. 1147: M. Wschebor, Surfaces Aléatoires. VII, 111 pages. 1985.

Vol. 1148: Mark A. Kon, Probability Distributions in Quantum Statistical Mechanics. V, 121 pages. 1985.

Vol. 1149: Universal Algebra and Lattice Theory. Proceedings, 1984. Edited by S. D. Comer. VI, 282 pages. 1985.

Vol. 1150: B. Kawohl, Rearrangements and Convexity of Level Sets in PDE. V, 136 pages. 1985.

Vol 1151: Ordinary and Partial Differential Equations. Proceedings, 1984. Edited by B. D. Sleeman and R.J. Jarvis. XIV, 357 pages. 1985.

Vol. 1152: H. Widom, Asymptotic Expansions for Pseudodifferential Operators on Bounded Domains. V, 150 pages. 1985.

Vol. 1153: Probability in Banach Spaces V. Proceedings, 1984. Edited by A. Beck, R. Dudley, M. Hahn, J. Kuelbs and M. Marcus. VI, 457 pages. 1985.

Vol. 1154: D.S. Naidu, A.K. Rao, Singular Pertubation Analysis of Discrete Control Systems. IX, 195 pages. 1985.

Vol. 1155: Stability Problems for Stochastic Models. Proceedings, 1984. Edited by V.V. Kalashnikov and V.M. Zolotarev. VI, 447 pages. 1985.

Vol. 1156: Global Differential Geometry and Global Analysis 1984. Proceedings, 1984. Edited by D. Ferus, R.B. Gardner, S. Helgason and U. Simon. V, 339 pages. 1985.

Vol. 1157: H. Levine, Classifying Immersions into $\mathbb{R}^4$ over Stable Maps of 3-Manifolds into $\mathbb{R}^2$. V, 163 pages. 1985.

Vol. 1158: Stochastic Processes – Mathematics and Physics. Proceedings, 1984. Edited by S. Albeverio, Ph. Blanchard and L. Streit. VI, 230 pages. 1986.

Vol. 1159: Schrödinger Operators, Como 1984. Seminar. Edited by S. Graffi. VIII, 272 pages. 1986.

Vol. 1160: J.-C. van der Meer, The Hamiltonian Hopf Bifurcation. VI, 115 pages. 1985.

Vol. 1161: Harmonic Mappings and Minimal Immersions, Montecatini 1984. Seminar. Edited by E. Giusti. VII, 285 pages. 1985.

Vol. 1162: S.J.L. van Eijndhoven, J. de Graaf, Trajectory Spaces, Generalized Functions and Unbounded Operators. IV, 272 pages. 1985.

Vol. 1163: Iteration Thoory and its Functional Equations. Proceedings, 1984. Edited by R. Liedl, L. Reich and Gy. Targonski. VIII, 231 pages. 1985.

Vol. 1164: M. Meschiari, J.H. Rawnsley, S. Salamon, Geometry Seminar "Luigi Bianchi" II – 1984. Edited by E. Vesentini. VI, 224 pages. 1985.

Vol. 1165: Seminar on Deformations. Proceedings, 1982/84. Edited by J. Ławrynowicz. IX, 331 pages. 1985.

Vol. 1166: Banach Spaces. Proceedings, 1984. Edited by N. Kalton and E. Saab. VI, 199 pages. 1985.

Vol. 1167: Geometry and Topology. Proceedings, 1983–84. Edited by J. Alexander and J. Harer. VI, 292 pages. 1985.

Vol. 1168: S.S. Agaian, Hadamard Matrices and their Applications. III, 227 pages. 1985.

Vol. 1169: W.A. Light, E.W. Cheney, Approximation Theory in Tensor Product Spaces. VII, 157 pages. 1985.

Vol. 1170: B.S. Thomson, Real Functions. VII, 229 pages. 1985.

Vol. 1171: Polynômes Orthogonaux et Applications. Proceedings, 1984. Edité par C. Brezinski, A. Draux, A.P. Magnus, P. Maroni et A. Ronveaux. XXXVII, 584 pages. 1985.

Vol. 1172: Algebraic Topology, Göttingen 1984. Proceedings. Edited by L. Smith. VI, 209 pages. 1985.